Asia's Innovation Systems in Transition

NEW HORIZONS IN THE ECONOMICS OF INNOVATION

Founding Editor: Christopher Freeman, *Emeritus Professor of Science Policy, SPRU – Science and Technology Policy Research, University of Sussex, UK*

Technical innovation is vital to the competitive performance of firms and of nations and for the sustained growth of the world economy. The economics of innovation is an area that has expanded dramatically in recent years and this major series, edited by one of the most distinguished scholars in the field, contributes to the debate and advances in research in this most important area.

The main emphasis is on the development and application of new ideas. The series provides a forum for original research in technology, innovation systems and management, industrial organization, technological collaboration, knowledge and innovation, research and development, evolutionary theory and industrial strategy. International in its approach, the series includes some of the best theoretical and empirical work from both well-established researchers and the new generation of scholars.

Titles in the series include:

Systems of Innovation and Development
Evidence from Brazil
Edited by José E. Cassiolato, Helena M.M. Lastres and Maria Lucia Maciel

Innovation, Competence Building and Social Cohesion in Europe
Towards a Learning Society
Edited by Pedro Conceição, Manuel V. Heitor and Bengt-Åke Lundvall

The Dynamics of Innovation Clusters
A Study of the Food Industry
Magnus Lagnevik, Ingegerd Sjöholm, Anders Lareke and Jacob Östberg

Technological Systems and Intersectoral Innovation Flows
Riccardo Leoncini and Sandro Montresor

Inside the Virtual Product
How Organisations Create Knowledge Through Software
Luciana D'Adderio

Embracing the Knowledge Economy
The Dynamic Transformation of the Finnish Innovation System
Edited by Gerd Schienstock

The Dynamics of Innovation in Eastern Europe
Lessons from Estonia
Per Högselius

Technology and the Decline in Demand for Unskilled Labour
A Theoretical Analysis of the US and European Labour Markets
Mark Sanders

Innovation and Institutions
A Multidisciplinary Review of the Study of Innovation Systems
Edited by Steven Casper and Frans van Waarden

Innovation Strategies in Interdependent States
Essays on Smaller Nations, Regions and Cities in a Globalized World
John de la Mothe

Internationalizing the Internet
The Co-evolution of Influence and Technology
Byung-Keun Kim

Asia's Innovation Systems in Transition
Edited by Bengt-Åke Lundvall, Patarapong Intarakumnerd and Jan Vang

Asia's Innovation Systems in Transition

Edited by

Bengt-Åke Lundvall

Professor of Economics, Department of Business Studies, Aalborg University, Denmark and Special Invited Professor, School of Economics and Management, Tsinghua University, Beijing, China

Patarapong Intarakumnerd

National Science and Technology Development Agency (NSTDA), Thailand

Jan Vang

Copenhagen Institute of Technology, Aalborg University, Denmark and Centre for Innovation, Research and Competence in the Learning Economy (CIRCLE), Sweden

NEW HORIZONS IN THE ECONOMICS OF INNOVATION

Edward Elgar

Cheltenham, UK • Northampton, MA, USA

Published by
Edward Elgar Publishing Limited
Glensanda House
Montpellier Parade
Cheltenham
Glos GL50 1UA
UK

Edward Elgar Publishing, Inc.
136 West Street
Suite 202
Northampton
Massachusetts 01060
USA

A catalogue record for this book
is available from the British Library

Library of Congress Cataloguing in Publication Data

Asia's innovation systems in transition/edited by Bengt-Åke Lundvall,
Patarapong Intarakumnerd, Jan Vang.
p. cm. — (New horizons in the economics of innovation)
Includes bibliographical references and index.
1. Technological innovation—Economic aspects—Asia. 2. Technology transfer—Asia. 3. Technological innovation—Government policy—Asia. 4. Organizational change—Asia. 5. Organizational learning—Asia. 6. Knowledge management—Asia. I. Lundvall, Bengt-Åke, 1941– . II. Intarakumnerd, Patarapong. III. Vang, Jan. IV. New horizons in the economics of innovation.
HC412.A78325 2006
338'.064095—dc22 2005031680

ISBN-13: 978 1 84542 713 9
ISBN-10: 1 84542 713 0

Printed and bound in Great Britain by MPG Books Ltd, Bodmin, Cornwall

Contents

Figures

Tables

Contributors

Tilman Altenburg is Head of the 'Private Sector Development and State Reform' Department at the German Development Institute (GDI), the government-funded think tank for development policy issues in Germany. Before joining the GDI, Mr Altenburg was research fellow of the Latin American Institute of the Free University Berlin and the Philipps University Marburg. He received his doctorate in Economic Geography from the University of Hamburg, Germany in 1991. Since 1986 Mr Altenburg has done empirical research on different issues of economic development in Latin America and Asia, with a focus on industrial policy and SME promotion.

Erman Aminullah is a director of the Centre for Science and Technology Development Studies (Pappiptek), at the Indonesian Institute of Sciences (LIPI), Jakarta. His current research activities are on innovation policy and economic complexity.

Erik Baark is Associate Professor at the Division of Social Science, Hong Kong University of Science and Technology. He received a Ph.D. in Information and Computer Science from the University of Lund in 1986, and a Dr.phil. from the University of Copenhagen in 1998. His primary interests are related to innovation systems and policies in China and other East Asian countries. Recent research projects have involved analysing innovation in service industries and software development in China and India.

Martina Fromhold-Eisebith is an economic geographer and holds a full professorship at the Department of Geography, Geology and Mineralogy at the University of Salzburg, Austria. Her research interests cover various issues of technology-oriented regional development, in particular relating to university–industry interaction, concepts of collaboration-based development, and related policies. Her empirical research has covered regions in Europe (Germany, Austria) and Asia (India, Indonesia).

Peter Gammeltoft is Assistant Professor at the Copenhagen Business School, Department of International Economics and Management. He acquired his Ph.D. in International Development Studies on the basis of a study of the Asian electronics industry, with a specific focus on Southeast

Asia. Prior to pursuing a Ph.D. he worked as a Senior Consultant with Accenture, primarily developing and deploying ICT systems in the Nordic health care industry. His current research interest is international management of technology and innovation, with a specific focus on the development of the software industry in Asia and foreign software R&D labs in China.

Shulin Gu is Visiting Professor, Tsinghua University, China and Honorary Senior Research Fellow, UNU/INTECH (United Nations University Institute for New Technologies). Research interests are in innovation and development, innovation systems in developing countries, reforms of S&T systems in China, agricultural technology development and rural/regional development in China.

Patarapong Intarakumnerd is the Project Manager of Thailand's National Innovation System Study at the National Science and Technology Development Agency (NSTDA), Thailand. His research interests include national innovation systems in Thailand and newly-industrializing countries in Asia, industrial and knowledge-based clusters, technological capabilities and evolution of latecomer firms, financial incentives for enhancing innovation capabilities and competitiveness, roles of Research Technology Organisations (RTOs) in industrial clusters and national innovation systems, science and technology (S&T) indicators for developing countries, and the co-evolution of the intellectual property regime and technological development in developing countries.

K.J. Joseph is Professor at the Centre for Development Studies in Trivandrum, Kerala, India. He has been Visiting Professor at Jawaharlal Nehru University, Visiting Senior Fellow at RIS, New Delhi and a consultant to UNESCAP. His research interests include issues relating to the national innovation system, informational technology and regional economic integration.

Nagesh Kumar is Director-General of RIS, New Delhi. Dr Kumar has previously served the United Nations University-Institute for New Technologies, Maastricht, the Netherlands, directing its research programme on Globalization, FDI and technology transfers. He has also been a consultant to the World Bank, ADB, UNCTAD, UNIDO, UN-ESCAP and ILO, among others. Recipient of the Exim Bank's first Trade Research Award and a GDN Research Medal, Dr Kumar has written extensively on the developmental impact of FDI, regional economic integration in Asia, WTO and development, new technologies and development, among other themes, and is a co-editor of *South Asia Economic Journal*.

Kong-Rae Lee is currently a research fellow of the Science and Technology Policy Institute (STEPI), Seoul, Korea. He is serving as the president of the Korean Society for Technology Management and Economics, and member of the Korea Academy of Science and Technology. He received his Ph.D. from SPRU, University of Sussex, UK. He served as policy advisor to the Minister of Science and Technology, in the Republic of Korea. His research areas are national, regional and industrial innovation systems, particularly multidisciplinary R&D activities in innovative clusters. He has published a book, *The Source of Capital Goods Innovation* at Harwood Academic Publishers.

Bengt-Åke Lundvall is Professor of Economics at Aalborg University, Department of Business Studies, and Special Invited Professor at Tsinghua University, School of Economics and Management, Beijing, China. His current research is on economics of knowledge and innovation in relation to economic development. He is former Deputy Director of OECD Directorate for Science, Technology and Industry and initiator of the global network of innovation scholars GLOBELICS (www.globelics.org).

Hiroyuki Odagiri is a Professor of Economics at Hitotsubashi University, Japan. His research interests are in the theory of the firm, industrial organization, and economic studies of innovation. He has written numerous books and academic papers in English and Japanese. Among the books published in English are *The Theory of Growth in a Corporate Economy*, *Growth through Competition, Competition through Growth*, and *Technology and Industrial Development in Japan*.

Mikkel Lucas Overby is a doctoral student at the Copenhagen Business School (CBS). His research interest focuses on firm boundary decisions with particular attention to intermediate forms or organization such as strategic alliances and outsourcing relationships. He has co-authored a book on this topic along with a number of book chapters and papers. Mikkel Lucas Overby has been a Visiting Scholar at the Indian Institute of Management Bangalore (2003) and Stanford University (2005).

Naubahar Sharif is a Research Assistant Professor in the Division of Social Science, The Hong Kong University of Science and Technology. He received his Ph.D. degree from the Department of Science and Technology Studies (S&TS) at Cornell University. In his current research program he is investigating the history of the innovation systems concept within a broader sociology of technology framework and applying the empirical results of his research to Hong Kong.

Jan Vang is Assistant Professor at the Copenhagen Institute of Technology, Aalborg University in Copenhagen, Denmark as well as researcher at CIRCLE, Lund University, Sweden. His research focuses on globalization of innovation systems; special attention is paid to Asia and Scandinavia and to so-called creative industries. He has written and (co)-edited numerous books, special issues and papers. He is book editor of the journal *Science, Technology and Society*.

Henry Wai-chung Yeung, Ph.D., has been Associated Professor in Economic Geography at the Department of Geography, National University of Singapore since January 2000. His research interests broadly cover the geography of transnational corporations, Asian firms and their overseas operations and Chinese business networks in the Asia-Pacific region. Professor Yeung has published widely on transnational corporations from developing countries, in particular Hong Kong, Singapore and other Asian Newly Industrialized Economies. He is the author of three books and the editor or co-editor of four other books. He has had over 60 research papers published in internationally refereed journals and 20 chapters in books. He is Editor of *Environment and Planning A*, *Economic Geography*, and *Review of International Political Economy*, and Asia-Pacific Editor of *Global Networks*.

Preface

This book is the outcome of collaboration among scholars working on innovation systems and related themes of competence-building systems in Asia. This collaboration was developed in connection with the worldwide network GLOBELICS and the region-wide network ASIALICS. The book is the second in a series of regional studies of innovation systems and a follow-up of the book *Africa First* (2004) edited by Mammo Muchie, Peter Gammeltoft and Bengt-Åke Lundvall.

Most of the papers included in the book were presented at the First Asialics Conference on Asian Innovation Systems and Clusters held in Bangkok in Spring 2004. The conference was organized by the National Science and Technology Development Agency (NSTDA) and King Mongkut's University of Technology Thonburi, Thailand with the support of the Japan International Cooperation Agency (JICA). At this conference it became clear – not surprisingly – that there are important differences among the Asian countries in terms of institutional set-ups supporting innovation, government policies and industrial structures. But what was exciting was that almost all of the national innovation systems were introduced as being in a process of 'transition'. They were all trying to get on to a new trajectory more intensive in learning and use of knowledge, including scientific knowledge.

This commonality, which has inspired the title, organization and structure of this book, reflected, among other factors, the aftermath of the financial crisis in the middle of the 1990s which hit most of the Asian economies as a shock. Even if the core of this crisis was financial speculation and the bubble economy – together with the later global crisis of the 'new economy' – it also raised critical questions about the sustainability of the trajectories followed in the most successful of the Asian economies.

1. Asia's innovation systems in transition: an introduction

Bengt-Åke Lundvall, Patarapong Intarakumnerd and Jan Vang

INTRODUCTION

Economies in transition has become the standard term for former centrally planned economies in Eastern Europe, the former Soviet Republics and Asia. This reflects the general connotation of *transition* as referring to a process where there is a change of an object, concept or system *from one state to another*. We have borrowed this term for the title of this book – *Asia's Innovation Systems in Transition* – to capture what is going on in a number of Asian countries, well aware that most of them have never been centrally planned economies. We did so because we think that *transition* captures quite well what is going on in the different national innovation systems analysed in this book.

The transition we refer to is a process where one constellation of institutions is turning into a different constellation of institutions. With institutions we mean norms, rules and conventions, formal and explicit as well as informal and implicit. Institutions are layered into organizations and into the minds of people, and they are slow to change. In our context the institutions in focus are those related to the production, diffusion and use of knowledge. The transition may, as in the standard use of the term, involve a different balance between market, state regulation and collective versus private property.

But they may also reflect a change in the relationship between knowledge producers and knowledge users or the emergence of a new mode of innovation. In some of the national cases presented below, this kind of change has already taken place over the last decade; in others they are on the policy agenda and attempts are made to move further ahead. But there are also cases, such as Indonesia, where the need for a transition is obvious but where it remains blocked by established power structures and by institutional inertia.

We see a specific value in combining *transition* with an innovation system perspective both in general, and in the specific context of the Asian economies. In general, it is a way to enrich our understanding of innovation systems. The focus is not on how systems are structured at a specific moment but rather on how they go through qualitative change when exposed to a combination of external transformation pressure and growing internal contradictions.

Second we believe that the transition perspective may be seen as a necessary complement to prevailing ideas of *catch-up economic growth*. Catching-up refers to quantitative growth and it may hide the fact that in order to keep growing, national systems from time to time need to go through a process of qualitative change affecting institutions, organizations and relationships between organizations. If this is not well understood it remains a mystery why certain countries, such as Japan, may be caught in growth traps or run into serious sustainability problems, as China, when investing more resources within their old growth model.

Third, and specifically for Asia, the transition perspective challenges the idea that some Asian systems are to be seen as 'models' that can be used as benchmarks for copying by other developing countries. If anything, they are moving targets and what will determine their future performance is not so much their current characteristics, it is rather their capability to make the transition to a new state. We will also show that while they have in common the fact that they are in transition, the transition challenge is quite different in, for instance, Japan, China and Thailand.

Not all the chapters are explicit on all these dimension of the transition process but they give interesting examples that help to understand how transition problems in innovation systems work themselves out in real life. Thus apart from, in their own right, being interesting empirical contributions to understanding what is happening in Asian innovation systems, the chapters in this book also help to fill a gap in the theoretical literature.

INNOVATION SYSTEMS

The combination of elements forming the national systems of innovation (NSI) concept makes it highly 'dialectical'. Innovation signals discontinuity while 'system' tends to be associated with a stable structure. Some have argued that the most dubious element of the concept is 'national' since it brings in, *ex ante*, a level of analysis that might not be the most adequate for understanding the process of innovation. In contrast to this we believe that it has become even more important to be explicit about the national dimension as 'globalization' becomes a major trend. To cope with the problems connected with globalization calls for an understanding of the historical role of national systems. The analysis of how various countries differ in terms of institutional set-ups supporting innovation and learning is important in this context.

Rather we see as more problematic the standard use of the term 'system'. The system appears in different social and academic discourses. There is a tendency to think about a system as a stationary self-reproducing set of elements with interrelationships. This is reflected in much of the literature on innovation systems. The result is that there is much focus on prevailing institutions and structures and less on qualitative change in the structure and in the institutional set-up. Empirical analysis is often an attempt to describe the system in terms of structure, institutions and organizations as well as the interrelationships between organizations with little concern for how it changes.[1]

INSTITUTIONS AND DEVELOPMENT

That the institutional set-up is a fundamental dimension of the national innovation system is not a controversial issue and from the very beginning the literature on 'innovation systems' takes this as fundamental starting point (Freeman, 1982; Lundvall, 1985).

Some of the first attempts to link explicitly the economic literature on institutions to innovation systems were by Björn Johnson (1988 and 1992). One of his basic points was that the uncertainty that characterizes innovation makes it even more necessary to include institutions in the analysis when the focus is on innovation and innovation systems. Rational choice referring to well-defined alternatives cannot explain what comes out of a process where outcomes are by definition unknown. Therefore institutions understood as rules, norms and habits are crucial for the outcome of what individuals decide and do in relation to innovation.[2]

There is a new tendency in international organizations that work on development issues to focus on institutions as perhaps the most important development factor. This tendency is interesting and useful but the focus remains narrow and one may wonder if the relatively narrow spectrum of institutions, which have been in focus, really can explain so much of the development process as it is claimed.

According to the World Bank (2002: 8), institutions have three main objectives: they channel information about market conditions, goods and participants; they define and enforce property rights and contracts; and they regulate competition. Within this framework, transaction costs that determine market opportunities typically stem from insufficient information, incomplete definition and enforcement of property rights, and barriers to entry to markets. These problems, as excessive transactions costs in general, have to do with inadequate institutions. Improved institutions that prop up market exchange and raise returns would support development.

In recent publications also IMF emphasizes the importance of institutions for growth. Sometimes institutions are even referred to as 'root causes' of economic development. Sometimes the power acknowledged to institutions is quite impressive. IMF (2003) for example concludes that if the quality of institutions in sub-Saharan Africa were to 'improve to the levels in developing Asia', per capita income would rise by 80 per cent, and if its institutions 'rose to world average levels' the average per capita economic growth rates would become 2 per cent higher. Like the World Bank, the IMF focuses on a narrow range of market-supporting institutions related to the security of property rights, good governance and measures to restrict corruption.

Well-functioning markets are important for development and so are uncorrupted civil servants and efficient regulation procedures in the economy. The problem is that this narrow view does not consider the crucial question of how institutions may support learning and innovation (except for the role of the formal school system, which belongs to the 'established' growth factors). The impact on learning and innovation of, for example, labour market institutions, financial institutions, economic policy regimes and a host of norms supporting (or undermining) a learning culture need to be analysed. We believe that in the current context of the globalizing learning economy this is a serious limitation (Arocena and Sutz, 2000).

UNDERSTANDING THE TRANSITION OF INNOVATION SYSTEMS

In what follows we will define transition as *radical institutional change*. To understand the mechanisms that increase the pressure for transition of innovation systems we need to find a way to link changes that take place outside the system to what goes on inside. There are two mechanisms that work simultaneously that together may undermine the performance of innovation systems.

One mechanism refers to situations where the environment changes so that the prevailing institutional set-up becomes ill-suited for the problems that the environment raises: we might refer to this as emergence of contingency mismatch.[3] The other mechanism refers to situations where endogenous economic growth within the system makes it reach limits for further economic growth – we might refer to this as emergence of inherent limits to growth. The successful catching-up process may be a process outgrowing the old national institutional set-up. To overcome mismatch and limits to growth radical institutional change (transition) is necessary.

One example of the first type could be the success and stagnation of the British economy as the technological environment changed, so that the most promising technological opportunities changed from mechanical engineering toward chemistry and electrical engineering. The old system was not geared to the new technological opportunities and characteristics of the US and the German systems were more in tune with the new opportunities. Japan may be taken as an example of how limits to growth are inherent in a successful growth model. The Japanese model was highly successful in catching up but it is less successful in operating at the frontier of science-based technologies. A different type of inherent limits to growth linked to sustainability is now becoming increasingly visible in China's innovation system.

In both cases the key barrier is 'institutional'. Even when major policy efforts are made and new organizations – such as science parks and technological institutes – are established, the institutional setting may be slow to follow the changes taking place. The recent public policy efforts in Thailand may be confronted with such problems. Old ways to do things do not disappear just because the formal organizational framework is modernized.

We use the innovation system concept in this book because we believe that studying transition through this 'focusing device' is useful for theoretical as well as practical purposes. It helps us understand why transition is difficult; it has to involve changes, not only in a single institution but rather in constellations of interconnected institutions sometimes supporting each other.

THE GLOBALIZING LEARNING ECONOMY AND THE EXTERNAL TRANSFORMATION PRESSURE

All national systems are exposed to changes in the environment. These changes relate to growing importance of knowledge and learning on the one hand, and the increasing international interdependence on the other hand.

There is growing agreement that knowledge is now at the very core of economic welfare and development. Nations, regions, industries and firms with a faster rate of growth are those that successfully manage to generate and apply knowledge. The OECD, for example, has consistently stressed the move toward a *knowledge-based economy* (Foray and Lundvall, 1996).

However, we have preferred to refer to the 'learning economy' (Lundvall and Johnson, 1994) because we believe that this captures better the dynamics of our age.

The concept is based upon the hypothesis that over the last decades an acceleration of both knowledge creation and knowledge destruction has taken place. Individuals and institutions need to renew their competencies more often than before, because the problems they face change more rapidly. And at the same time the segments of society affected by accelerating change have grown considerably. Therefore, in a wide set of economic activities, what constitutes success is not so much having access to a stock of specialized knowledge. The key to success is, rather, rapid learning and forgetting (when old ways of doing things get in the way of learning new ways).

Another major trend is so-called globalization. In recent years the interconnections between geographically different parts of the world have considerably increased and this has also multiplied the learning opportunities. But globalization is not a completed process. In some areas such as markets for financial assets it has gone a long way while in others related to competence building and innovation, national borders still remain. While some parts of the economy are at the core of the current trends, others have been marginalized. We have therefore preferred to refer to a 'globalizing' rather than to a 'global' economy, to stress that the current state of the world remains far from one characterized by a truly global economy and society.

It is important to emphasize that the 'learning economy' and the 'globalizing economy' are interconnected. A circular process has taken place. On the one hand, the development of an integrated world economy has allowed agents to acquire information, expertise and technology at a faster pace and often at lower costs than in the past. On the other hand, the current globalization has been nurtured by a generation of new technologies. The major technological advances of the last quarter of a century have in fact occurred in fields that allow the production, communication, transmission and storage of information. ICTs have in other words acted as the material devices to allow globalization to occur. Finance, production, media and fashion would not be as global as they are today without the generation of new technologies. In this sense, the 'learning' and 'globalizing' dimensions of the world economy strongly reinforce each other (see Archibugi and Lundvall, 2002).

An important element in this new context is that competition, as well as learning, has become more global and more intense in most parts of the economy. This is true especially in markets related to information technology: they are at the same time the carriers for the transmission of new knowledge, those where the rate of change is faster and those where competition has become extreme. But the production of traditional manufactured products such as textiles, toys and ships has also experienced a more intense competition, and substantial parts of these industries have moved out of Europe, Japan and North America to other parts of the world. Service-related activities such as shipping and software engineering are getting more and more exposed to global competition. Now, also traditionally protected and regulated areas (telecommunication, collective transport, public utilities, health and education), are becoming strongly exposed to competition.

IMPLICATIONS FOR EMERGING ECONOMIES

The globalizing learning economy offers both new opportunities and new threats for emerging economies. Several of the Asian economies have become more integrated in the global value chains that constitute an important characteristic of the globalizing learning economy and this has helped them to achieve high growth rates. This contrasts with the majority of the Latin American economies where the opening up of the national economies has seen more destruction of capabilities than stimulation of economic growth in its wake. Africa has largely been left out of the globalization game and this has gone hand in hand with stagnation or growing poverty.

There are many possible explanations for these major regional differences in the response to 'globalization'. A simple explanation is that there are certain prerequisites that need to be present in order to benefit from being integrated in the world economy. The first is skilled people and technological capability. The second is a certain degree of political control over the process of internationalization. The third is coherence in society with acceptance of certain rules of the game so that not everyone in society goes after immediate private benefit – legally or illegally. These factors may not be sufficient but they seem to be necessary.

Not all of the Asian countries studied here have all these prerequisites. Indonesia is not strong in any of the three dimensions. But it is interesting to note that those countries that have established the necessary prerequisites have responded differently to the opportunities. To some degree this reflects different levels of economic development but there are also other factors at play. The size of the economy, the economic system as well as unique historical factors have defined different transition challenges for each single national system.

In Japan the transition challenge is to build new institutions that support radical innovation in new advanced fields of science such as bio- and nanotechnology. In China it is to build institutions that help firms out of an imitative mode in many fields of technology but also to build institutions that make growth sustainable in social and ecological terms. In India the challenge is to transfer the successful institutional set-up in software and information technology to other sectors.

The switch might be incremental or radical (sometimes referred to as discontinuous) but where institutional economics usually stresses the incremental nature of institutional transitions, several Asian countries are united in having experienced radical institutional transition several times in the last couple of decades. Firstly, this was either during the rapid transition from predominantly agricultural economies to manufacturing economies (and in some cases even high-tech economies) or during the transformation from closed to rather open economies. Secondly, the recent financial crises in Asia also initiated a wave of institutional transitions.

THE STRUCTURE OF THE BOOK AND THE CONTENT OF THE CHAPTERS

The book may be seen as constructed from three different parts. The first part, is about 'Asian Countries entering knowledge-based competition'. It shows how national systems through different mechanisms have been able to upgrade their production structure and move away from low-wage competition. All the three chapters emphasize the spatial dimension of innovation and knowledge. Chapter 2 looks at the need to build and build upon specialized capabilities at the regional level when it comes to enter knowledge-based competition. Chapter 3 analyses the role of the Indian diaspora in attracting FDI to the Indian ICT sector. Chapter 4 is explicitly about the need for strategies to bridge international, national and regional systems of innovation.

While the three chapters in the first part all draw upon empirical illustrations the next four chapters about 'Asian innovation systems in transition', are in-depth case studies of four innovation systems in Asia. Chapter 5 on Thailand gives a rather optimistic picture of government attempts to establish the prerequisites for a transition to a new trajectory. Chapter 6 looks at the historical roots of the Hong Kong innovation system and shows how the current problems and opportunities of transition can be traced far back. Chapter 7 on the Indonesian system brings out many of the factors that make transition especially difficult – segmentation of society, economics and politics and the ethnic issues are among the most serious problems. Chapter 8 is about the Korean system and especially about why it has been successful for so long. But also in this case there is a growing need for transition toward an innovation mode that puts a premium on creativity.

The third part of the book, 'Science-based innovation in Asia and the need for policy learning', brings together three national studies that focus on science-based innovation and a chapter on policy learning in China. Chapter 9 by Odagiri uses biotechnology to illustrate the kind of transition toward a more science-based innovation mode that now takes place in Japan. Chapter 10 gives a detailed presentation of how the institutional framework supported the growth of the Indian software industry and discusses how far lessons can be drawn for the new ASEAN countries (Vietnam, Laos, Cambodia and Myanmar). Chapter 11 shows how the public authorities in interaction with transnational companies have shaped a framework that has fostered the growth of the high-technology industry in Singapore. Finally, Chapter 12 is about the use of the innovation system concept as a support for policy learning, and it uses experiences from China to show how central governance needs to be complemented with regional authority and governance.

ASIAN COUNTRIES ENTERING KNOWLEDGE-BASED COMPETITION

Chapter 2, by Tilman Altenburg, 'Opportunities for Asian countries to catch up with knowledge-based competition', uses case study material from electronics and shrimp farming to illustrate how the less developed economies in Asia gradually can move from low-wage and natural resource competition to knowledge-based competition. He argues that this kind of transition calls for a coordinated effort of a multitude of private and public agents in the national innovation system, with central and regional government playing important roles. A common effort to build local specialization in terms of production and capabilities is crucial. Building infrastructure and attracting foreign direct investment supporting the specialization is important and so is a strong focus on the absorption of knowledge from abroad in the strategic areas. Altenburg points out that the barriers to entry in terms of investment in R&D are very high in most high-technology sectors and this is one reason why he sees an incrementalist strategy where the starting point is the prevailing specialization pattern as the most realistic for countries such as Thailand, Malaysia, Indonesia and Burma.

But he mentions the software sector as one possible exception, and this connects his chapter to Chapter 3, 'Transnational communities, offshore outsourcing and offshore subsidiaries: the case of the Indian IT service industry' by Jan Vang and Mikkel Lucas Overby, where the Indian ICT industry is used as an illustrative case. The chapter is about the role transnational communities play as agents that can promote economic development in their country of origin. The authors challenge the brain drain hypothesis as well as the interpretation of transnational communities as equalizing a 'brain gain'. They develop a dynamic interpretation, suggesting that transnational communities are crucial for reducing uncertainty related to foreign investments in their country of origin especially in the initial phases of industrialization. However, the advantages diminish over time due to improved institutional regulation and management competencies of the firms in both the host and home countries of the transnational communities. The contribution is illustrated by a case analysing how the importance of the Indian transnational community as agents of economic development in their home country changes during the course of evolution of the Indian IT service and software industry.

Chapter 4, 'Effectively linking international, national and regional systems of innovation: insights from India and Indonesia', by Martina Fromhold-Eisebith, is about how innovation systems on different geographical scales can be interlinked and coordinated in order to promote economic development. The chapter first depicts major conceptual features of the NSI, RSI and ISI approaches. Then it develops ideas about how different system scales could logically be linked in order to constitute a scale-bridging systemic complex that may be termed 'National Supersystem of Innovation' (NSSI) which combines ideas of the three notions in complementary ways. Then the arguments are illustrated by examples from India and Indonesia. India – more implicitly than explicitly – selectively profits from scale-crossing system constellations; Indonesia hardly does, although possessing some good potential. The Indonesian case shows particular deficits in utilizing its RIS potential in ways that would be necessary for a functional NSSI.

ASIAN NATIONAL SYSTEMS IN TRANSITION

Chapter 5, 'Thailand's national innovation system in transition', by Patarapong Intarakumnerd, is about how a latecomer country's innovation system can be transformed from being weak and fragmented to become stronger and more coherent. It is argued that Thailand's national innovation system is in transition. Passive firms characterized by slow technological learning, ineffective and incoherent government policies, isolated education and training institutes, technologically unsupportive and risk-averse financial institutions, low-capacity trade/industry associations and an unfavourable institutional context have been perpetuated for the past fifty years of Thailand's industrialization. These have now begun to change due to two key factors. First the new government has initiated a major policy shift and implemented a management style significantly different from those in the past. Second, the recent economic crisis has had a strong and sometimes positive impact on key actors in the NIS. But the author also recognizes that the transformation is slow and difficult. Above all, it is difficult to change the mindsets and routines of some actors. A longer time-frame is needed for serious examination regarding whether the extent of these changes is large enough to make significant impacts on Thailand's innovation capabilities and long-term competitiveness.

In Chapter 6, 'Hong Kong's innovation system in transition: challenges of regional integration and promotion of high technology', by Erik Baark and Naubahar Sharif, it is shown how a system of innovation has been emerging in Hong Kong during the past century. Hitherto, however, technological innovation has not been regarded as an important element of Hong Kong's developmental experience. Technological innovation has only recently started to attract serious attention in Hong Kong, where the government in 1998 launched a new strategy in pursuit of knowledge-intensive economic growth. The authors also point to a need for a clear awareness of the opportunities and limitations of a more active role of the government in the shaping of future innovation in Hong Kong. They identify a range of areas where private and public initiatives to develop innovative capabilities coexist. In some cases, these initiatives appear to be mutually supporting. In other cases, public policies seem to be preoccupied with serving narrow business interests or even contradict the professed ambitions of promoting innovative industries. In other words, the role of government needs to be more solidly grounded in the principles of comprehensive and coherent policy-making that have been informed by innovation systems research (see, for example, OECD, 1997; 1999 and 2002).

In, Chapter 7, 'The Indonesian innovation system at a crossroads', by Peter Gammeltoft and Erman Aminullah, it is shown that one of the most conspicuous features of the Indonesian innovation system as it evolved under the 32-year rule of President Soeharto, was its segmentation. The end of Soeharto's rule in 1998 brought a series of economic, political and social reforms. The chapter shows that Indonesia is at a crossroads, having to come to grips with two major impetuses for reform: one is the multi-dimensional domestic reform process, the other is the one induced by what is commonly referred to as 'globalization'. The authors point to the need to avoid excessive (borrowed) capital investment and to shift focus from mere economic growth to techno-economic development. They point to the corporate structure as a serious constraint and they argue that future

Indonesian economic development requires a shift in the mindset of industrial leaders from mere economic to techno-economic perspectives. This will require a stepping up of industrial R&D, and government needs to engage in: restoration of the technological infrastructure, strengthening the linkages between public research institutions and industry, and better enforcement of competition in the economy.

Chapter 8, 'Performance and sources of industrial innovation in Korea's innovation system', by Kong-Rae Lee, makes an attempt to isolate the crucial factors behind the success of the Korean innovation system but also the weak elements that call for a re-orientation of the system. At the macro level factors such as hardworking people and aggressive learning, export-oriented strategy, sequential capability, heavy investment in R&D activities and the active role of government have been important. But there is also a unique strategy of 'crisis learning'. At the meso level the crucial success factors vary by industry. Exploitation of 'economy of speed' may be the most important in the semiconductor industry – R&D support of GRIs for the mobile telecom service industry and aggressive learning and in-house development of production technology for the automobile industry. Although Korea's innovation system has showed good performance in industrial innovation, it also has weaknesses calling for shifts in the strategy. Among the obstacles to strengthen innovation of Korea's industry are identified: low social trust, labour unrest, lack of flexibility in the education system, immobility of professional manpower, unbalanced regional innovation, underdevelopment of innovation management and rigid management of public R&D institutes. For Korea it is a challenge to overcome these drawbacks in order to strengthen the creative capability of the national innovation system.

SCIENCE-BASED INNOVATION IN ASIA AND THE NEED FOR POLICY LEARNING

Chapter 9, 'Advance of science-based industries and the changing innovation system of Japan', by Hiroyuki Odagiri, also indicates a transition from one trajectory to another. With the decline in demand in existing industries, intensifying technological competition on a global scale, and the rapid progress of scientific knowledge, Japan now aims at advancing *science-based industries*. Accordingly, Japan's national innovation system is changing. In part, it is a spontaneous change occurring in response to changing market needs. But it is also a consequence of conscious policy efforts to change the institutional, legal and policy frameworks so they become better adapted to these kind of industries. This chapter, taking biotechnology as a case, shows how technological changes, socio-economic changes, and institutional changes interact with each other, creating a new and yet path-dependent national innovation system. The financial system of Japan, characterized by a close bank–firm relationship and the presence of stable shareholders, has been complementary to the labour system characterized by a long-term worker–employer relationship. And this system has been conducive to the accumulation of firm-specific human skills and the close intra-firm (and intra-group) information sharing, which made cumulative technological innovation easier. Still, to promote new industries and new firms, the economy needs to foster reallocation of talented people through external markets (as opposed to internal labour markets) and the supply of more venturous funds (for which banks lack comparative advantages).

Chapter 10, 'National innovation systems and India's IT capability: are there any lessons for ASEAN newcomers?', by Nagesh Kumar and K.J. Joseph, traces the factors that have led to the build-up of substantial IT capability by India. It is shown that the national system of innovation supporting IT capability may be seen as an outcome of government policies. These included development of a system of higher education in engineering and technical disciplines, creation of an institutional infrastructure for S&T policy-making and implementation, building centres of excellence and numerous other institutions for technology development. In addition, the institutional interventions like the setting up of software technology parks were highly helpful for IT exports. The chapter also draws lessons from the Indian experience for the new members of ASEAN, viz. Cambodia, Laos, Myanmar and Vietnam in their attempt to establish national capabilities in ICTs. The chapter shows that in these countries, with the possible exception of Vietnam, the prerequisites for building such capabilities are still very weak. But the author also argues that much could be learned from India when it comes to facilitating their leapfrogging. The chapter underscores the opportunities and potential for cooperating with India in developing their IT capabilities.

In Chapter 11, 'Innovating for global competition: Singapore's pathway to high-tech development', Henry Wai-chung Yeung examines the experience of Singapore, a city-state in Southeast Asia, in harnessing the positive benefits from cluster development for high-tech industrialization. Specifically, it provides empirical evidence from several surveys and interviews with foreign and local firms and institutions conducted between 1999 and 2000 to show how the state in Singapore has been highly active in developing R&D capabilities through science parks, harnessing the SME advantage through reverse technology flows to foreign TNCs and nurturing cluster development in the chemical industry. Taken together, these cases point to the importance of state institutions and foreign TNCs in chartering a peculiar pathway to high-tech development in Singapore.

In Chapter 12, 'Policy learning as a key process in the transformation of the Chinese innovation systems', by Shulin Gu and Bengt-Åke Lundvall, the idea of innovation systems is linked to economic development and especially to policy learning in a developmental context. The authors emphasize the evolutionary and systemic foundation of the innovation system's approach. In the second part of the chapter these general points are illustrated in relation to three important policy themes in China: regional development, university–industry linkages and transformation of innovation systems. If anything, policy learning is more demanding in development countries than it is in the rich parts of the world. A characteristic for the economic development process is that from time to time a new trajectory needs to be opened up. This calls for initiatives at the central level but in order to be successful there must follow a rather long period of adaptive learning at all levels of the innovation system. Neither over-centralized systems that leave no autonomy at lower-level policy learning nor decentralized ones that lack the central governance mechanism necessary to initiate radical change will be able to cope with this double challenge.

THE CHALLENGE OF TRANSITION IN THE DIFFERENT INNOVATION SYSTEMS

As can be seen, there has been radical institutional change in several of the innovation systems studied in this book. Today there are new transition challenges that are different for each system. For India it is a major challenge to replicate the success story in software in other sectors in order to get the system on to a catch-up trajectory where economic growth rates become higher than population growth. The growth in China is still very much based upon massive investment, foreign technology and low-wage competitiveness. A change toward indigenous technological capabilities and innovation remains to be realized. While the concentration of political will-power at the central level may be seen as something that makes transition more realistic, the uneven regional development in both of these big countries calls for more efficient and stronger regional governance. Another big country, Indonesia, seems to have even bigger problems with governance. The segmentation of society, politics and economics and the negative impact of ethnicity contribute to making the innovation system incoherent and to undermining the coordination necessary to establish a common development trajectory. Here the necessary transition is difficult to manage.

Japan is already a highly developed economy and its transition problems relate to what should come after a successful catch-up strategy has been realized. The institutional characteristics that served the Japanese innovation system so well for several decades – long-term inter-firm relationships, patient capital and long-term employment contracts – are now increasingly seen as hindering the growth in some of the new science-based sectors. These institutions are now under pressure from markets and also critically reassessed and reformed by government. It is not the first time that former success makes the transition especially difficult. Freeman (2002) gives a series of historical examples of which the historical loss of industrial leadership for England in the twentieth century is of course the most obvious.

For the smaller countries in the region that have not yet come very far in terms of catching up and where knowledge-intensive production remains marginal (Malaysia, Vietnam, Cambodia) it is still somewhat of an open question as to what kind of transition it is possible to realize. Some contributions in this book give examples of successful incremental change where the existing specialization in low-wage products has been taken as a starting point but then has been combined with an upgrading through the build-up of a national and regional knowledge infrastructure. If this can be combined with a kind of leapfrogging in fields such as software industry, remains to be seen.

Thailand is an interesting intermediate case where the government under rather adverse conditions has made efforts to promote knowledge-based economic development through a set of coordinated initiatives. It will be interesting to see whether or not these initiatives are sufficient to open up a new trajectory. If they were to succeed, lessons could be drawn for other countries in the region who are in similar conditions.

Hong Kong and Singapore are small countries that have witnessed very rapid growth over the last decade. But even these small Asian tigers may have to enter into a transition process. The focus has been more on successful commerce than on promoting innovation, and this focus might not be sustainable in the context of the globalizing learning economy. Korea has a stronger knowledge base and has been successful in catching up on the basis of technology efforts. But in some respects it may be moving into a situation similar to that in Japan. There is a need to adopt new institutional frameworks that promote creativity and indigenous innovation.

Even these countries have different transition experiences; a tentative conclusion can be drawn as follows. Two major factors can contribute in upgrading a national innovation system. Internal factors, namely improving capabilities of key actors of the system and changing patterns of interaction between actors, can lead and sometimes pressure other actors in the system to change accordingly. For example, the governments of Hong Kong,

Thailand and Singapore spearheaded the transformation to the knowledge economy and encouraged other actors in the systems to follow. However, the systemic natures of development processes emphasized in the innovation systems approach means that the weakest node in the system can result in systemic failures.

Hence, for example, a government initiated procurement project is unlikely to result in the expected goals unless firms in the private sector are prepared to undergo transformation in the internal organization and external collaboration in ways matching the requirement of a modern learning economy. In the case of China, authoritarian Chinese leadership styles can prevent the needed change. Second is the change, in the external environment that would affect all actors in the system to change, such as Thailand's economic crisis in 1997 and Japan's decade-long recession. But while externally generated 'creative destruction' of inefficient innovation systems can provide the needed 'kick' to change the path of the innovation system in the right direction, it can also result in a vicious circle, resulting in increased poverty, de-industrialization and political conflicts.

LESSONS TO BE LEARNT

Another reason to apply the innovation systems perspective is that it helps to avoid naïve borrowing of 'best-practice' policy across national borders. What seems to work well in one systemic context might not do so in another. Asian NIEs' experiences with high growth have been used to launch a general campaign for attracting foreign investment. In this campaign, too little attention has been given to how the incoming foreign capital has been linked up with local strengths and needs in the different cases. Some of the important national differences in this respect can be understood by reading the chapters in this book.

While there are few specific policies and institutions that can be copied across national borders, there are still lessons that can be learnt from the experiences brought together in this volume. One lesson is that the debates about whether or not state intervention should play a role when it comes to promoting transition may be closed. When it comes to successful transition it is difficult to find one single case in this book where the public sector was not actively involved in institution building and policy. Public intervention is a necessary element both in establishing and renewing innovation systems.

This is not to say that state intervention has always been for the best. There are several examples in the book where it is clear that governments have slowed down or derailed necessary reform. Sometimes this has happened because the state was too much influenced by partial interests, and sometimes incompetence and corruption within the public sector has been part of the problem. But, even so, it is quite clear that in the absence of the state initiatives little successful transition would have taken place. So what is needed is to get the state to do the right thing. This has to do with social capital and/or the relative autonomy characterizing the so-called developmental state in the sense in which these concepts have been developed by Woolcock and Wade. What is needed, among other things, is a combination of autonomy and social integration among government officials.

There are other lessons that have to do with issues that seem to cut across national systems. In several of the countries studied in this book there seems to be a need for strengthening the regional dimension of economic development. This is strongly argued in the cases of Indonesia and Thailand. Also in China there seem to be strong needs to strengthen regional institutions and, not least, the governance capabilities of regional authorities. The balance between decentralization and centralization is, however, a delicate one. A high degree of decentralization carries the danger that regions compete against each other, and a 'race to the bottom' starts. Thus, instead of benefiting from the advantages of decentralization (for example local knowledge, increased incentives, and so forth) needed for industrial upgrading, the regions can engage in an even more ruthless price-competition.

Another crucial issue for countries aiming at knowledge-based economic development is how to promote the collaboration between universities and industry. This theme is addressed in several of the chapters. This focus is understandable in a situation where knowledge and competence is seen as a scarce resource. The example of India's software industry shows that investment in science and technology has a potential for stimulating growth. The unique Chinese pattern with new venture companies often owned by universities may be seen either as a model or as reflecting the weak absorptive capacity of the Chinese firms. It is important to take into account the different roles that universities play in the overall innovation system. Focusing on its capacity to give rise to concrete innovations is not a good idea if the most important role is to contribute to training and basic knowledge. Moreover, tying the universities too close to the immediate needs of the industry may prevent them from responding to the long-term needs of the industry.

A largely neglected dimension of development research conducted within the innovation systems approach has been the role that minorities and diasporas play as agents of economic development. Several of the chapters address these issues. In the chapter on Thailand and Indonesia the crucial role of the Chinese minority in business is of fundamental importance for understanding the innovation systems. In the chapter on Hong

Kong the origin of most of the population in mainland China is referred to as a key to explain the rapid growth of the Pearl River region. And in the chapter by Vang and Overby it is shown that Indian presence in the US IT business has been crucial for the location of transnational US firms in India.

THE END OF DEVELOPMENT RESEARCH?

But 'transition' is not a process relevant only for less developed countries in Asia or elsewhere. We believe that a similar volume entitled 'Europe's innovation systems in transition' might need to be written. And many of the basic concepts used in this book would be as useful for the one on Europe. Hereby we also imply that the idea that there should be a specific field of research called 'development research' might have become obsolete. The current ongoing rapid transformation of all economies, rich and poor, in the context of global competition undermines the assumption that there are mature modern economies where certain theories should be applied, and some others where underdevelopment calls for a different kind of theory. At first sight, this might sound like a threat for scholars who have been insiders in development research for many years, but actually they should rejoice because what is happening is actually that many of their tools and methods are now becoming as needed for developed economies as they are for less developed economies.

When analysing the economy of a rich country there has been a tendency to assume that institutions and social aspects could be neglected because markets and competition had been fully developed into 'maturity' leaving us with the option to use the tools of 'pure economics'. Only when we look at less developed countries would there be a need to reflect upon 'institutions' and 'culture' – and then as a kind of 'friction' that needs to be taken into account when designing strategies for economic development. We think that this view of the world is fundamentally wrong. In the current context of globalization where knowledge and learning becomes increasingly important, 'transition' and 'transformation' of socio-economic systems are ubiquitous processes to be found in the North and West as well as in the South and East.

NOTES

1. This is true both for many of the national case studies presented in chapters in Nelson (1993). It is also to some degree true for the more analytical chapters in Lundvall (1992).

Freeman's historical work on innovation systems is much more about transition (Freeman, 1997; 2002).

2. Later Johnson and Edquist developed the distinction between organizations and insttutions in relation to innovation and innovation systems (Edquist and Johnson, 1997). While specific organizations may be seen as 'incarnations' or 'containers' of institutions – such as patent offices incarnating intellectual property rights or as universities housing a specific knowledge production mode – they should not be defined as institutions.
3. This concept is inspired by the organizational and managerial literature on organization situational and contingency fit and misfit (Donaldson, 2001; Gresov, 1989; Burton and Obel, 1998). Similar ideas are developed in the work by Christopher Freeman (1997 and 2002) where he discusses the emergence of mismatch in historical perspective.

REFERENCES

Archibugi, Daniele and Bengt-Åke Lundvall (eds) (2001), *The Globalising Learning Economy: Major Socio-Economic Trends and European Innovation Policy*, Oxford: Oxford University Press.

Archibugi, D. and B.-Å. Lundvall (eds) (2002), *The Globalizing Learning Economy*, Guildford and King's Lynn: Oxford University Press.

Arocena, Rodrigo and Judith Sutz (2000), *Interactive Learning Spaces and Development Policies in Latin America*, DRUID working papers, No. 00-13.

Burton, R.M. and B. Obel (1998), *Strategic Organisational Diagnosis and Design; Developing Theory for Application*, Boston, MA: Kluwer Publishers.

Donaldson, L. (2001), *The Contingency Theory of Organisations*, Thousands Oaks, CA: Sage Publications.

Edquist, C. and B. Johnson (1997), 'Institutions and organisations in systems of innovation', in C. Edquist (ed.), *Systems of Innovation: Technologies, Institutions and Organisations*, London: Pinter.

Foray, D. and B.-Å. Lundvall (1996), 'The knowledge-based economy: from the economics of knowledge to the learning economy', in D. Foray, and B.-Å. Lundvall (eds), *Employment and Growth in the Knowledge-based Economy*, Paris: OECD Documents.

Freeman, C. (1982), 'Technological infrastructure and international competitiveness', Draft paper submitted to the OECD ad hoc group on Science, technology and competitiveness, August, mimeo. Now as Freeman, C. (2004) 'Technological infrastructure and international competitiveness', *Industrial and Corporate Change*, **13** (3).

Freeman, C. (1995), 'The national innovation systems in historical perspective', *Cambridge Journal of Economics*, **19** (1).

Freeman, C. (1997), 'Innovation systems: city-state, national, continental and sub-national', mimeo, paper presented at the Montevideo Conference, University of Sussex, SPRU.

Freeman, C. (2002), 'Innovation systems: city-state, national, continental and sub-national', *Research Policy*, **31**.

Gresov, C. (1989), 'Exploring fits and misfit with multiple contingencies', *Administrative Science Quarterly*, **34** (3), 431–54.

IMF (2003), *World Economic Outlook: Growth and Institutions*, April, Washington: International Monetary Fund.

Johnson, B. (1988), 'An institutional approach to the small country problem', in C. Freeman and B.-Å. Lundvall (eds), *Small Countries Facing the Technological Revolution*, London and New York: Pinter Publishers.
Johnson, B. (1992), 'Institutional learning', in B.-Å. Lundvall (ed.), *National Innovation Systems: Towards a Theory of Innovation and Interactive Learning*, London: Pinter Publishers.
Lundvall, B.-Å. (1985), *Product Innovation and User–Producer Interaction*, Aalborg: Aalborg University Press.
Lundvall, B-Å. (ed.) (1992), *National Systems of Innovation: Towards a Theory of Innovation and Interactive Learning*, London: Pinter Publishers.
Lundvall, B.-Å. and B. Johnson (1994), 'The learning economy', *Journal of Industry Studies*, **1** (2), 23–42.
Nelson, Richard R. (ed.) (1993), *National Systems of Innovation: A Comparative Study*, Oxford: Oxford University Press.
OECD (1997), *Industrial Competitiveness. Benchmarking Business Environments in the Global Economy*, Paris: OECD.
OECD (1999), 'The knowledge-based economy: a set of facts and diagrams', Document presented at the 1999 Ministerial Meeting on Science and Technology Policy, Paris: OECD.
OECD (2002), *Dynamising National Innovation Systems*, Paris: OECD.
World Bank (2002), 'Building institutions for markets', World Development Report 2002, Washington, DC: Oxford University Press.

2. Opportunities for Asian countries to catch up with knowledge-based competition

Tilman Altenburg

INTRODUCTION

In the worldwide division of labour the production factor 'knowledge' is growing increasingly more important. Price proportions for the traditional factor costs of labour, capital and land are no longer sufficient to fully explain the specialization patterns of industrial locations. The ability to generate innovations is becoming the crucial factor for competitiveness.

Innovations are for the most part the result of systemic-interactive processes involving a large number of specialized private- and public-sector actors, and they require complex infrastructural and economic policy inputs. Innovation dynamics should therefore develop in particular in locations endowed with highly qualified personnel, competitive and diversified firms and effective institutions. With the exception of some highly developed Asian countries and regions (such as South Korea, Singapore, Taiwan and Hong Kong), these favourable conditions only rarely exist in Asia's developing and emerging economies and where they exist they are limited to very specific sub-sectors (for example, India's satellite industry). Asia's developing countries are thus faced with the risk of being cut off from the self-reinforcing processes involved in competition for innovations. This confronts us with the question of how Asia's newly industrializing nations may best master the transition from labour-cost-based and resource-based to knowledge-based competitive advantages.

The present chapter offers some conceptual thoughts on this and points out, using practical examples from the region, how it is possible for countries to catch up with knowledge-based competition and start to build the necessary national innovation systems. In the first section I will argue that there is a strong rationale for strengthening the knowledge base of the Asian economies, and I will summarize some findings from recent publications on innovation that demonstrate its systemic character and then go on to discuss the implications for latecomer industrialization. Section 2 deals with concrete opportunities for countries to catch up technologically even though they have not yet developed mature and differentiated national innovation systems. The section discusses opportunities in activities that currently owe their competitiveness to low labour costs as well as resource-based activities. In both cases the section discusses general trends in the global division of labour and provides empirical evidence from the region. As far as empirical evidence is concerned, the section draws on experiences from the electronics industry (which in most countries of the region is largely restricted to simple assembly and testing operations) and the shrimp-farming industry, a resource-based sector with a potentially very high knowledge content. Section 3, finally, formulates some policy-related conclusions.

THE ROLE OF KNOWLEDGE AND INNOVATION IN BUILDING COMPETITIVE ADVANTAGES

The Rationale for Strengthening the Knowledge Base of the Economy

Competitive advantages resulting solely from prices for traditional factors of production are invariably threatened by factor movements, which lead to price adjustment and, accordingly, to declines in the profit rate. Firms – or in aggregate terms: locations – are, however, able in part to elude this price competition if they succeed in creating knowledge-based competitive advantages that are more specific in nature and thus difficult to replicate. Since knowledge cannot be codified, transferred and traded completely and without transaction costs, an innovative producer will be able to use his edge as a temporary market-entry barrier for competitors or he will in some cases even formalize this edge by applying for patent protection. The higher such barriers, the more profitable production may be: producers who manage to erect technological barriers that shield them from competition may pocket innovation rents, that is their capital invested yields above-average returns. If, on the other hand, we look at knowledge-*ex*tensive products without any relevant market-entry barriers, we find that the worldwide oversupply of low-skilled workers exerts strong downward pressure on returns. For technological latecomers this translates into deteriorating terms of trade.[1]

The Systemic Character of Innovation

But what precisely are the conditions under which firms, locations or societies are able to generate innovations systematically? Innovations require increasingly complex knowledge. In developing new industrial products today, it is as a rule necessary to combine up-to-date knowledge from different fields, that is knowledge on the properties of certain materials, on mechanics and electronics. Many products are for this reason jointly developed in a systematic process involving customers and complementary producers. Moreover, the products finally sold are often not merely standardized physical goods but (customized) 'solutions' that involve coupling products with knowledge-intensive services (design, maintenance, marketing, financing). This requires integration of specialized service providers. Another factor is the growing speed at which innovations prevail in the market (and are rendered obsolete by the next cycle). The aim is therefore not only to possess specialized knowledge and to be able to tap complementary stocks of knowledge and combine these innovatively with one's own knowledge, but also, and in addition, to master this task faster than competitors.

The following features outline the systemic character of innovation:[2]

1. In today's economy innovations are not extraordinary occurrences which cut through and abruptly transform otherwise 'innovation-free' everyday business routines. Rather, innovation is a continuous, gradual process, one that takes place constantly and everywhere – although by no means with the same dynamism. This process results in constant changes in products, production techniques, organizational processes, marketing, and so on.

   There are many areas in which innovations are systematically pursued. This is particularly evident in cases where specialized R&D departments are set up. But systematic pursuit of innovations can also be observed frequently in the routine operations of companies and institutions. Modern concepts of organizational development aim to create 'learning organizations' with clearly defined goals and indicators that are used to constantly monitor goal attainment and ensure that no time is lost in effecting process adjustments as soon as discrepancies are noted between targets and actual results. Today, mechanisms designed to check performance against defined goal parameters are firmly established in many organizations. Employees are given incentives to be on the constant lookout for possible improvements. In business practice this becomes evident in the widespread adoption of concepts like continuous improvement processes. Under framework conditions in which technological parameters and markets are changing at an ever-increasing pace, it is furthermore becoming more and more important for companies to be able to go beyond fixed parameters ('single-loop learning') and to establish procedures aimed at regularly and systematically questioning these parameters ('double-loop learning').[3]

2. The innovation process is increasingly moving away from the linear course of the past towards a circular-cumulative process involving numerous feedback loops. These loops occur not only within individual firms, as described in the above point, but also between the stages of a value chain. In the early phases of industrial development the phases 'invention', 'innovation' (that is, further development of an invention to the point of marketability) and 'diffusion' (establishment in the market) usually followed one another in clear succession. What we have today is an iterative process in which innovations are continuously tested and adapted. Users' requests, for example, are incorporated in the development process at an early stage, and established products or processes are systematically reviewed and modified.

   In this way innovation becomes an interactive process in which numerous actors work collectively to produce reciprocal external effects. Thus reality is increasingly taking leave of the picture drawn in Schumpeter's[4] early work, where innovations were to a large extent the individual achievements of creative individuals. Interaction takes place principally between firms in up- and downstream stages of the value chain (that is synchronized product development involving parts suppliers), but it can also be observed among firms at the same stage of the value chain (where it serves to achieve economies of scale) and between firms and scientific, research, training, business-promotion and other institutions. The importance of intensive cooperation with suppliers and research institutions has been stressed for decades now. More recently, interactions between manufacturers and demanding lead users have also been accorded great importance. Lead users frequently create incentives or apply pressure to induce producers to improve products. Porter refers, over and above this, to the innovation-driving effects created by challenging and differentiated demand.[5]
3. Knowledge can never be 100 per cent codified; it is, qua experience, always bound in part to people and institutions (tacit knowledge) and as such is not readily transferable. Furthermore, a wholly private appropriation of the outcomes of investments in new stocks of knowledge is seldom possible. Spillovers, that is unintended transfers to third parties, are more or less unavoidable. The production factor 'knowledge' is in this sense highly vulnerable to market failure. This, too, is of relevance to the question of business locations. Transfer of tacit knowledge hinges on interpersonal contacts; specialists are not totally mobile and specialized regional pools of skilled workers are therefore at times essential. This means that production processes cannot be broken down at will and distributed across business locations with factor-cost advantages.

4. The interactive character of innovation processes implies a great need for coordination of the various actors involved. The amount of information needed on product features, markets, potential cooperation partners, technological options, organizational forms and the like is constantly increasing, making decision-making more and more complex. The increasing specialization and differentiation of value-added processes leads at the same time to the creation of new interfaces between subsystems. New, knowledge-intensive forms of moderation are called for to ensure that this wealth of information is properly structured and communicated between the various actors involved, without this leading to exploding transaction costs. These 'interfacing services' include, for example, the assessment, evaluation and legal formulation of impending make-or-buy decisions, mergers or acquisitions, the coordination of logistic subsystems, establishment of quality standards along the value chain, moderation of communication processes in multicultural teams and among business partners, to mention but a few.
5. Locational specialization is shaped by historical developments and is to this extent path-dependent. Locations which are still in the early stages of the profile-building process have, initially, a multitude of specialization options – viz. all options which offer them comparative advantages based on cost factors. However, any initial specialization calls for specific investments, for instance, in relevant training programmes. Since innovations are of a cumulative nature and build on existing stocks of knowledge, constellations of actors, preferences and interactions, this initial specialization inevitably pre-shapes the further path of development. Possible economies of scale and external effects must be considered in the future allocation of scarce resources, which means that alternative patterns of specialization for which no initial investments were made are, comparatively, less profitable and are therefore abandoned.
6. High levels of private and public investment are required to create efficient, specialized business networks with high synergy potentials. Many inputs for knowledge-based clusters have, at least in part, the character of public goods, particularly in the fields of R&D, training and regional strategy formulation. Such fields are in need of public institutions and policy instruments if a socially optimal outcome is to be achieved. The more target-oriented and specific these inputs are, the greater the path dependence of the regional specialization pattern.
7. The systemic character of innovation and networking among firms and between firms and the public sector requires development of information and communication technologies. These technologies have

dramatically reduced the costs of storing, handling, moving and combining information and have made different patterns of national and international networking possible.[6] This has increased the possibilities to come up with new combinations of codified and tacit knowledge and interactive learning but at the same time puts pressure on producers to keep up with the most innovative competitors. While most earlier episodes of technical change centred on particular products or industrial sectors, information technology is generic. It impacts on every element of the economy, both goods and services, as well as on R&D, production, marketing and distribution.

Challenges for Latecomer Industrialization

In view of these characteristics a given industrial location today has little chance to catch up in global innovation competition if it fails to meet the requisite system conditions – that is highly qualified manpower, a differentiated company structure, a dense network of effective business-related institutions and an adequate set of economic and social rules. While cooperation and network synergies may emerge across spatial barriers, the existence of relevant networks in a given industrial location makes many things much easier. Spatial proximity may entail a number of localization economies;[7] it may for instance lower search costs, accelerate the flow of information that is not fully codifiable, and give rise to specialized local labour markets and complementary company structures.

The fact that innovations are becoming increasingly relevant to competition, and the dynamics of their emergence is linked to increasingly complex system conditions work in favour of concentration of innovative activities in locations that are already especially well endowed. This would lead us to expect a spatially polarized development that would have particularly negative effects on technological latecomers like Asia's developing countries and emerging economies, since the latter are able to meet at best only a small part of the necessary locational prerequisites.

However, the polarizing tendency of agglomeration advantages are subject to two qualifications: factor-cost (mainly labour cost) differentials and the uneven distribution of natural resources.

The first factor relates to factor-cost differentials. These run counter to the polarizing tendency because in agglomerations certain production costs are driven up by strong demands – this goes in particular for wages, incidental wage costs and real estate prices. Traditional neoclassical economic theory assumes that such differentials tend to balance one another out because they create incentives to invest in locations with low factor prices. While the network dependency of innovation processes sets limits to these levelling factor movements, companies nevertheless have incentives to cash in on factor-cost differentials. This would lead us to assume that firms

- will locate knowledge-intensive activities at sites with highly specialized, differentiated system conditions;[8]
- will, however, outsource production processes that are largely unreliant on knowledge-based locational synergies. These will mainly be routine operations for which factor costs are an important competitive aspect, and knowledge-based externalities ('spillovers') are unimportant.

It is in this way that complementary locations may emerge on the periphery.

The second factor relates to the uneven distribution of natural resources. Geography matters, because natural resources – fertile soil, oil and mineral resources, naturally beautiful areas with tourist potential, strategic trade locations, and so on – are unevenly distributed. A favourable resource endowment is an important initial locational advantage. Some of Asia's emerging economies are blessed with fertile soil, abundant water, plentiful marine resources or richness in mineral resources, and this enables such countries to become exporters of primary products. Others, such as Singapore, benefit from their advantageous location. Such geographic factors bring about economic rents. However, what seems to be a blessing may also create economic problems – for example, the Dutch disease-effect that drives the value of the local currency up and diminishes export competitiveness, but also political problems related to rent-seeking behaviour.[9] In fact, research on comparative economic history reveals that economic performance does not correlate with natural resource endowment.[10] The leading nations of the Asian Miracle, South Korea and Taiwan, were highly successful despite being poorly endowed with natural resources. Hence, though geography matters, it would be a mistake to see it as destiny.

Based on this brief discussion of the relevance of clustering, factor costs and natural resources, we can identify analytically three ideal types of location:

- knowledge-based locations whose competitive advantages and specialization patterns rest on an accumulation of highly specific production factors and interactions and can therefore not be explained adequately by classical trade theory;[11]
- labour-cost-based locations whose competitive advantages rest mainly on price advantages from (unskilled) labour;
- natural-resource-based locations.

The competitive advantages of developing countries – in Asia as elsewhere – are based almost exclusively on either natural resources (agriculture, fishery, oil and mining, tourism) or labour-cost advantages. Given the general trend toward falling terms of trade for most products of these categories, the challenge is how to move toward more knowledge-based advantages. This is a task that must be accomplished despite the weakness of existing innovation systems: most Asian countries still rank low on international comparisons as far as their technological capabilities and the performance of their public institutions are concerned.[12]

CATCHING UP DESPITE WEAK INNOVATION SYSTEMS

Viewed from the perspective of development policy, the question is how locations whose competitive advantages have until now been based almost exclusively on wage-cost advantages and/or favourable natural-resource endowments can, under today's conditions, be further developed with a view to knowledge-based specialization. The following sections will discuss this issue for both product groups, citing a number of concrete success stories.

Upgrading of Labour Cost-based Advantages: Labour Cost-based vs Knowledge-based Locations

The integration of the countries of Southeast and East Asia in the world economy in particular rests largely on factor-cost-based competitive advantages. In many countries the latter have long since ceased to be restricted to the classic labour-intensive light industries (garments, toys, footwear); indeed they also account for many investments that have been made in the automotive and electronics industries. This is why the export statistics of emerging economies like Malaysia and Thailand now often display patterns of industrial specialization that are usually classified as 'human-capital-intensive'. But looked at more closely, they turn out to be mainly factor-cost-based specialization patterns in that only certain activities with a low innovation content (such as assembly and testing) are conducted in the countries concerned. The crucial factor is that these locations as a rule lack the necessary systemic conditions typical of knowledge-based clusters in industrialized countries.

We have argued that knowledge-based competitive advantages are suited to creating innovation rents and thus higher factor incomes. Due to the cumulative character of innovations, however, this situation is far more typical of very well endowed locations in OECD countries. But these locations may be forced to contend with shortages of and price increases for different factors of production (such as specialized manpower, unskilled workers, real estate). The pull of localization economies may be countered by potential localization diseconomies. What this means is that certain production functions that can, technically, be outsourced or relocated and are at the same time dependent on factor costs, will be relocated at other sites.

These new 'complementary locations', which start out competing on the basis of factor prices, are faced with a challenge, namely that their terms of trade deteriorate in the division of labour with knowledge-based locations. They therefore have no choice but to seek to upgrade their locational advantages. The question for these locations is whether and how they will, under today's conditions, be able to assert themselves technologically against the centripetal forces, the localization economies, of established well-endowed locations.

Building Knowledge-based Advantages

The following section will, in three basic lines of argumentation, demonstrate that it is possible to master the transition from simple, wage-cost-based specialization to innovation-based specialization.

1. Transnational corporations are more and more systematically subdividing their functions, reorganizing their internal corporate structures, concentrating on core competences and outsourcing marginal tasks and functions. One effect of trade liberalization and new transportation, information and communication technologies is to facilitate the spatial division of value-added processes. Relocation of company branches or outsourcing may mean chances for non-cluster locations with knowledge-intensive infrastructure.
2. In practice the distinction made above between factor-cost- and knowledge-based competitive advantages lacks discriminating power. To this extent there is no 'locational dualism' in which it would be possible to clearly delineate 'centre' and 'periphery'; what we have is instead more a complex spatial pattern with manifold agglomerations that combine, in different ways, cost- and knowledge-based factors and whose 'degree of knowledge intensity' could be plotted along different points on an axis. These different levels of knowledge intensity reflect different levels of 'maturity' of locations; that is to say, it is conceivable for locations to gain international profile on the basis of simple factor-price advantages with a view to attracting a critical mass of investments and then to gradually deepen the emerging specialization

patterns by taking on more demanding tasks, creating complementary system components and developing more complex interaction patterns. In other words, the complex system characteristics required to compete for innovations can be acquired in an incremental process.

3. One factor that facilitates the transition is that today the individual stages in global value-added processes are highly interlinked in functional terms and certain, in part demanding, standards are often enforced through the overall value chain – that is also for simpler processes for which factor costs are still a key locational criterion. This ensures that outsourced goods and services always involve certain best practices as regards technical and management aspects. It is now the learning capacity of the actors in new locations, and/or the endowment of local and national innovation systems, that decide on the extent to which this information is in fact harnessed to move up in the 'locational hierarchy'.

In what follows, these three lines of argumentation will be discussed in more depth and underpinned with examples from practice.

Outsourcing and relocation as opportunities

Globalization entails a continuous increase in the competitive pressure to which national markets are exposed. Many large corporations in industrialized countries have responded to this situation by engaging in two different kinds of restructuring.

First, they increasingly specialize in those areas in which they have marked competitive advantages. It has often proven reasonable to concentrate a corporation's resources on a limited number of areas which promise economies of specialization or scale and to outsource areas of production and services that do not constitute part of their core competences. In view of rapidly changing market conditions, these 'make-or-buy' decisions have to be made again and again for various areas of operation. Identifying core competences, deepening and rounding them off by developing complementary fields of activity (or by acquiring companies that fit the need), and outsourcing non-essential areas is therefore not a one-off step taken to adapt to a new techno-organizational paradigm, they are continuous processes.

The core competences of leading firms often consist in knowledge-intensive value-added stages up- or downstream of industrial production; these would include R&D, design, establishment of brand names and coordination of logistic tasks in the value chain. These intangibles are often predicated on the ability to acquire or generate and to innovatively combine up-to-the-minute knowledge in many areas (new technical possibilities, demand trends, markets, logistic and financing concepts and the like). This kind of ability is difficult to copy, and it may therefore pave the way to innovation rents.[13] On the other hand, innovative, specialized firms are often inclined to outsource standardized services that are more or less easy to come by in the market. These often include industrial mass production and many non-specific services, from gardening/landscaping services to knowledge-intensive logistic and IT services.

Second, production processes are structured and divided up spatially so as to be able to utilize specific locational advantages for every stage of the value-added process. This need not necessarily mean outsourcing, it can also take on the form of expansion of a company's branch network. For example, R&D tasks may be concentrated at locations with favourable human capital endowments and potential research synergies, while simple, labour-intensive processes will tend to be drawn to locations with low labour costs.

Both processes, outsourcing and relocation of a company's production facilities at locations with specific factor endowments, open up opportunities for developing countries and countries in transformation or other peripheral locations. Furthermore, another marked tendency favouring this trend may be seen in a decline in both the costs involved in overcoming space (such as shipping rates) and in non-spatial transaction costs (e.g. those bound up with information and communication technologies, trade liberalization and harmonization of international investment law). In other words, many services can be purchased across national borders and across major distances without incurring any substantial additional costs.

However, outsourcing and relocation presuppose that it is possible, in both organizational and spatial terms, to remove value-added stages from their current production context. In view of the above-outlined systemic-interactive, network-bound character of many production processes, this situation is by no means given everywhere. Certain characteristics must exist if this is to succeed.[14]

- processes must be readily (re)structurable;
- the relevant know-how should pose no major codifiability problems;
- a low level of interdependence with complementary goods and services;
- a low level of dependence on local institutional inputs;
- a low level of dependence on interaction with customers;
- optimally low costs for information transfer and movement of goods.

In the case of the garment industry, the wage-intensive sewing process poses no technical problems to separation from the upstream stages (yarn/thread and cloth production), it is largely unreliant on specialized services and it is easy to learn. The same applies for the assembly and testing of many electronic devices and the digitalization of standardized data or the use of call centres to process relatively homogeneous customer inquiries. However, more and more production-relevant knowledge is so geared to certain specific customers or market segments that the companies concerned are bound to the locations of these customers or of complementary specialized firms and institutions.

In other words, numerous industry specifics have to be taken into account when companies decide on the relocation of value-added stages. Whether or not a location in a developing country will be able to profit from these opportunities furthermore depends on local framework conditions. The crucial factors here include not only factor costs (that is labour and real estate) but also political stability, legal security for investments, an appropriate and reliable infrastructure (telecommunications, electricity, roads, airports) and well-trained specialists. Locations that fail to meet minimum general or industry-specific requirements will as a rule not qualify for international investments.

Gradual transitions between labour-cost- and knowledge-based competitive advantages

Approaches based on the paired opposites 'factor-cost-based' vs. 'knowledge-based' lack discriminating power. Factor-cost- and knowledge-based cost advantages must instead be seen as poles of a continuum. The way in which we assess the production factor 'labour' will, for instance, depend on the factor's productivity. The more knowledge is incorporated in the factor, the higher will be its potential productivity. True, knowledge-based competitive advantages may, temporarily, prove largely independent of factor prices – because, say, no competitor offers a comparable product and a supplier is thus in a position to earn innovation rents. But as soon as the first imitators appear on the scene the situation will be defined by price competition. In addition, in every stage of production it is essential to bear in mind a number of different production factors that entail different effects of scale, externalities and transaction costs, and this means that in selecting locations it is necessary to take into account a variety of different elasticities and to come to terms with goal conflicts.[15] Locational decisions are to this extent based on a number of effective contexts that are far more complex than the dichotomy presented in the above heading would seem to indicate. In reality this is a matter of 'overtones' nuances: in a given production operation, what is the share of *easily substituted factor costs* in relation to the *harder-to-substitute knowledge* share? Or, viewed in dynamic terms: whether an innovation leader succeeds in renewing or patenting his knowledge advantage before his competitors have a chance to catch up.

The fact that the transitions between factor-cost-based and knowledge-based competitive advantages are fluid makes it possible for locations to gradually improve their competitive position. In other words, locational upgrading is a gradual and incremental process that rarely involves clearly identifiable individual technological thresholds which, once crossed, would open the way to a fundamental breakthrough.

Every production process has capital requirements of its own, calls for particular manpower skills profiles and the availability of certain complementary services in situ. To this extent there is for every process a specific combination of optimal locational factors. Proceeding on the assumption of reduced investment barriers and declining costs for overcoming space (that is shipping costs), companies are freer to choose locations and subdivide their value-added processes in such a way as to ensure that, in the ideal case, they will be able to select for every value-added stage a particular location that meets their specific needs.

One of the reasons for locational differentiation is that *different* elements of the value chain are broken down spatially in terms of their specific factor-endowment requirements (Figure. 2.1); another is the different quality requirements encountered at *one and the same* value-added stage. Figure 2.2 shows how one single electronics corporation distributes its operations across four Asian countries, seeking to achieve the optimal combination of different locational factors needed for each area of production.

Embedded knowledge: every investment transports know-how

Today the individual stages in global value-added processes are largely functionally interlinked. Information flows and constant recombination of stocks of knowledge in the value chain are what give rise to innovations. This systemic-interactive character of innovation was discussed in an earlier section.

Both information exchange with up- and downstream processes and interactive learning are becoming increasingly important even for the more simple value-added stages for which locational choices are still largely dependent on wage costs. The reasons for this are:

1. *The need for closer coordination of logistics and other process standards*. The competitiveness of individual firms is coming more and more to depend on whether up- and downstream value-added stages are structured efficiently. These conditions force firms seeking to realize competitive advantages to focus more and more on developing unutilized efficiency potentials *in the supply chain*. This is visibly transforming competition between companies into competition between

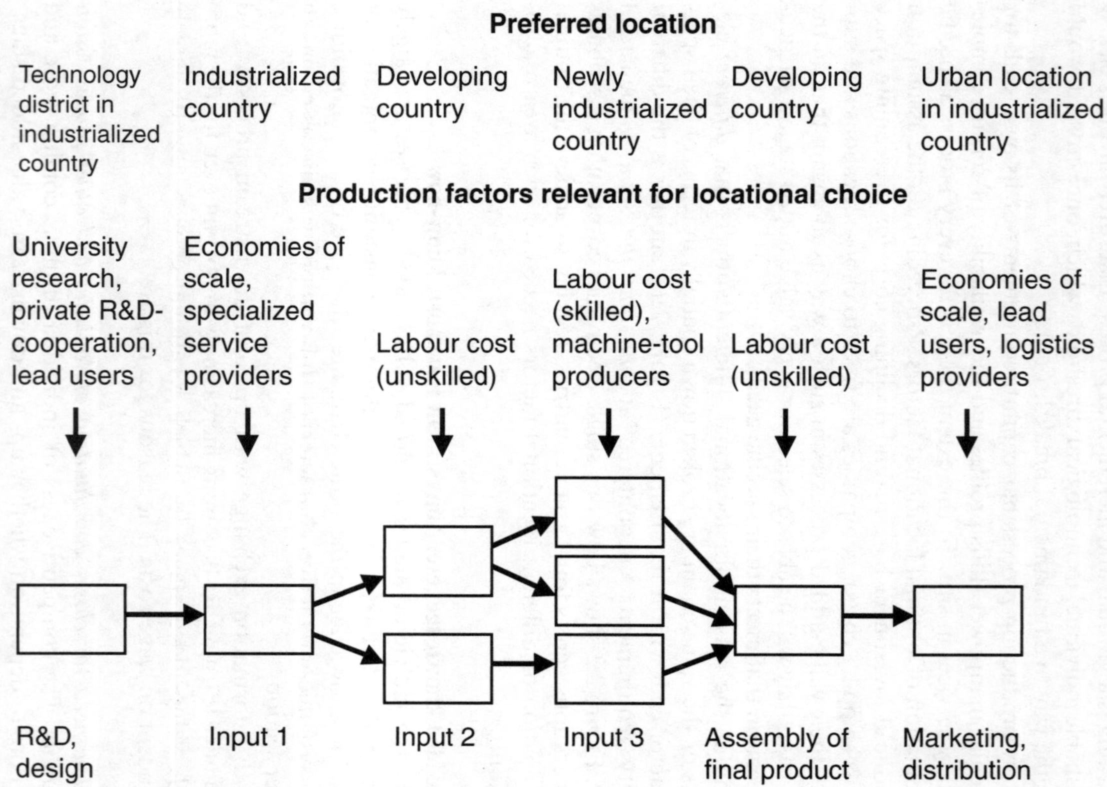

Figure 2.1 Locational choices along value chains: a stylized example

Location	Activity		Cost of direct labour	Labour cost engineer
Singapore	IC design Testing of highly sophisticated processors Failure and device analysis	↑	390	1290
Penang, Malaysia	Developing design packages for advanced logic and memory devices Assembly and testing of memory and logic devices	Increasing knowledge content	205	680
Bangkok, Thailand	Assembly of plastic parts Testing, marking and packing of logic and memory products		170	425
Suzhou, China	Assembly and testing of cost-sensitive high-volume devices Distribution		75	170

Source: http://www.amd.com; interview with Yuthana Hemungkorn, Managing Director, AMD Thailand, 31 March 2003; labour cost data: Panichapat (1999).

Figure 2.2 Regional distribution of production facilities of the American transnational Advanced Micro Devices (AMD), 2003, and labour costs in the hard disk drive industry in US $/month, 1999

value chains. Here success depends in large measure on the quality of supply chain management. The main concerns here are cost efficiency (in particular the simultaneous need to avoid warehousing costs and supply bottlenecks), speed and supply reliability. Even simple wage-cost-based activities must be integrated into the value chain in order to avoid supply bottlenecks and unnecessary inventories. This calls for electronic inventory control systems that are compatible with those used by chain partners. Likewise other standards, such as the ISO standards, are gradually gaining ground and they call for closer cooperation with firms positioned up- or downstream in the value-added process.

2. *Technological indivisibilities*. In some cases a value-added stage A may be highly wage-cost-intensive, while value-added stage B is knowledge-or capital-intensive and both may be so closely dovetailed in the production process that they cannot be separated spatially. In this case the investor is faced with a decision: if he chooses a location featuring large numbers of skilled workers and specialized firms, he will as a rule be forced to pay relatively high wage costs even for simple, standardized labour-intensive processes; but if he chooses a typical low-wage location, he will have to invest there in the development of specialized manpower and suppliers. In this case the investor will have to supply substantial know-how, even for allocation decisions for which factor-cost advantages are the deciding factor. In the example in Figure 2.3 the investor, who is actually interested only in low wage costs for semi-skilled workers, will have to ensure that the subsidiary trains skilled workers and promotes machine-tool suppliers for input 3, since the production of input 3 cannot be separated spatially from up- and downstream wage-cost-intensive processes. The electronics industry in Penang (Box 2.1) offers a good example of a case of this kind.

If the direction of technological development indicates that innovation is becoming a cross-cutting function inherent (though in different intensity) in every stage of production, the line separating knowledge- and factor-cost-based locations will tend to blur. If at the same time there is a need for a more comprehensive, rapid and reliable flow of information between individual value-added stages, this will mean that certain value-added stages will have to be brought back closer together in spatial terms. This explains the concentration encountered in many industrial and service clusters.

For locations that have in the past found a specialized place in international value chains on the basis of their low factor costs, this may open

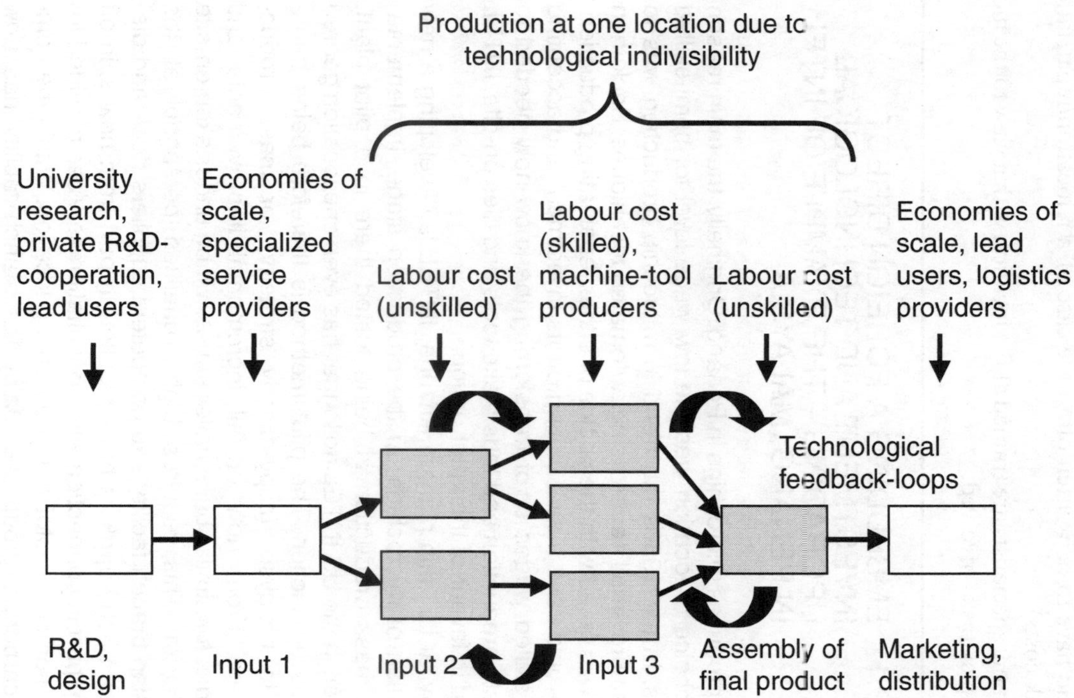

Fgure 2.3 Indivisible stages of the value chain

up chances for upgrading, though it may also spell the end of the location's viability. A location's chances for upgrading are better

- the higher the investments already made – since these investments would have to be written off in the case of a relocation/repatriation (sunk costs);
- the more specialists are trained at the location with a view to the more complex tasks involved;

BOX 2.1 EMBEDDING A FOREIGN DIRECT INVESTMENT AND TECHNOLOGICAL UPGRADING – THE EXAMPLE OF INTEL IN PENANG, MALAYSIA

INTEL produces microchips in Penang. Originally, the main reason for its choice of location was the low wage level for semi-skilled workers. In the 1980s the trend in microchip production was to manufacture small series using automated techniques. Back then INTEL was faced with the choice of either repatriating production, which would have meant writing off its investment and accepting higher skilled wages or of transferring the know-how needed to automate production and thus make tools and dies on-site. INTEL decided in favour of the latter option.

Today, at its main plant in Arizona, INTEL is developing a new generation of microchips up the production stage. Implementation of mass production is being tested there in a pilot plant. However, in the US INTEL no longer has even one location geared for mass production. The pilot method is therefore being transferred to Malaysia (copy-exactly strategy) for mass production there. From now on all incremental improvements and deviations for similar but smaller series will be undertaken on-site in Malaysia. This requires highly qualified personnel at the Malaysian branch facility and competent suppliers (tool- and die-makers). Such suppliers have resulted in part from new spin-off firms, in part in connection with consultation services provided for a large number of SMEs at the Penang location. Within the combined corporate group the Malaysian branch facility has now become an important technology provider which provides consulting services in developing new locations. Indeed, some local machine tool-makers have now themselves become transnational corporations.

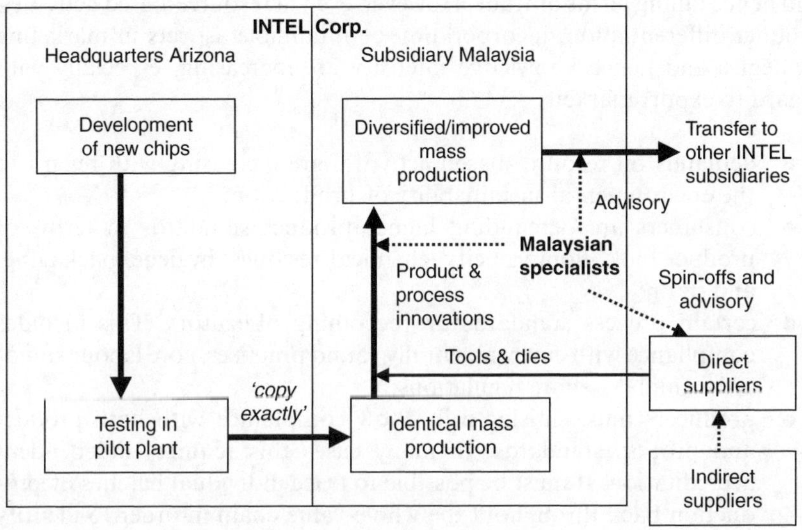

Figure 2.4 Technological upgrading at the location of assembly plants: the example of INTEL Corp., Malaysia (microchips)

- the more complementary suppliers and institutions that have been established at the location.

In this sense every local investment in one of the areas mentioned serves to increase the degree of embeddedness[16] of the outsourced value-added stages.

The example of the INTEL corporation in Malaysia shows how developing countries can make use of these trends toward outsourcing, relocation of production facilities in low-wage countries, and integration of value-added processes to build their own core competences, 'embed' foreign investments and stabilize a location in spite of drastic wage increases (Box 2.1).

Upgrading of Resource-based Advantages: New Challenges in Resource-based Activities

Besides labour-cost advantages, natural-resource-based activities are a second important pillar of Asia's economies. These activities are usually regarded as being less knowledge-intensive and as having less potential for creating innovation rents. In fact, barriers to entry for the production of most agricultural commodities are low, frequently leading to oversupply and hence falling terms of trade. However, even in resource-based activities, product differentiation, incorporation of intangible aspects in marketing strategies and hence knowledge intensity are increasing, especially with regard to export markets:

- demands on resource-based activities are increasing with regard to the environmental sustainability of production;
- consumers are demanding higher product standards in terms of product look, homogeneity, chemical residues, hygiene, packaging, and so on;
- certain process standards are becoming obligatory. This includes compliance with environmentally sound practices, core labour standards, animal welfare regulations;
- producers must satisfactorily show compliance with both product and process standards. In many cases this requires independent accreditation. It must be possible to trace individual batches of production back throughout the whole value chain in order to identify sources of irregularities. This may require complex systems to document any product movement, use of identification tags, often including bar-coding technology;
- this in turn requires that all actors involved in a value chain agree upon certain procedures including information, communication and accreditation standards. In some cases the data to be handled may be quite complex, requiring specialized electronic data interchange software;
- processing, sophisticated packaging and other value-adding activities are becoming more important;
- as specific consumer preferences gain importance, product differentiation increases, intangible characteristics (like 'organically grown', 'farm grown' or 'traded fairly') become distinguishing features that add value to the product.

In sum, markets become more demanding. Barriers to entry for simple commodity producers tend to rise and new opportunities for product differentiation emerge. Tapping into these markets requires increasingly complex institutional back-up, including market information, laboratories, improved logistic concepts, certification bodies, and so on.

In what follows I will present a more detailed analysis of shrimp farming in Thailand in order to exemplify the complex challenges and the multitude of knowledge-creating and -diffusing actors within a resource-based activity.

Building an innovation system for a resource-based activity: the example of shrimp farming in Thailand[17]

Thailand is the world's largest producer and leading exporter of farm-raised shrimp, supplying 35 per cent of the world market for Black Tiger shrimps.[18] In 2001 exports of farmed shrimp earned Thailand US$2.2 billion.

This economic success is being achieved at high environmental costs. For some years, shrimp farming was very profitable as environmental costs could easily be externalized. Today, however, due to high stocking rates and discharge of wastewater into irrigation canals, diseases have spread throughout the whole country. Virus infections have become a serious threat to shrimp cultivation, leading to frequent crop failures and substantial economic losses. Short-term solutions have focused on the heavy usage of a wide range of industrial chemicals, mainly antibiotics, to control diseases, aquatic vegetation and to disinfect water. These chemicals not only have a negative impact on the quality of soil and water but also threaten exports as importing countries impose increasingly rigid conditions. In particular the European Union has recently introduced a zero-tolerance policy for antibiotic residues and now rejects contaminated shrimp.

In addition, in the past many shrimp ponds were established in mangrove areas and other wetland ecosystems. The destruction of these ecosystems has had major negative impacts.[19] For instance, due to overexploitation and deforestation of mangroves, wild-caught Black Tiger broodstock[20] has become extremely rare. Revenues from shrimp farming decrease as broodstock caught from the sea becomes scarcer and smaller, a factor which reduces its economic value.

All these developments threaten Thailand's competitiveness in the world shrimp market. Environmental problems have thus grown to such an extent that they challenge the viability of the sector itself. In a worst-case scenario, shrimp farming might even collapse, as it did in Taiwan in 1988, as a result of overly intensive shrimp production at high environmental costs.[21]

In addition, shrimp farming has some negative spillovers to other economic activities. Deforestation of mangroves has destroyed nursery areas for the larvae of many species of fish and other marine resources, making fishery less profitable, and impoverishing entire coastal communities which depend on fishing.[22] Drainage of wastewater from shrimp ponds, containing chemicals and feed residues, as well as the practice of transporting saline water in enormous amounts to inland shrimp farms seriously affects the profitable use of agricultural land in adjacent areas.

To cope with the above-discussed environmental and economic challenges, knowledge-intensive innovations are called for at different levels. To develop eco-efficient farming systems, it is necessary to take the following steps:

1. Intensify existing research cooperation between industry and science. For many of the above-mentioned problems experienced by Thailand's shrimp aquaculture, technological solutions have to be developed and these in some cases require a considerable research effort. Among the most pressing research issues are domestication of broodstock;[23] detection and treatment of diseases; genetic improvement; and secure and efficient pond management techniques, including the development of low-salinity culture techniques. Shrimp aquaculture is one of the most prominent examples in Thailand of how research and technology institutes can pool resources and cooperate with the private sector to create knowledge-based competitive advantages. Research is being conducted to address the main challenges facing the sector, and linkages between science and industry are more intense than in any other Thai industry. Several universities[24] and the Department of Fisheries have placed emphasis on shrimp research and the national research organization BIOTEC funds a special Shrimp Biotechnology Programme, reflecting a national research priority in this sector. Furthermore, several private-sector companies are highly committed to shrimp-related R&D. In some cases, concerted efforts are being undertaken to advance research. The Centre of Excellence for Shrimp Molecular Biology and Biotechnology, for instance, is a joint research centre of Mahidol University and BIOTEC, and has close research contacts with the private sector. In addition, the Shrimp Culture Research & Development Co. Ltd has been set up as a public–private cooperation project aimed at the domestication of broodstock. Considerable public research is directed toward the achievement of eco-efficiency.
2. Enhance the regulatory capability of national institutions and enforce land-use and related environmental regulations. The enforcement of land-use and related environmental regulations is often problematic, because the same government agencies are responsible for in part contradictory tasks, that is enforcement of environmental policy on the one hand and industrialization and export promotion on the other hand.[25] Adding to this complexity is the fact that shrimp culture is affected by a large number of laws and regulations, including land laws, water laws, environmental laws, fishing laws and so on. Therefore it will be necessary to improve the coordination of departments in charge of different issues related to shrimp production needs. Several regulations have been successfully implemented, such as expansion of

inland shrimp-farming was banned in 1998, because it increases salinity levels in soil and groundwater, and this has led to a considerable reduction of inland shrimp-farming. Other regulations, for example those concerning the drainage of saltwater into public freshwater systems or farming areas and further deforestation of mangroves, have not been fully enforced. It appears that visible aspects such as sedimentation ponds and the ban on inland shrimp-farming have by and large been enforced, whereas regulations concerning the reduction of antibiotics and regulation of effluent discharges are more difficult to put into effect.

3. Improve pond and wastewater management systems. Pond and wastewater management refers to the removal of uneaten food and other waste products from a pond by exchanging its water. Water management is a very critical process, since some farmers do not change the water of their ponds as frequently as they should, because they are afraid that diseases might spread to their ponds with new water, which may be contaminated by neighbouring farms. Improved water management includes treatment of wastewater in sedimentation ponds before it is released into the drainage system. Furthermore, construction of sufficient pumps and aerators, careful dosage of feed and adequate stocking rates of shrimp larvae are crucial to maintaining good water quality. All this shows that pond management is highly knowledge-intensive and requires cooperation among farmers.
4. Reduce the use of chemicals. Although awareness has increased considerably, the use of antibiotics and other chemicals is still widespread among shrimp farmers. The search for less intensive but still economically viable farming methods is being hampered by feed and chemical firms, which, while they are the main source of information for farmers, are interested in maximizing inputs such as feed and chemicals. Given the prominent role of feed and chemical corporations as advisers to shrimp farmers, the public sector should become more active in the promotion of less intensive farming methods, e.g. by supporting research on viability, by providing testing services or by exploring market opportunities for organically grown shrimp.
5. Promote good cultivation practices. Taking into account that many thousand farms throughout the country are engaged in shrimp farming, attempts to enforce command-and-control mechanisms for environmental protection would far exceed the capacities of Thailand's public administration. It is therefore necessary to complement government regulation by means of industry self-regulation based on standards and codes of conduct. Such standards not only contribute to internalizing environmental costs of shrimp farming, they are also an

important means of regaining consumer confidence, differentiating the market and thus increasing the competitiveness of the shrimp sector. If seafood companies are to gain access to international markets and also to maintain their competitiveness, they are going to have to comply with international quality standards in the global market. This requires the establishment of certification and auditing systems for different standards, ranging from basic hygiene standards such as HACCP to labels for organically raised shrimp. In addition to promoting such internationally recognized standards, in 1999 Thailand's Department of Fisheries initiated a national Code of Conduct for shrimp farms that provides guidelines for the development of voluntary management systems to reduce negative social and environmental impacts. Thus far, however, self-regulation by the shrimp industry has not been very effective. Codes of Conduct and quality labels will only gain general acceptance among farmers if such initiatives on the marketing side prove to be successful. If they are to gain consumer confidence, moreover, it will be necessary to establish certification bodies and a transparent tracking and tracing system along the whole value chain to enable authorities to trace a batch of shrimp back to the producer.

6. Increase the value-added of shrimp exports. As an additional measure aimed at enhancing its competitiveness in the world market, Thailand has opportunities to increase the value-added of its shrimp exports by focusing on its advantages in the processing business, that is its knowledge of export markets and the specific processing skills and techniques that are required especially by the Japanese market.[26] Product diversification includes high-grade processed foods, that is microwave (ready-to-eat), 'eco-shrimp' and sushi, as well as sophisticated packaging for demanding customers. The government initiative to promote Thailand as a 'kitchen of the world' may help to tap into international high-end markets.

Altogether, shrimp farming is faced with some quite complex challenges. Substantial technological and social innovations are needed that call for systemic solutions involving both private and public efforts.

Large companies which supply feed and chemical inputs are the main providers of farming technology, including know-how concerning stocking rates, providing the optimal amount of feed to achieve fast growth without overfeeding and putting the pond ecosystem at risk, the use of aerators and pumps, the use of chemicals, and so on. The CP Group, a Thai-based agro-business transnational, alone employs 400 consultants to provide advisory services. CP's operations range from feedmills, hatcheries and demonstration farms, laboratory testing and diagnostic services for shrimp farmers to shrimp-processing plants. Several extension centres equipped with laboratories and staffed with experienced biologists provide water-quality testing, disease diagnosis and seminars. Furthermore, CP is engaged in research activities on the domestication of broodstock, improvement of feed and genetics.[27] Private-sector engagement will be crucial for further improving pond and wastewater management systems and establishing good cultivation practices.

On the public side, a large number of institutions have been created to promote and regulate shrimp farming, minimize ecological damage, conduct research, provide extension services, set and implement standards, market products internationally, and so on. Certain environmental and land use regulations need to be better enforced. Furthermore, taking into account that the leading private corporations are biased toward intensive farming systems which are not ecologically sustainable, there is a need for public institutions to promote less intensive eco-efficient farming methods and to support the adoption of a Code of Conduct. Finally, several industry associations represent shrimp farmers' interests and provide technological information. Together with government institutions, these industry associations should help to develop a shared vision of the industry's future trajectory and to improve the coordination of a sector-specific innovation system.

CATCHING UP IN TECHNOLOGY AS A TASK OF ECONOMIC GOVERNANCE

To sum up, it can be said that in the framework of the international division of labour, technological catching-up processes are still possible for firms and locations which belatedly seek integration and have nothing else to offer in the initial phase but comparative cost advantages in unskilled labour or natural resources. This positive message, however, has to be put into perspective: technological upgrading is not easy to achieve, and it is evident that only a few countries and regions manage to progress significantly. The successful experiences presented in this chapter benefited from some especially favourable conditions.

In the case of Penang's electronics industry, for instance, the cluster has been one of the early movers in the microprocessor industry, and at the time when automation of chip manufacturing advanced significantly, companies had already made substantial 'sunk investments' in training personnel and a sizeable cadre of engineers and technicians had been built up. Consequently, the emerging cluster was less footloose than it may have been in other locations which entered the industry later on. Multinational corporations therefore preferred to invest in the technological advancement of Penang's industry rather than relocating to lower cost countries and writing off their investments. In the case of Thailand's shrimp industry, the country benefited from very favourable agro-ecological conditions and the existence of a large national agribusiness conglomerate with substantial capital, technological expertise and market experience.

Nevertheless, it would be a mistake to see such specific conditions as destiny – competitive advantages can deliberately be created if local agents take strategic action. Both the theoretical arguments and the empirical observations demonstrate that catching up is possible if firms or locations pursue an explicit strategy with clearly defined goals and at the same time make flanking investments in national innovation systems. One of the central challenges involved is to integrate both the central private-sector change agents and a broad range of public-sector institutions (not only research and technology organizations but also regulatory bodies, extension services, and so on) in a common strategy. There are four key fields of intervention here.

Agreement on Technological Goals

The cumulative and path-dependent character of innovation processes calls for creation of corridors of specialization. Knowledge-based competitive advantages are reliant on special inputs from a national innovation system, that is research, training and infrastructure facilities adapted to a location's pattern of specialization. No location can afford to provide a very great range of special inputs. To this extent it is important to invest scarce resources in a promising specialization pattern.

In other words, it is essential for latecomers to be able to recognize what technological developments are to be anticipated in different industries, what short-term mix of available factors they will need to seek integration in the existing division of labour, what learning paths are accessible to them in view of this initial specialization, and how high the market-entry barriers are that hinder technological upgrading. In some industries economies of scale or technological complexity could well mean that market-entry barriers are today too high for Asia's emerging economies.[28] This is why – with the exception of a few new branches of industry, such as software – it is essential to start out chiefly with incremental improvements in wage-cost- and resource-intensive industries.

In a complex market economy in which competitiveness is the outcome of systemic interaction processes, it is extremely difficult to anticipate the technological perspectives of alternative paths of specialization. In particular, it would be unrealistic to seek to achieve this using a top-down planning approach. Today – even in countries with a highly centralist and hierarchic tradition, such as South Korea[29] – the trend is more in the direction of participatory, publicly moderated search processes aimed at more closely defining the rough direction of technological specialization. The instruments of technology foresight (flanking research on technical and demand trends, systematic benchmarking against other locations, and so on) can be used to ground and systematize this search process.[30] As soon as the matter at hand is fine-tuning, the focus will turn more and more to linking promotion with competition, self-initiative and self-risk on the private-sector side, that is in the form of co-financed institutions or by awarding support funds in connection with competition procedures. Only in this way is it possible to ensure that technological initiatives will seek their bearings in the market.

Especially in federal systems, this participatory and subsidiary approach presupposes complex coordination processes involving local and regional authorities. Some relevant decisions will have to be made at the national level, some at the supranational level. This goes for the development of efficient institutions of market regulation and commercial law, but also for national dialogue on the rough direction of social transformation projects. At the same time, however, regional-level governance competence assumes an important function here, because it is at this level that production networks with a high communication density and a variety of non-market, trust-based forms of cooperation emerge.

Further Development of Special Locational Advantages

The more a location develops in the direction of knowledge-based competitive advantages, the more high-grade and special services it will have to offer in fields like research, training and infrastructure. For peripheral regions the most important element of specialization is in most cases likely to be the training of specialists. Many of the promising clusters in developing countries initially sought integration on the basis of wage-cost advantages for unskilled or semi-skilled workers. In connection with automation, increasingly demanding company organization models and special customer wishes, however, nearly all industries develop a need for qualified workers. This in turn implies a need for more on-the-job training and the establishment of training centres. Locations such as Bangalore/India (software), Penang/Malaysia (electronics) and Tijuana/Mexico (automotive parts, TV sets) today distinguish themselves mainly by their supply of *specialist manpower*.[31] An important role is likewise played by physical infrastructure endowments – that is modern industrial parks, ports and airports; for electronic processes, telecommunications is also important.

The challenge for public or parastatal institutions of the national innovation system, including universities, training centres, research and test laboratories, is often not solely, or indeed not even mainly, the quality of their physical and manpower resources but the need to improve their forms of internal organization, the value systems of their employees and their incentive systems. Many institutions are not equipped to identify private demand, to cooperate with the private business sector and to generate marketable innovations.[32]

Strategic Acquisition of Foreign Direct Investments

Precisely on the periphery, internationally efficient clusters rarely emerge exclusively on the basis of local firms. Internationally active corporations that are thoroughly familiar with industrialized-country markets and hold relevant market shares, that are leaders in R&D, that have mastered logistic concepts in the value chain as well as other standards can seldom be replaced as agents of innovative knowledge and as entry points to markets.[33] Targeted efforts to attract these corporations are for this reason highly important. This goes in particular for those corporations that Best[34] refers to as 'developmental enterprises', noting that their corporate philosophy is keyed to promoting innovations and strategic capabilities in both their own operations and in their environment, and in this way to generating new niche markets and major local spillovers.

The marketing of locations should not emphasize only traditional factor-cost advantages but at the same time underline potential industry-specific economies of scale and externalities, low transaction costs and other 'soft' locational factors. It is in this way possible to focus mainly on those investors who are capable of enhancing a location's profile. Furthermore, location marketing and business promotion should be closely coordinated with a view to ensuring that investment projects are optimally 'embedded' in their regional context.[35]

Promotion of the Transfer of Knowledge and Technology

As noted above, competition is increasingly a matter of giving a structure to a corporation's upstream and downstream relationships that is more efficient than the value chain of competitors. Corporations are therefore interested in seeing the development in their region of an efficient structure of businesses and institutions that are in a position to provide them with complementary services. It is for the most part too expensive and time-

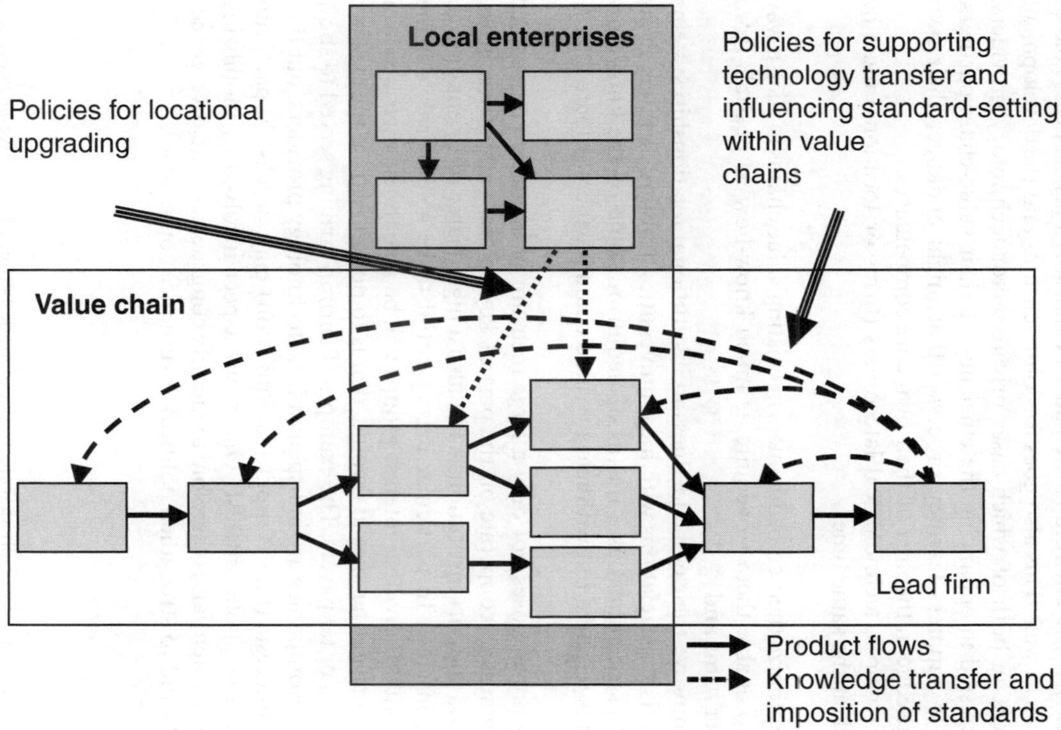

Source: O'Connor (1993), UNCTAD (2001).

Figure 2.5 Enhancing complementary technological learning through locational and value chain policies

consuming for them to import specialists, subcontracted products and services.

However, it is likewise too costly for a corporation to develop or initiate on its own all of the complementary structures required, especially in view of the fact that the environment created in this way may be accessible to competitors as well. This also goes for joint ventures and the development of suppliers, in both of which cases private-sector technology providers, while interested in enhancing the efficiency of their value-chain partners, will seek to suppress learning processes that might endanger their own knowledge edge in the area of their own core competences.

In other words, technological latecomers (firms or locations) must be concerned, at the same time:

- to optimize their embeddedness in existing value chains in such a way as to enable them to acquire as much know-how as possible from other firms and
- to improve their embeddedness in institutional relationship systems (that is their relations with local universities, training facilities, firms and associations) as a means of acquiring some of the know-how that is required but withheld by value-chain partners (Figure 2.5).[36]

All this presupposes not only a large measure of strategic competence and willingness to cooperate on the part of key actors – ministries, firms, research and support institutions – but also a high degree of orientation to the public interest. In particular it is essential to be able to rule out the possibility that individual interest groups will be able to use their political connections to garner special favours, that is to pocket rents at the expense of consumers or taxpayers. The example of shrimp farming served to show that private companies are indispensable as technology providers, but that their own interests (that is input maximization) may, at least in part, run counter to the public interest. One main aspect involved in mobilizing endogenous potentials is therefore to identify congruencies and divergences in interests and to strengthen political governance competence.

NOTES

1. It can be shown empirically that the terms of trade between knowledge-*in*tensive and knowledge-*ex*tensive goods is shifting to the disadvantage of the latter. This can be observed not only in the trade of natural resources for industrial goods but also in the decline of prices for industrial goods exports from developing countries (which are as a rule less complex in technological terms) in relation to industrial goods exported by highly developed countries (Kaplinsky, 2000).

2. Based on Lundvall (1992); Nelson (1993); OECD (1999).
3. Morgan (1998), pp. 79ff.
4. Schumpeter (1934).
5. Porter (1990), pp. 109ff.
6. Kline/Rosenberg (1986), p. 280.
7. Lösch (1954).
8. Audretsch/Feldman (1996) show that the knowledge intensity of branches of industry correlates positively with their propensity to form local clusters.
9. See Kruger (1974).
10. E.g. Landes (1998).
11. In particular, the Heckscher-Ohlin trade theorem.
12. See the Global Competitiveness Rankings of the World Economic Forum: www.weforum.org.
13. Kaplinsky and Morris (2001), pp. 25ff.
14. Gassmann (1997), p. 148.
15. Storper (2000), pp. 252ff.
16. See Grabher (1993).
17. For details, see Altenburg et al. (2004), pp. 64ff.
18. Patmasiriwat et al. (1998), p. 30.
19. Lebel et al. (2002); Flaherty et al. (1999), pp. 2050ff.
20. Gravid female brooders are called broodstock. At present it is not commercially viable to reproduce Black Tiger shrimp in captivity.
21. Kagawa (2003), p. 23.
22. http://www.american.edu/TED/THAISHMP.HTM.
23. Not only because sea-caught broodstock is becoming increasingly rare, but also because only domesticated broodstock can guarantee disease-free parent generations.
24. E.g. Chulalongkom University; Mahidol University, Kasetsart University, Prince of Songkhla University.
25. Flaherty et al. (1999), p. 2053, refer to Kaosa-ard et al. (1995).
26. Kagawa (2003), p. 83.
27. Patmasiriwat et al. (1998), p. 11.
28. In the auto industry, for instance, no country has managed – despite costly industrial policy projects – to build a sustainably competitive national industry since South Korea's success.
29. Hanna et al. (1996).
30. See Johnston (2003).
31. For Bangalore see Lateef (1997); for Penang see Box 2.1; for Tijuana: Carrillo and Mortimore (1998).
32. See OECD (2002).
33. See Knorringa/Schmitz (2000) for the important role of global buyers. Even Asia's emerging economies are still highly dependent on foreign transnational customers. Shrimp farming in Thailand is one exception where the lead firm is local.
34. Best (1999), pp. 4f.
35. Lall (1995) refers to this as 'target and guide policy', mentioning Singapore as an example. Thailand has a programme designed to promote cooperation between suppliers and local SMEs; the programme is run by the Board of Investment.
36. This combined use of externalities in vertical value chains and externalities in given locations can be observed in all successful cases of latecomer development – e.g. South Korea, Ireland, Singapore, Taiwan. See e.g. Battat et al. (1996).

REFERENCES

Altenburg, T. et al. (2004), 'Strengthening knowledge-based competitive advantages in Thailand', Bonn: German Development Institute, Working Paper 1/2004.

Audretsch, D. and M.P. Feldman (1996), 'R&D spillovers and the geography of innovation and production', *American Economic Review*, **86** (3), 641–52.

Battat, J., I. Frank and X. Shen (1996), 'Suppliers to multinationals. Linkage programs to strengthen local companies in developing countries', Foreign Investment Advisory Service, Occasional Paper 6, Washington, DC.

Best, M.H. (1999), 'Cluster dynamics in theory and practice: Singapore/Johor and Penang Electronics', Cambridge: The Judge Institute of Management Studies, *Research Papers in Management Studies*, WP 9/1999.

Carrillo, J. and M. Mortimore (1998), 'Competitividad en la indústria de televisores en México', *Revista Latino-americana de Estudos do Trabalho*, **4** (6), 79–100.

Flaherty, M., P.P. Vandergeest and P. Miller (1999), 'Rice paddy or shrimp pond: tough decisions in rural Thailand', *World Development*, **27** (11), 2045–60.

Gassmann, O. (1997), 'F&E-Projektmanagement und Prozesse länderübergreifender Produktentwicklung', in A. Gerybadze, F. Meyer-Krahmer and G. Reger (eds), *Globales Management von Forschung und Innovation*, Stuttgart: Schäffer Poeschel, pp. 133–73.

Grabher, G. (ed.) (1993), *The Embedded Firm*, London: Routledge.

Hanna, N. et al. (1996), 'The East Asian miracle and information technology. Strategic management of technological learning', World Bank Discussion Papers No. 326, Washington, DC.

Johnston, R. (2003), 'Tools for managing the future: a portfolio approach', APEC Centre for Technology Foresight, Bangkok, 17–21 February.

Kagawa, M. (2003), 'Trade linkages in shrimp exports: Japan, Vietnam, and Thailand', unpublished manuscript, Auburn, Alabama.

Kaosa-ard, M. et al. (1995), *Natural Resources Management in Mainland Southeast Asia*, Bangkok: Thailand Development Research Institute.

Kaplinsky, R. (2000), 'Globalisation and unequalisation: what can be learned from value chain analysis?', *Journal of Development Studies*, **37** (2), 117–46.

Kaplinsky, R. and M. Morris (2001), *A Handbook of Value Chain Research*, University of Brighton.

Kline, S.J. and N. Rosenberg (1986), 'An overview of innovation', in R. Landau and N. Rosenberg (eds), *The Positive Sum Strategy: Harnessing Technology for Economic Growth*, Washington, DC: National Academy Press, pp. 275–306.

Knorringa, P. and H. Schmitz (2000), 'Learning from global buyers', *Journal of Development Studies*, **37** (2), 177–205.

Krueger, A. (1974), 'The political economy of the rent-seeking society', *American Economic Review*, **64**, 291–303.

Lall, S. (1995), 'Industrial strategy and policies on foreign direct investment in East Asia', *Transnational Corporations*, **4** (3), 1–26.

Landes, D. (1998), *The Wealth and Poverty of Nations*, New York: Norton.

Lateef, A. (1997), 'Linking up with the global economy: a case study of the Bangalore software industry', Geneva: International Institute for Labour Studies (IILS), Discussion Papers.

Lebel, L. et al. (2002), 'Industrial transformation and shrimp aquaculture in Thailand and Vietnam: pathways to ecological, social, and economic sustainability?', Stockholm: The Royal Swedish Academy of Sciences, *Ambio*, **31** (4).

Lösch, A. (1954), *The Economics of Location*, New Haven: Yale University Press.
Lundvall, B.-A. (ed.) (1992), *National Systems of Innovation. Towards a Theory of Innovation and Interactive Learning*, London: Pinter.
Morgan, G. (1998), *Images of Organization: The Executive Edition*, San Francisco: Berrett-Koehler Publishers.
Nelson, R.R. (1993), *National Systems of Innovation. A Comparative Study*, Oxford: Oxford University Press.
O´Connor, D. (1993), 'Electronics and industrialisation. Approaching the 21st Century', in K.S. Jomo (ed.), *Industrialising Malaysia. Policy, Performance, Prospects*, London/New York: Routledge, pp. 210–33.
OECD (1999), 'Managing national innovation systems', Paris: OECD.
OECD (2002), 'Benchmarking industry–science relationships', Paris: OECD.
Panichapat, Ch. (1999), 'Hard disk drive industry in Thailand', http://www.boi.go.th/ english/focus/ hddAug.html.
Patmasiriwat, D., O. Kuik and S. Pednekar (1998), 'The shrimp aquaculture sector in Thailand: a review of economic, environmental and trade issues', CREED Working Paper 19.
Porter, M.E. (1990), *The Competitive Advantage of Nations*, New York: The Free Press.
Schumpeter, J.A. (1934), *The Theory of Economic Development: An Inquiry into Profits, Capital, Credit, Interest and the Business Cycle*, Cambridge, MA and London: Harvard University Press.
Storper, M. (2000), 'Globalization, localization, and trade', in G.L. Clark, M.S. Gertler and M.P. Feldman (eds), *The Oxford Handbook of Economic Geography*, Oxford: Oxford University Press, pp. 146–65.
UNCTAD (2001), 'Promoting linkages', World Investment Report 2001, New York and Geneva.

3. Transnational communities, offshore outsourcing and offshore subsidiaries: the case of the Indian IT service industry

Jan Vang and Mikkel Lucas Overby

INTRODUCTION

Globalization and increased emigration have recently spurred a surge of interest among researchers and policy-makers in the role transnational communities play as agents of development in their home countries. This research finds that transnational community networks are becoming increasingly important in fostering economic development. Despite the fact that multinational corporations increasingly rely on outsourcing and offshoring to realize cost and innovation advantages through globally distributed resources and competences, traditional theorizing on the issue mostly neglects the insights from this literature. In this chapter we use the term 'offshore outsourcing' to denote situations in which a firm contracts out or sells its assets to a third party supplier located in a different country for an agreed time period, whereas 'offshoring' refers to situations where a firm establishes a subsidiary in a foreign country. Transaction cost economics, which has emerged as the dominating theory for explaining whether firms choose to outsource or organize activities in-house, is silent on the offshore aspect. Transaction cost theory focuses on how economic hold-up threats limit the firm's propensity to outsource but disregards the different challenges firms face when outsourcing to or establishing subsidiaries in institutionally distant contexts.

Development researchers, in contrast, deal with the offshore aspect through their focus on migration. Development researchers have traditionally treated transnational communities – globally dispersed migrant networks[1] – as synonymous with emigration of skilled labour as expressed in the 'brain drain hypothesis'. According to the brain drain hypothesis, emigration is costly for the country that has invested in the emigrants' education or carrying opportunity costs as the country loses particular skilled employees and thus the opportunities for reaping the associated economic benefits. Hence, development studies have concentrated on the negative features of the emigration trend but have been relatively silent on the potential positive feedback aspects.

The aim of this chapter is to use insights from the recent literature focusing on transnational communities to contribute to the theory on offshore outsourcing and offshoring as well as elaborating on shortcomings in this literature. We investigate the role that members of transnational communities play as agents of development in their home countries by reducing problems of institutional distance, which multinational corporations in developed countries face when outsourcing or establishing a subsidiary in a developing country. The chapter contributes to the ongoing literature on outsourcing and offshoring on three main accounts. First, it develops a critical alternative or supplement to transaction cost economics by pointing to the importance of transnational communities in reducing uncertainty through bridging institutional gaps between countries. Second, we challenge the brain drain hypothesis, not by falsifying it but rather by emphasizing the positive aspects of the 'brain drain' for economic development in the transnational communities' home country. Finally, we extend the rather static contemporary research on transnational communities, by emphasizing how the role of the transnational communities as agents of economic development changes over time as firms gain experience with handling outsourcing/offshoring processes and the formal institutional regulations in the 'host' country improve. We illustrate our insights empirically through a case study on the role of the Indian transnational community in facilitating the rapid growth of the Indian IT service and software industry. This industry presents an ideal case for studying the role played by a transnational community throughout the industrial evolution from the development phase to the maturity phase and the recently initiated restructuring phase.[2]

The chapter proceeds as follows: first, we present transaction cost economics as the dominating theory for explaining make-or-buy decisions. This is followed by a critical discussion of the theory that identifies its limitations and elaborates on the role of transnational communities in the theoretical framework. Second, we argue for an institutional perspective on outsourcing/offshoring decisions that acknowledges the role of transnational communities and how this role changes over time. We hereafter present a case study on the role played by the Indian transnational community in the US in forming the domestic Indian IT service and software industry. Finally, we discuss and conclude on our findings and suggest fruitful paths for future research.

A CALL FOR TRANSCENDING TRANSACTION COST ECONOMICS

Transaction cost economics has emerged as the dominant theory for explaining outsourcing decisions. In this section we investigate the merits and flaws of the theory in terms of explaining economic organization across national and institutional borders. We first provide a brief introduction to transaction cost theory. This is followed by a section illustrating the limitations of the theory. Finally, we contend that acknowledging the role of transnational communities in reducing transaction costs, drawing on insights from institutional theory, may overcome some of these limitations.

Make-or-buy Decisions in Transaction Cost Economics

Despite its reputation in development studies for having too narrow a focus, transaction cost economics has been the main source of inspiration for assessing offshore make-or-buy decisions. In Williamson's version of transaction cost economics, focus is on efficiency considerations and incentive structures aimed at dealing with hold-up problems. Hold-ups may take place in situations where an agent considers making an investment, which only at high costs can be used in another relationship. Agents are considered to be boundedly rational, which implies incomplete contracts that make sequential decisions processes, in terms of specifying the exact prices, conditions of delivery and so forth, necessary. Renegotiations will be conducted in situations different from the initial contract, since at least one party subsequently has made relation-specific investments. When the relation-specific investment has been made the 'fundamental transformation' occurs and the state of affairs changes from a large number to a small number bargaining situation. This makes the party making the investment, the subcontractor, vulnerable to opportunistic behaviour as the exchange partner knows the profit-maximizing subcontractor will accept an offer, which is significantly lower than initially agreed. The price merely needs to exceed what the subcontractor would be able to charge in second-best use. Since the subcontractor can predict this scenario, he will not make relation-specific investments. He might offer to make an investment in a general-purpose machine or similar, which is less efficient than a machine tailored to the specific relationship. The general-purpose investment will either not be able to produce the needed quality or only at higher costs than what can be achieved by a specific investment. Hence, to ensure efficiency, incentives must be created for the subcontractor to make the relation-specific investment by reducing hold-up risks through credible commitments, or equity investments. Nevertheless, at times the risk of hold-up cannot be efficiently mitigated and integration becomes the solution of last resort. Integration eliminates the risk of hold-up as incentives theoretically become fully aligned.[3]

Limitations of Transaction Cost Economics

Despite transaction cost theory's valuable insights it is incapable of explaining the empirical increase in outsourcing transactions involving asset-specific investments. Moreover, it neglects the significant barriers for firms establishing offshore subsidiaries. In particular, we find three problems with the approach. First, although the traditional hold-up problem is relevant for outsourcing decisions, the Williamsonian framing is insufficient. Institutional differences between countries are neglected in the transaction cost framework, which means that the theory cannot identify the added challenges a firm faces when outsourcing or making offshore investments to a foreign country characterized by a different institutional setting. This omission also prevents transaction cost theory from analysing whether outsourcing or offshoring is the most efficient solution for coping with institutional differences.

Second, by adopting the transaction as the unit of analysis, transaction cost economics disregards the impact of repeated games and learning effects. Williamson frames the hold-up situation as a one-shot game. Yet, in reality most contracts involve repeated transactions. Therefore, firms will not renegotiate on the initial contracts since the value of future collaboration will be larger than the immediate value realized from a hold-up. In addition, it may be costly for a firm to act opportunistically if it consequently gets a reputation for being unreliable. The one-shot, opportunism-attentive approach is relevant for firms entering a 'new' country, where the formal institutions are weak and the firms face uncertainty about which subcontractors they can trust. However, as the firms working in the country gain knowledge on whom to trust, and norms and values affecting collaboration, they will gradually be able to navigate paying less attention to hold-up problems. In other words, as firms learn to navigate in another institutional context they learn how to overcome hold-up problems and other challenges coming from opportunistic behaviour. Learning effect can also impact the degree of asset specificity. While asset specificity is high initially when an investment is made, this can change over time as firms learn about potential alternative partners and alternative uses. Decreases in asset specificity over time implies that activities which might be organized in offshore subsidiaries in time T_0 may be more efficiently outsourced in T_1 as the firm learns to cope with the challenges.

Third, transaction cost theory only focuses on threats of partner opportunism but disregards third party hold-up risks. While government bodies, just as contractual partners, learn over time, for instance in terms of building a formal institutional setting reducing the risk of opportunistic behaviour in their country, such bodies may also act opportunistically. Hold-ups by governments are relevant for firms as government behaviour in many developing countries actually may be less predictable than the behaviour of subcontractors. Potential hold-ups by governments press towards outsourcing instead of offshoring as the former solution typically requires investments with a lower degree of asset specificity.

While the general reasoning of transaction cost economics has proved valuable in several cases it is insufficient to explain make-or-buy decisions in cross-institutional settings. Differences in norms and values cannot be reduced to assumptions of narrow economic (bounded) rationality. In some institutional contexts it is more generally accepted to behave opportunistically than in others. Hence, the reputation effects will be minor in comparison to other contexts as some institutional contexts make actors focus on short-term profits. In addition, there are often contractual 'grey-zones' where it is difficult to prove that a partner's opportunistic behaviour is actually opportunistic and not merely a fair result of changed circumstances. Hence, firms cannot always rely on the disciplining effect of the value of repeated interaction, thus theorizing has to take norms and values into account. In sum, transaction cost theory identifies critical issues pertaining to make-or-buy decisions and the theory holds some relevancy for an initial outsourcing/offshoring phase. Yet, it lacks institutional-sensitivity and dynamics and consequently overlooks that learning over time makes firms better equipped to cope with the uncertainty and differences in institutional environments.

Transnational Communities and Transaction Cost Economics

Recent years have witnessed a surge of interest among scholars in transnational communities. The literature argues that transnational communities provide a linkage between the different institutional contexts, and that members of transnational communities are important agents for creating economic development in their home country. Saxenian sums it up:

> these immigrants [emigrants from developing countries, ed.] can provide the critical contacts, information, and cultural know-how that link dynamic – but distant – regions in the global economy. They can create social networks that enable even the smallest producers to locate and maintain mutually beneficial collaborations across great distances and facilitate access to foreign sources of capital, technical skills, and markets. The proliferation of such relationships over time can result in the creation of a transnational technical community – one that can transfer the market and technological know-how needed to support a dynamic of industrial upgrading (2002b: 186).

What is the role of transnational communities in a transaction cost economic perspective? Transnational communities have only received rudimentary treatment in the transaction cost (inspired) literature. Yet, some contours can nevertheless be identified. Transnational communities, it is argued, can reduce uncertainty, search costs and hold-up problems surrounding cross-border contracting and investing by using networks in their home countries. Hereby, the transnational communities' home country network allows firms in the developed world to utilize globally distributed competencies they otherwise would not. In this way, transnational communities can serve as strategic assets to firms in the developed world while simultaneously acting as agents of economic development in the developing world. Thus, in situations where the traditional Williamsonian theory would suggest the activity to be made in-house or that the subcontractor should refrain from undertaking the investment, outsourcing via a transnational community may actually be possible, because of the role played by the community members. Members of transnational communities can choose collaborative partners, whom they know personally and believe they can trust, and hereby reduce hold-up risks. Furthermore, transnational community networks are often primordial in nature as they build on family relations, kinship and altruism, and agents in such networks usually do not hold each other up. This reduces contracting costs, gives better opportunities for using incomplete contracts, and so forth. In short, transnational communities may influence make-or-buy decisions in ways that are not accounted for by transaction cost theory. Firms relying on members of transnational communities can make relational-specific investments leading to either outsourcing or establishment of foreign subsidiaries, which they would otherwise not have been able to undertake. By not only representing a human resource outflow, but also facilitating investment flows from developed to developing countries, transnational communities challenge the net loss contention of the 'brain drain' hypothesis. Yet, to develop a more fine-grained understanding of the role that transnational communities play over time, as agents of economic development, we need to appreciate the institutional effects of increased outsourcing/offshoring.

AN INSTITUTIONAL PERSPECTIVE ON OUTSOURCING/OFFSHORING

The previous sections stressed the need for a theoretical alternative to transaction cost economics in order to more adequately explain cross-border outsourcing and subsidiary formation. In this section we present an institutional alternative to the transaction cost theory that highlights central evolutionary aspects related to outsourcing and offshoring to developing countries. We focus on the following questions: how do institutions affect firms' outsourcing/offshoring decisions, how do these decisions change over time, and what is the role of transnational communities in bridging institutional differences?

Institutions and Outsourcing/Offshoring Decisions

In institutional economics, embracing the literature on national systems of innovation, business systems and neo-institutionalism, firms are not fully informed entities acting according to profit maximizing principles. Rather, firms are embedded in the national institutional context from which they originate. Peng explains: 'Since no firm can be immune from the institutional frameworks in which it is embedded, there is hardly any dispute that institutions matter' (2002: 251). Hence, institutions influence firms' choices and behaviours in a non-deterministic way, as well as their perception of other institutions in different contexts. They are 'the rules of the game in a society or, more formally, are the humanly devised constraints that shape human interaction', and in addition they shape the agents' cognitive schemas. Institutions cover both formal and informal aspects. The formal aspects refer to the legal regulation of behaviour, while the informal institutions refer to norms, values and non-legal sanctions guiding the agent's behaviour and perceptions of the world. Formal institutions by definition are highly codified, but they rest on a tacit foundation. Informal institutions might to some extent be codified but tend to be tacit and taken for granted in the institutional context. Thus, for example, US firms are shaped by the country's particular history, the national labour market regulation, what it means to be a legitimate firm in the US, and how culturally distanced other countries are perceived to be.

What are the implications for understanding outsourcing and establishing foreign subsidiaries in an institutional perspective? The 'traditional' conclusion drawn is that, contrary to what transaction cost theory would suggest, outsourcing patterns have less to do with microeconomic issues than with the institutional context in the country where the firm has its origin. This has at least two implications. First, firms' behaviour is shaped by formal and informal drivers towards structural isomorphism in their country of origin. Secondly, 'when organizations attempt to expand beyond their national boundaries they implicitly take with them their nation's history of socioeconomic choices. On the international scene, they become representatives of their home nation' (Whitley, 1994). We contend that the greater the institutional distance – not to be conflated with geographical distance – the higher costs and difficulties associated with both outsourcing and establishing foreign subsidiaries.

Institutional Learning and Outsourcing/Offshoring Decisions

Thus, in situations where firms do not have experience with cross-border outsourcing or offshoring, they face additional challenges in identifying the activities that can benefit from being geographically repositioned. The firms face uncertainty on whether to offshore or outsource production, which in turn depends on the institutional distance. Large institutional distances makes it difficult to find the right subcontractor, that is a subcontractor with high degrees of security, who respects property rights, who delivers the requested quality in time, who is reliable when it comes to promises, who has an organization structure that matches their own, and so forth. It also makes interactive learning difficult since as Lundvall (1992) argues 'when cultural differences between users and producers are present, certain types of messages will be difficult to transmit or decode . . . Cultural distance between users and producers may block the interaction'. The formal institutional distance is also crucial. Among other things it refers to property rights enforcement, labour market law, accounting principles and government hold-ups. The lack of formal institutions results in greater uncertainty for firms and thus more reluctance when considering outsourcing and offshoring decisions.

However, firms are not 'victims' of institutions or the lack hereof. That is, while the organization structures and the firms' perceptions of the world might be influenced by the original institutional context of the firms, they are nevertheless capable of learning from outsourcing/offshoring experience in a particular country and collaborating with a particular firm. Formal and informal distances are important but firms learn and tend to follow diverse strategies for coping with the uncertainty. To minimize risk firms may experiment with different solutions with limited commitment of resources in the initial phases in order to gain experience on how to overcome problems emanating from formal and informal institutional distances. That is experimenting with building trust relations, coping with cultural differences, how property rights might be a problem, how labour market regulation functions, and so forth.

Moreover, outsourcing across borders cannot be seen solely as a dyadic relationship between two firms. Several firms participate in similar exercises, which creates basis for knowledge sharing, knowledge spillovers (through consultants, publications, collaboration) and also in some cases the establishment of industry-specific institutions. These elements assist in gradually reducing formal and informal institutional distance between firms originating in different countries. Government bodies in developing countries are also capable of learning from interaction with foreign firms and adjusting the formal institutions to the needs of the foreign firms (to the extent the political climate allows for it). They can reduce the bureaucratic red tape, create economic incentives and transparency, enforce property rights, and so forth. Developing appropriate formal institutions requires learning and experimentation in government bodies. Hence, adopting an institutional evolutionary perspective acknowledging learning effects both in firms and government bodies, suggests that firms will be more inclined to establish offshore subsidiaries and outsource over time as the associated costs and difficulties decrease.[4]

Institutions, Transnational Communities and Outsourcing/Offshoring

As the industry evolves, contracting parties in both the developed and the developing world learn and gain experience. In parallel the regulatory authorities learn how to lower unwanted barriers to trade and improve legal regulation. This co-evolution of firm and government learning reduces uncertainty. What are the implications of this causality for the role of the transnational community? If firms are inexperienced and there are few cases to learn from, the firm moving an activity to a country characterized by a large institutional distance faces the above-mentioned challenges. In such situations transnational community members may play a crucial role and serve as boundary spanners or bridge builders between the firms involved, because of their unique home–host country cultural and social capital.[5] Cultural capital consists of knowledge and tools about culture and language that allow them to navigate in their home country as well as their host country. Social capital in terms of a network in their home country allows them to identify collaborating partners in their home country. The identified collaborative partners are typically persons whom the members of the transnational community have reasons to trust. They trust them either because they are members of a kinship group, because they – or their families – have known each other for a long period, or because they can decode the local signals on whom to trust. Trust here refers not only to an expectation of non-opportunistic behaviour but also open and honest communication, thus reducing the problems associated with lack of transparency in competences, quality delivery on time, constraints on learning, and so on. Moreover, in some cases members of the transnational community have networks reaching into the government bodies, which modify difficulties stemming from the formal institutional regulation. In this way transnational communities are crucial in facilitating outsourcing and offshoring decisions and their implementation despite great institutional differences.

However, as the institutional distance decreases over time, as explained above, the firms become less dependent on the competences of transnational community members. The firms will have developed more broad competences and their own cultural and social capital as more employees will acquire cross-cultural communication and collaboration skills. This makes the role of the transnational communities less significant for firms contemplating outsourcing or offshoring. In addition, if the government learns to accommodate the needs of the foreign firms and in general adjusts the formal institutions to industry requirements this also reduces the institutional distance, and hence reduces the importance of the transnational community. We do not claim that uncertainty disappears, eliminating the advantages of transnational communities completely. Members of transnational communities might still embody cultural and social capital valuable for the foreign firms they work for. They might also have developed first-mover advantages, which in the long term will provide comparative advantages in comparison to their western partners. However, in general their importance will decrease.

To sum up: in this section we have argued that the greater the institutional distance between a firm's home country and the country within which it considers organizing a particular activity, the higher the costs and difficulties. Departing from the static efficiency approach of transaction cost theory, we add a dynamic dimension by suggesting that the co-evolution of institutional development (a function of learning in government bodies) and firms' outsourcing/offshoring competencies (a function of learning in the firms in both the developed and developing countries) decreases the importance of transnational community members as bridge-makers between firms having their origin in different institutional contexts. This does not eliminate the advantages of the members totally, but the initial advantages are reduced.

INDIAN TRANSNATIONAL COMMUNITY AND THE INDIAN IT SERVICE AND SOFTWARE INDUSTRY

In this section we empirically illustrate the role transnational communities may play in facilitating outsourcing and offshoring decisions within firms in their host countries. Thus, we focus on their role as agents of economic development in their home country and elucidate how the role changes over time. We pay special attention to the following aspects: how transnational communities can bridge institutional distances and reduce hold-up threats, how the use of transnational communities as vehicles for outsourcing and offshoring produces learning effects, and how these effects impact the importance of the communities over time.

We use the Indian IT software and service industry as an empirical case. The case is interesting as the market for IT services is one of the fastest growing. In 1995 it amounted to US$76 billion while in 2002 it has grown to US$140 billion. We focus on the Indian IT service industry. It has evolved from being almost non-existing two decades ago to being a billion dollar industry today. It is also an industry where the Indian transnational community in the US has played, and continues to play, an important role. The importance of the Indian community is indicated by the stylized fact that in Silicon Valley alone more than 750 IT firms have a CEO with an Indian background (2001 numbers), Indians received around half of the H1-B visas (special visas for experts) and half of them (135 000 in 2001) work in the IT industry (www.northsouth.org). Moreover, members of transnational communities are also returning to India. This trend was reinforced after 9/11, where a hostile US environment caused the number of Indian IT professionals returning from the US to India to increase to 35 000–40 000 (*Businessworld India*, 2003).

The Development Phase

The IT software and service industry has since its emergence been dominated by US organizations as it was driven by interaction between national US security institutions and universities. Until the 1980s, production of IT services was still predominantly a US phenomenon (subsequently an OECD country phenomenon) and outsourcing of IT services mainly occurred in Silicon Valley, while the East Coast IT firms were vertically integrated. From the late 1980s and onwards, the industry gradually globalized. In the developing world the vast majority of the IT based business was located in India. The first major multinational corporation entering India was Digital Equipment in hardware. Soon after, Texas Instruments, Motorola and General Electric followed.

The main reasons for choosing India were cost reduction, the existence of excess capacity of engineers in India, time zone difference, and widespread English skills. Still, not everything pointed to India as a natural location for outsourcing or offshoring of these services. Like several other developing countries India was about to dismantle the ISI strategy but had not yet developed an internal institutional structure adapted to the IT industry. Moreover, the local Indian capacity within the field was limited as few Indian firms at that time had significant IT competences. Rather the majority of firms were situated in the low end of the IT service industry. At the same time, most US firms only had limited experience with outsourcing to developing countries. This combination of few highly skilled Indian firms and little experience in transacting within developing countries generated a highly degree of uncertainty for the US firms around issues such as which subcontractors had the appropriate competences, which subcontractors were trustworthy and which bureaucratic and cultural obstacles they would face.

The differences between Western economies and India are well established in the sociological literature. Weber pointed to the radical differences between Hinduism and Christianity. Some of the concrete challenges that firms offshoring to India face are related to the caste system (whom to hire, what it means if you hire a low caste in a higher position), language (Indians speak English fast and their body language is significantly different from westerners' body language), clarity of arguments (Indians tend not to like straight answers) among others. The business psychologist Geert Hofstede has tried to systematize cultural differences and has measured the difference between India and the US. He divides culture into five dimensions: Power distance, Individualism, Uncertainty avoidance, Masculinity and Long-term orientation[6] (see Table 3.1).

According to Hofstede the major differences between India and the US are in the degree of power distance, where India is a society characterized by a high degree of power distance, which is only moderate in the US. US is a very individualistic country, which is not the case for India. Finally, Indians tend to favour long-term commitments as opposed to short-term commitments in the US. These findings indicate a large institutional distance. Moreover, the bureaucracy modelled on the UK-colonial administration combined with widespread corruption and incompetence that characterize many developing economies work to increase the initial distance.

In short, the US firms experienced a high degree of uncertainty and large institutional distance, both formal and informal, which should create non-trivial cost and difficulties of transacting. The institutional differences hereby constrained US firms' propensity to outsource to and establish subsidiaries in India. From a transaction cost perspective, we would expect that

Table 3.1 Cultural differences between India and the US

	Power distance	Individualism	Uncertainty avoidance	Masculinity	Long-term orientation
India	77	48	40	56	61
USA	40	91	46	62	29

Source: www.spectrum.troyst.edu/~vorism/hofstede.htm.

the firms choosing to outsource would expect significant hold-up threats, due to high uncertainty and small number bargaining situations.

The multinationals appreciating the opportunities in India arguably tried to lower these transaction costs. Initially, for instance, the US firms only moved rather simple and trivial activities to India such as maintenance of existing code or reengineering code from one programming language. These activities did not involve any high degree of asset specificity, and hence they did not expose the firms to great hold-up risks. Also US firms recognized three reasons for Indian firms not to engage in opportunistic behaviour; first the value of future collaboration might exceed the value of reneging on current contracts, second the need for reaching minimum efficient scale, and finally the importance of reputation in the industry. Yet this is not enough to explain the initial uptake in outsourcing and foreign subsidiary establishment in India. As co-founder of Infosys, one of India's leading technology firms with revenues of $754 in 2002–2003, explains:

> In the early '90s, when we went to the United States to sell our services, most chief information officers, didn't believe that an Indian company could build the large applications they needed . . . We realized that there was a huge gap between, on the one hand, how prospective Western clients perceived Indian companies and, on the other, our own perception of our strengths.

To adequately explain the increase in capital flows from the US to India it is necessary to understand the role of members of the Indian transnational community in the US. Several members of this community held important positions in US firms. These members played a significant role in shaping the outsourcing and offshoring decisions in the US firms as the following examples illustrate.

Large institutional distance and significant uncertainty prevented US-based Motorola from utilizing the advantages of India. In 1991, Motorola established MIEL, a software subsidiary in Bangalore. Despite the obvious cost advantages no product sector within Motorola was willing to risk sourcing its software needs from MIEL. Ramachandran and Dikshit (2002) explain: 'The first breakthrough came when Arun Sobti, an Indian who was a senior manager in Motorola's Land Mobile Product Sector in Florida, USA, decided to give MIEL a chance'. They also did some internal marketing with other divisions in their companies even though this was not part of their formal roles. According to Ramachandran and Dikshit, although the first project from Sobti was successful, Sobti was unable to give any more projects to MIEL, because he faced budgetary cuts in his division. However, Sobti continued to help: he put Shrikant Inamdar, the then General Manager (Operations) in MIEL on to the Cellular sector, and he personally lobbied with the sector's management and helped MIEL get its second contract for a Motorola product called CT2. Since the work was in the cellular domain, it afforded MIEL an opportunity to learn about the wireless technology that Motorola was famous for.

Social networks were also important when Texas Instruments (TI) set up its first international IT subsidiary in Bangalore, India in 1985. The establishment was made possible, because the Indian TI vice president, Mohan Rao, utilized his professional position in the US and his knowledge of the Indian political bureaucratic system to facilitate TI's entrance into India. Rao used this combination to get access to top-level people in the Indian government, which in turn allowed him to push the ideas of building an Indian IT industry and to establish a TI plant in India. In other words, his knowledge of the Indian political culture allowed him to reduce the bureaucratic uncertainties and deal directly with top-level politicians in the Indian government. The bureaucracy also ran more smoothly because TI bought the most modern IT equipment and gave it to the Indian government.

Hence, in line with Saxenian, we argue that, 'As they [Indians in the US] gained seniority in US companies in the 1990s, many non-resident Indians (NRIs) were instrumental in convincing senior management to source software or establish operations in India to take advantage of the substantial wage differentials for software skill' (Saxenian, 2002b, 192). Hereby the Indian transnational community in the US played a crucial role in the development phase of the Indian IT service industry.

The Maturity Phase

Since the reform of the Indian economy in 1991 and onwards the Indian IT industry has undergone a remarkable transformation with a large growth in the numbers of Indian IT service firms. US-based IT firms such as Accenture, EDS and CSC are increasingly locating subsidiaries in India. In addition, firms from other high-tech industries, such as the pharmaceutical industry and the banking industry, have started to outsource IT services to the country, and non-IT firms in US and Europe have also started exporting support activities to India such as data entry, payroll processing, and call centres. By 1999, about 185 Fortune 500 companies were sourcing software from Indian vendors. The majority of Indian firms remain in the lower end of the IT service industry, but a growing number have managed to move up the value ladder in terms of the services they produce.

What facilitated this increase in outsourcing and offshoring? Unquestionably, the liberalization of the Indian economy played a significant role. Also, Indian firms spend substantial resources on obtaining quality certificates. Certificates function as signals that reduce uncertainty. They provide important reputation effects that complement recommendations through word-of-mouth and testimonials from satisfied customers. Thus, the top Indian IT service companies make a great effort to obtain the highest certification (ISO to CMM to CMMi to PCMM). However, other factors were influential as well; in particular the evolutionary development of firms and institutions. First, both US and Indian firms developed competencies in bridging the institutional distances. Through working together they learned about each other's differences. It can be argued that this trend has been more pronounced among the larger Indian firms, because of their predominance in the contracting schemes and their ability to attract Indian workers previously employed by US India-based subsidiaries. In contrast, the small and medium sized firms still lacked competences to reduce institutional distances. Second, the co-evolution of Indian institutions played an important part. As the liberalization was initiated in 1991 state bodies started learning about how to facilitate the inflow of investments. The Indian IT service industry organization (National Association of Software and Service Companies), and the Indian government, together with individual Indian firms launched a campaign to increase foreign awareness of the business potential in India and to facilitate the required transactions.

How much can then be attributed to members of transnational communities? In general these developments indicate that the community's assistance was not strictly required any more. This does not mean that the transnational community became insignificant. Social and cultural capital could still reduce the uncertainty, but the role played by community members changed. Some community members that had success in the US as technology entrepreneurs now became venture capitalists. They typically funded ventures that have a front end (sales and marketing) in the US and a back end (software development) in India. During the dotcom-boom, anecdotal evidence suggested that a firm's chances of getting VC funding improved if it had an Indian CTO/CIO. (According to Saxenian, Indians were in charge of about 10 per cent of the technology companies started in Silicon Valley between 1995 and 2000.) Other members of the Indian transnational community have become intermediaries/sales people for software companies, either as consultants or employees. A few have also opened companies in India leveraging their relationships in the US (for example Pradeep Singh of Aditi, a former Microsoft employee). Organizations such as TiE (The IndUS Entrepreneurs) started spanning both the US and India and creating a global network. This raised the image of India as a source of high technology and indirectly promoted software contracts to Indian firms. Note that Microsoft's co-founder was a member of the Indian transnational community and the same was the case with Hotmail founder Sabeer

Bhatiya and Sun Microsystems' co-founder Vinod Khosla. Also, managers, who have worked for other multinationals in India, play a key role. For example, Raman Roy, who set up GE Capital's back-office in Gurgaon (near Delhi), was later persuaded by a leading venture capital firm to set up his own company to provide third-party services. His company, Spectramind, was subsequently acquired by Wipro, a large Indian IT services company.

To sum up, the facts presented herein illustrate that in the early phase of the Indian IT service industry's life cycle the transnational community played a crucial role in the development of the industry through influencing investments decisions in the US. It did so because of the role it played in reducing the formal and informal institutional distances and other uncertainties. This was made possible because of the cultural and social resource endowments of the communities. Hence, the transnational community was central in attracting foreign firms in the early phase, thus of generating economic development in their home country.[7] As the industry matured, both Indians and US firms improved their competences in handling offshore outsourcing to and offshoring in India, building up cultural competencies and creating their own local networks. Firms also learned from each other's experiences. The formal institutional regulation was also greatly improved. All this implied that the central ability of the transnational community to act as bridge builders became less important. Their advantages gained from being first movers, however, led the community members to continue to play an important, but less central and different, role in the Indian IT software and service industry (a role that the recent restructuring might challenge).

DISCUSSION

While transaction cost economics provides a means for identifying a particular problem surrounding an investment decision, the approach is insufficient in the particular context of explaining US firms' offshore outsourcing and offshoring decisions in regard to India. Despite the fact that the industry is characterized by complexity and measurement difficulties (among other reasons because of the intangible nature of the 'product'), which potentially opens it up for opportunistic behaviour, the industry has not been burdened by a high level of hold-ups. One reason for this is that in the Indian IT services industry, reputation in the market is vital. The unit of analysis should not be single transactions but rather repeated interactions so the value of future transactions and learning effects are taken into account. In other words, as both transnational communities and institutional evolutionary development reduce hold-up problems in the Indian IT software industry, opportunism-driven explanations for governance decisions are insufficient.

There is little doubt that India has experienced a significant brain drain. In 1998, about 30 per cent of the graduates from the famed Indian Institute of Technology and 80 per cent of the graduates in computer science went to the US to continue their careers. However, the brain drain has had positive effects. These 'brain gains' have been emphasized in the literature, and we find in the context of the Indian IT service industry that people in the transnational community have played important roles for economic development. They have worked as bridge builders between western and Indian organizations to overcome problems of great institutional distance. A recent study found that the greater barriers to trade in India were the multiplicity of regulations governing product markets, distortions in the market for land, and widespread government ownership of businesses. Members of transnational communities have used their cultural and social knowledge to reduce the uncertainty for US firms associated with these problems.

Hence, in relation to the 'brain drain' hypothesis it is important also to acknowledge the positive effects of emigration. Our case study illustrates how the Indian transnational community has been central for building the Indian IT service industry through their influence on outsourcing and investment decisions in the US. A McKinsey Global Institute study shows that for every dollar previously spent on business processes in the US, which now goes to India, India earns a net benefit of at least 33 cents. This includes taxes, wages paid by US companies, and revenues earned by Indian vendors and their suppliers. Notably we find that the role of transnational communities has changed over time from them being vital for economic development to assuming moderate importance. We believe that this insight not only pertains to the Indian transnational community in the US but to transnational communities in general.

Interestingly, it appears that the role of the transnational community may currently be undergoing a change once again. Industrial transformations in the IT service industry have led strong US IT service firms to introduce shorter product cycles and hence shorter lead-times. This has been a deliberate strategy to maintain market position by inhibiting competitors from profitably imitating or copying their products and services. The shorter lead-time has also created a new demand for India to participate in a more active way in development, and Indian firms do occasionally face problems related to this. As a response Indian firms are increasingly locating subsidiaries in the US and Europe to cope with this challenge (and to be physically closer to sophisticated users which is important for entering the highest end of the value-added ladder). This geographical restructuring of the IT service industry also challenges the first mover advantages gained by the members of the Indian transnational community.

CONCLUSION

Our tentative discussion suggests that the Williamsonian version of transaction cost theory ignores important aspects surrounding offshore outsourcing and offshoring decisions. The theory focuses on individual transactions and opportunism-dependent explanations for governance choices, but it neglects the role played by transnational communities as agents of economic development. We have provided an initial step towards a conceptualization that actively links transnational communities, firm learning and institutional development to explain offshore outsourcing/offshoring decisions. We suggest that transnational communities have played important roles for economic development through reducing institutional distances between organizations in the different institutional settings in the early phase of the development of the industry. By facilitating the initiation of cross-border investment flows, knowledge transfer, and so forth, the transnational communities are central in starting the process of institutional development and firm learning. However, because of these community-initiated evolutionary developments, the importance of the transnational communities decreases over time.

NOTES

1. Anthropologists and sociologists have developed sophisticated definitions of the construction and transformation of identity among members of transnational communities, and so forth. The aim of this study differs from these, hence the more simple definition. What matters is that transnational community groups of entrepreneurs and professionals exist and that they have competencies they can use in an economic context. Moreover, we focus on the consequences of the acts of community members, not on the construction.
2. The recent restructuring of the Indian IT service industry permits studying the transformation of the role played by the Indian transnational community that would not have been possible earlier. The restructuring reflects fundamental economic transformations. It is almost unaffected by the burst of the dot.com bubble.
3. It is worth noting that Richardson (2002) points out a potential contradiction between what is efficient at the micro-level (static economic efficiency) and at the macro-level (dynamic efficiency of the system). He argues that general-purpose investments, in contrast to investments with high degrees of specificity, give the economy flexibility, which is central in responding to change. Thus, to be efficient at the macro-level investments may have a lower degree of specificity than is desired by the customer.

4. Governments might occasionally behave opportunistically to the detriment of foreign firms, but as developing countries increasingly rely on foreign direct investments for economic progress, such scenarios are less likely.
5. Obviously not all transnational community members possess this double-sided cultural and social capital.
6. Power distance is a measure of power and wealth social inequality; Individualism focuses on the degree the society reinforces individual or collective achievement and interpersonal relationships; Masculinity focuses on the degree the society reinforces, or does not reinforce, the traditional masculine work role model of male achievement, control, and power. The Uncertainty Avoidance Index (UAI) focuses on the level of tolerance for uncertainty and ambiguity within the society – that is unstructured situations, while Long-Term Orientation (LTO) focuses on the degree the society embraces, or does not embrace, long-term devotion to traditional, forward thinking values.
7. This being said it should also be 'remembered' that the 'Big 5' of the Indian software industry (Infosys, Wipro, TCS, HCL, and Satyam) were all started based on local effort with no obvious role of members of the transnational community; hence not everything was a function of the transnational community.

BIBLIOGRAPHY

Agrawal, V. and D. Farrell (2003), 'Who wins in offshoring', *McKinsey Quarterly*, Special Edition (4), 36–41.

Agrawal, V., D. Farrell and K. Remes (2003), 'Offshoring and beyond', *McKinsey Quarterly*, Special Edition (4), 24–33.

Appadurai, A. (1996), *Modernity at Large*, Minneapolis, MN: University of Minnesota Press.

Arora, A., V. Arunachalam, J. Asundi and R. Fernandes (2001), 'The Indian software services industry', *Research Policy*, **30** (8), 1267–87.

Balakrishnan, S. and B. Wernerfelt (1986), 'Technical change, competition and vertical integration', *Strategic Management Journal*, **7** (4), 347–59.

Business World India (2003), An ABP PvT Ltd Publication, India.

Christerson, B. and R. Appelbaum (1995), 'Global and local subcontracting: space, ethnicity, and the organization of apparel production', *World Development*, **23** (8).

Ciborra, C. (1993), *Teams, Markets and Systems*, Cambridge: Cambridge University Press.

Cohen, R. (1997), *Global Diaspora: An introduction*, Seattle: University of Washington.

Creehan, S. (2001), 'Brain strain: India's IT crisis', *Harvard International Review*, **23** (2), 6–7.

Devan, J. and P. Tewari (2001), 'Brains abroad', *McKinsey Quarterly*, Special Edition: Emerging Markets (4), 51–60.

DiMaggio, P. J. and W. Powell (1983), 'The iron cage revisited: institutional isomorphism and collective rationality in organizational fields', *American Sociological Review*, **48** (2), 147–60.

Freeman, C. (ed.) (1988), *Japan: A New National System of Innovation*, London: Pinter.

Gertler, M. (2004), *Manufacturing Culture: The Institutional Geography of Industrial Practice*, Oxford: Oxford University Press.

Gulati, R. and H. Singh (1998), 'The architecture of cooperation: managing

coordination costs and appropriability concerns in strategic alliances', *Administrative Science Quarterly*, **43** (4), 781–814.
Hazelhurst, S. (2001), 'Developing IT skills internationally: who's developing whom?', *Association for Computing Machinery. Communications of the ACM*, **44** (7), 27–8.
Joskow, P. (1985), 'Vertical integration and long-term contracts: the case of coal-burning electric generation plants', *Journal of Law, Economics and Organization*, **1**, 33–80.
Kotlin, J. (1993), *Tribes: How Race, Religion and Identity Determine Success In the New Global Economy*, New York: Random House.
Kristensen, P. H. (ed.) (2001), *The Craft Origins of Modern Management*, Bergen: Fakbokforlaget.
Kumra, G. and J. Sinha (2003), 'The next hurdle for Indian IT', *McKinsey Quarterly*, Special Edition (4), 42–53.
Lodovico, A., W. Lewis, V. Palmade and S. Sankhe (2001), 'India – from emerging to surging', *McKinsey Quarterly*, Special Edition: Emerging Markets (4), 28–50.
Lundvall, B.-Å. (1992), *National Systems of Innovation: Towards a Theory of Innovation and Interactive Learning*, London: Pinter.
Mahnke, V., M.L. Overby and J. Vang (2005), 'Strategic outsourcing of IT services: theoretical stocktaking and empirical challenges', *Industry and Innovation*, **12** (2), 205–53.
Masten, S., J. Meehan and E. Snyder (1991), 'The costs of organization', *Journal of Law, Economics, and Organization*, **7**, 1–25.
Monteverde, K. and D. Teece (1982), 'Supplier switching costs and vertical integration in the automobile industry', *The Bell Journal of Economics*, **13** (1), 206–13.
Nelson, R. (1993), *National Innovation Systems: A Comparative Analysis*, Oxford: Oxford University Press.
North, D. (1990), *Institutions, Institutional Change, and Economic Performance*, New York: Norton.
Osborn, R., J. Hagedoorn, J. Denekamp, G. Duysters and C. Baughn (1998), 'Embedded patterns of international alliance formation', *Organization Studies*, Fall (Special Issue: The Organizational Texture of Inter-firm Relations).
Patibandla, M. and B. Petersen (2002), 'Role of TNCs in a developing economy: the case of India's software industry', *World Development*, **30** (9).
Peng, M.W. (2002), 'Towards an institution-based view of business strategy', *Asia Pacific Journal of Management*, **19** (2/3), 251–67.
Peng, M.W. (2003), 'Institutional transitions and strategic choices', *Academy of Management Review*, **28** (2), 275–86.
Ramachandran, J. and P. Dikshit (2002), 'Motorola India Electronics Private Ltd', Case study, mimeo, Indian Institute of Management Bangalore.
Richardson, G.B. (2002), 'The organization of industry re-visited', *DRUID Working Paper* (02-15).
Saxenian, A. (1996), *Regional Advantage: Culture and Competition in Silicon Valley and Route* 128, Boston: Harvard Business School Press.
Saxenian, A. (2002a), 'The Silicon Valley connection: transnational networks and regional development in Taiwan, China and India', *Science, Technology & Society*, **7** (1), 117–49.
Saxenian, A. (2002b), 'Transnational communities and the evolution of global production networks: the cases of Taiwan, China and India', *Industry and Innovation* (Special Issue).

Silverman, B., J. Nickerson and J. Freeman (2003), 'Profitability, transactional alignment and organizational mortality in the US trucking industry', *Strategic Management Journal*, **18** (Summer Special Issue), 31–52.
Thompson, G., J. Rances, R. Levacic and J. Mitchell (1991), *Markets, Hierarchies and Networks: The Coordination of Social Life*, London: Sage.
Walker, G. and D. Weber (1984), 'A transaction cost approach to make-or-buy decisions', *Administrative Science Quarterly*, **29** (3), 373–91.
Walker, G. and D. Weber (1987), 'Supplier competition, uncertainty and make-or-buy decisions', *Academy of Management Journal*, **30** (3), 589–96.
Whitley, R. (1994), 'Dominant forms of economic organization in market economies', *Organization Studies*, **15** (2), 153–82.
Whitley, R. (1999), *Divergent Capitalisms: The Social Structuring and Change of Business Systems*, Oxford: Oxford University Press.
Williamson, O. E. (1975), *Markets and Hierarchies*, New York: Free Press.
Williamson, O. E. (1985), *Economic Institutions of Capitalism*, New York: Free Press.

4. Effectively linking international, national and regional innovation systems: insights from India and Indonesia

Martina Fromhold-Eisebith

INTRODUCTION

The notion of innovation systems aims at capturing the realization that innovation, regarded as a major driver of successful economies, is the outcome of the systemic interaction of various organizations and procedures, and of interconnected political, economic and social processes (Archibugi and Michie, 1997; Malecki, 1997; Hotz-Hart, 2000). These activities may relate to different spatial scales (Oinas and Malecki, 1999); this finds its expression in terminological distinctions that particularly address the national, regional and international levels. This chapter discusses how the three system scales could expediently be combined with respect to science, technology and innovation (STI) policies in Asian less developed countries, based on conceptual and empirical considerations.

First the idea of *national* systems of innovation (NSI) has appeared on the scene, promoted by economists like Freeman (1987; 1995) and Lundvall (1992a; 2003), and elaborated on by many others (see the contributions in Lundvall, 1992b; Nelson, 1993; Edquist, 1997). According to basic definitions, an NSI 'is constituted by elements and relationships which interact in the production, diffusion and use of new, and economically useful, knowledge . . . either located within or rooted inside the borders of a nation state' (Lundvall, 1992a; p. 2), which 'determine the innovative performance . . . of national firms' (Nelson and Rosenberg, 1993, p. 4). The NSI combines various agents (such as firms, public/policy agencies, education and research organizations), their modes of behaviour and relationships. In less developed countries foreign companies are particularly important elements (Chesnais, 1992), in addition to public players. It is assumed that systemic qualities and trajectories of technology development and learning are shaped by nation-specific socio-cultural and historic-genetic experiences and characteristics in dealing with issues of innovation and its promotion (Lundvall, 1992b; 2003; Nelson, 1993; Archibugi and Michie, 1997; Edquist, 1997; Lundvall and Maskell, 2000). Providing political guidance, some factors of success of NSIs have been identified, such as the creation of supportive institutions (key role of education) and of collaborations between R&D organizations and firms (Lundvall, 1992b; Nelson and Rosenberg, 1993; Freeman, 1995; Edquist, 1997). Yet, vast differences between NSI structures of various economically successful countries indicate that there is no universal 'best practice' recipe (Nelson, 1993; Hotz-Hart, 2000).

Second, the notion of *regional* systems of innovation (RSI) has been put forward by economic geographers and regional scientists (Cooke, 1992; 2001; Cooke et al., 1997; Braczyk et al., 1998), because the NSI concept neglects to explicitly highlight the crucial role of regionalized processes, locally bound assets and actor proximity for innovation-oriented collaboration.[1] A main argument is that regional institutions are actually the most suitable foundation for the evolution of innovative communities of firms and other organizations: they foster socially embedded collaboration and information exchange of co-located people based on trust, exchanges of 'tacit knowledge' and common socio-cultural values and friendships (Maskell and Malmberg, 1999).[2] Thus, the region 'may be the most appropriate scale for understanding the dynamics of, and organizing policy interventions directed towards, integrated innovation systems' (Bunnell and Coe, 2001, p. 575; drawing on Cooke et al., 1997). Defining the RSI as a complex 'in which firms and other organizations [. . .] are systematically engaged in interactive learning through an institutional milieu characterized by embeddedness' (Cooke et al., 1998, p. 1581) addresses a similar set of agents as in the case of NSI. Qualities of an RSI to function as an arena for STI policy making and the design and implementation of technology development and transfer strategies are also pointed out (Cooke, 1992; Cooke et al., 1997; Braczyk et al., 1998; Asheim and Cooke, 1999; Asheim, 2000).

Third, this chapter uses the term *international* system of innovation (ISI) in analogy to the other two notions, taking account of research that advocates an extended view as innovation increasingly depends on international information exchange and collaboration (Niosi and Bellon, 1994; Hotz-Hart, 2000; Bunnell and Coe, 2001). Internationalization proliferates among scientific communities, production systems and policy approaches, which has been underemphasized in the NSI and RSI literature so far. This view is particularly important when regarding less developed countries where inward technology transfer often prevails as the major way to acquire know-how (Chowdhuri and Islam, 1993; Mowery and Oxley,

1995; Enos et al., 1997). International systemic processes characterize various spheres of activity, from the R&D organization of Transnational Companies (TNCs) to the migration of human capital (Bunnell and Coe, 2001). In particular the cross-border mobility of qualified labour has been insufficiently conceived as a determinant of (local) innovativeness (Fromhold-Eisebith, 2002a).

Research has just started to take up the issue concerning how various spatial dimensions of innovation systems interact or could be combined (Hotz-Hart, 2000; Bunnell and Coe, 2001; Freeman, 2002; Bathelt and Depner, 2003), also including sector specificities (Malerba, 2002). On the one hand, attention should be paid to relationships operating between different levels since 'events do not occur exclusively at one particular scale but instead across various scales simultaneously, making it difficult to assign causal priority to one scale over the others' (Bunnel and Coe, 2001, p. 570), which involves less territorialized conceptions of innovation. On the other hand, multi-scalar approaches should explicitly take account of territorial, that is regional, national and international, specificities which still represent an important base for understanding innovation systems and for politically governing them. Policies have to consider differing prerequisites regarding those operational scales in promoting innovation. In this respect neither NSI, RSI nor ISI are expected to function well in isolation but mutually affect each other's performance. More analytically exploring these links may not only improve our understanding of innovation systems but also help design effective STI policies, which especially applies to less developed countries.

This chapter suggests how the national, regional and international levels of innovation systems could be combined, depicting policy options for newly industrialized countries (NICs) in Asia. They particularly need such a strategy due to the following scale-related STI characteristics: first, Asian 'latecomer' economies traditionally depend on international inputs of technology and know-how (Chowdhuri and Islam, 1993; Enos et al., 1997; Lall, 1998a). Second, catching-up in innovativeness is particularly hard for them as their role in the international division of labour is that of executors of standardized production routines using cheap and unqualified labour, rather than that of creators of their own 'brand new' products and processes for world markets based on indigenous R&D (Malecki, 1997; Dicken, 1998). Third, development policies in most Asian countries have so far either discounted issues of space and region-specificity (which especially applies to STI fields) or overly emphasized objectives of balanced regional growth and the reduction of interregional disparities. The high general importance of the latter task is beyond question as gradients of socio-economic wellbeing are usually very steep in less developed countries.

Yet, the logic of promoting peripheral areas contradicts the essentially metropolitan base of functional innovation systems (Koschatzky and Sternberg, 2000; Fritsch, 2002).

The next section develops ideas concerning how the three system scales could be linked in order to constitute a scale-bridging complex that may be termed 'national supersystem of innovation' (NSSI). Then, set against this normative framework, two country examples are discussed, India and Indonesia, showing to which extent their innovation-related strategies do, or do not, comply with recommended NSSI features. The conclusions point out what can be learnt for designing 'supersystemic' STI support in Asia.

OPTIONS FOR LINKING NATIONAL, REGIONAL AND INTERNATIONAL INNOVATION SYSTEMS

When the NSI, RSI and ISI concepts are regarded in conjunction several questions arise concerning how to effectively link processes on all spatial scales for policy purposes. Important issues are: in which respect do well-functioning RSIs constitute a successful NSI? Is a good NSI policy defined as (also) explicitly enabling the development of an effectual RSI? How could international impulses be integrated best, on the scale of NSIs or RSIs? The (inter)functioning of the system levels is explored from a theoretical but application-oriented perspective.

A major problem which obstructs consistent reasoning is that space-related requirements of innovative activities substantially differ between industries, or even companies (Koschatzky and Sternberg, 2000). They create scale-crossing sectoral systems of innovation that challenge the existence of NSIs (Nelson and Rosenberg, 1993; Malerba, 2002). For some industries international connections have top priority (such as electronics, pharmaceuticals), whereas others rather rely on a favourable national framework (for example biotechnology) (Bunnell and Coe, 2001), or the regional one (producer services). A functional NSI is especially important for complex high technology fields, such as aerospace R&D/production (Nelson and Rosenberg, 1993), while medium technology fields may, at least at times, not require that base. Additionally, the importance of different system scales varies over time relating to phases of the product life cycle.

To move ahead we largely abstract our view from sector specificities and look at general relationships between the major scales of innovation systems, keeping in mind the political interest of Asian NICs. A graph depicts relevant links between NSIs, RSIs and the wider ISI framework (see Figure 4.1), briefly explained in the following.

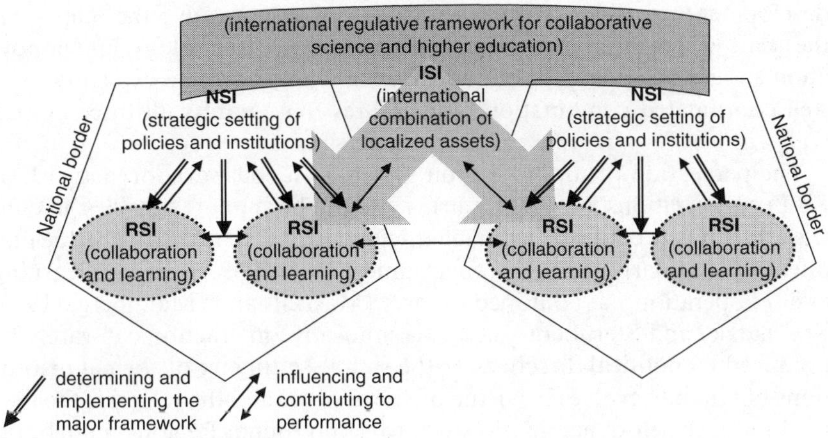

Source: Depiction by the author.

Figure 4.1 Interdependencies of national, regional and international systems of innovation (NSI, RSI and ISI)

For analysing NSI–RSI interdependencies arguments can be taken from the ongoing debate on which spatial level is more relevant for innovation. On the one hand, certain regional specificities support an alleged dominance of regional/local against national conditions (persistent industrial specializations, core-periphery differences, socio-cultural embedding of knowledge-intensive collaboration, particular regional government/governance structures; Braczyk et al., 1998; Cooke et al., 1998; Howells, 1999). On the other hand, local regulation rarely operates independently from the national framework, but usually builds upon nationally determined and constructed institutional or political settings (Bunnell and Coe, 2001; Freeman, 2002; Bathelt and Depner, 2003). Therefore the NSI, by virtue of political regulative and institutional influences, leaves its marks on innovation and technology developments in a country beyond the impact of RSIs (as empirically proven by Koschatzky and Sternberg, 2000), and despite a growing role of globalization and arrogating global corporations. This especially applies to most Asian countries with their tradition of centrally implemented and governed regional political structures and fairly centralist STI regulation.

Accordingly, the NSI, when interpreted as a strongly (but not only) politically determined context and set of strategic STI promotion instruments (Edquist, 1997; Lundvall, 2003), occupies a somehow superior role to the RSI (Figure 4.1). Setting the overarching institutional and legislative framework, it represents the superstructure that enables and promotes the development of RSIs (Freeman, 2002). Consequently, the nation is the scale where a 'supersystemic' promotion of technology and innovation should be anchored, but where agents ought to care explicitly for a well coordinated combination of measures that address all three system scales.

The perception of an innovation system as a collaboration network of R&D organizations, innovating firms and public support agencies, in contrast, can much better be associated with the RSI than NSI. The locality offers a particularly conducive environment for socially embedded learning and cooperation, as outlined above (Maskell and Malmberg, 1999; Koschatzky and Sternberg, 2000). As innovative interaction can rarely be produced by political directives or the simple setting up of formal institutions but mainly evolves by virtue of social processes, the region is the best level for such self-dynamic processes to gain ground. Regional collaboration, however, can only grow provided the necessary infrastructure and economic agents are locally present, which often relies on national allocation strategies and generally mainly applies to urban agglomerations (Fritsch, 2002). Hence, the RSI and its particular socio-economic network qualities are an important complement which depends on and contributes to the potential of the overlying NSI (Figure 4.1).

Consequently, a major role of the NSI should be to allocate relevant institutions and sufficient decision and financial powers to the right localities in order to enable the emergence of collaborative RSIs on this base, also catering to specific needs of localized sectoral innovation systems (Malerba, 2002). The nation, in turn, profits from evolving RSIs that become important constituents of the prospering NSIs, which may thus be conceived as incorporating several interlinked RSIs in a country. In this context an RSI is not operating as an autonomous, self-contained unit, but as a set of elements and connections that is linked with its NSI, other RSIs of its own nation and external ones, too.

International impacts on NSIs and RSIs, often perceived as a threatening factor that hollows out national powers (Niosi and Bellon, 1994), may constructively be included as a supporting factor that enriches existing strengths (Oinas and Malecki, 1999). Nations can profit from international connections by using the RIS scale (Figure 4.1), relying on the fact that ISIs are mainly driven by interests to exploit region-specific advantages for the sake of the entire cross-border structure. TNCs, for instance, that organize innovative activities in corporate networks, tap local (tacit) knowledge by locating in differently specialized regions and combine assets for their overall strategies (Bunnell and Coe, 2001). International procedures are therefore embedded in different NSIs via their RSI and link certain agglomerations, using their creative potential and at the same time contributing to it (Howells, 1999; Hotz-Hart, 2000; Malerba, 2002). Thus, it is the RSI that cares for an effective interplay of NSI and ISI. But the NSI sets the major institutional framework while the RSI provides the actual platform of the collaborative embedding of international agents, offering suitable seedbed structures that also draw on allocative decisions shaped by the NSI.

In this context national powers refer, first, to policies creating restrictions and incentives for foreign direct investors (admittance/location of firms in innovative sectors, 'local content' obligations, import–export requirements; see Marton, 1986; Dicken, 1998). Second, the NSI, by building up a domestic potential of R&D, education organizations and supply firms, determines the absorptive capacities of certain locations or RSIs for technological impulses and their attractiveness for innovation-oriented investors (Mowery and Oxley, 1995; Hotz-Hart, 2000). On this base TNCs can possibly be turned into a well-lubricated hinge between the international, national and regional levels, decisively strengthening all of them 'by acting as conduits for transferring tacit expertise between dynamic RSI' (Bunnel and Coe, 2001, p. 580), supporting industrial specialization (relating to the cluster idea) and providing access to state-of-the-art know-how. Some Asian NICs have already substantially improved their technological capabilities that way (Lall, 1998a), probably due to positive regional impulses of learning and innovation induced by TNCs where a corresponding, absorptive RIS pre-exists (Fromhold-Eisebith, 2002b). In turn, good RSI conditions also contribute to the performance of the ISI (Figure 4.1). Nations, third, are in charge of the financial promotion of international scientific and political interaction. Integrating certain countries (such as ASEAN or APEC states) into supranational structures of R&D support may even create international STI policy systems that interlink and strengthen the included RSI in different countries (as is aspired by the EU; Héraud, 2003). Fourth, crucial additional factors of an internationally integrated development of RSI, like qualified labour migration (Fromhold-Eisebith, 2002a), may be indirectly supported by national policies or, at least, monitored with respect to political consequences to be taken on national and regional levels.

On these foundations the features of a 'national supersystem of innovation' NSSI, can be outlined, in terms of a normative ideal and policy guideline for Asian NICs. The NSSI is mainly directed from the national scale, but purposefully creates and uses potential on the regional and international ones: central authorities explicitly act as 'masters of scales' in innovation promotion (which deviates from the original conception of NSI). Table 4.1 suggests some details on the division of tasks between the three spatial scales, which will be illustrated in the following Asian case studies.

Table 4.1 *Major features of a 'national supersystem of innovation' (NSSI) regarding functions at and links between spatial scales*

A) Tasks on the national scale

Major functions and focus	Link to the subnational regional scale	Link to the international scale
Determine the overall STI strategy of economic development and create a coherent institutional and legislative framework taking care of regional specificities and international contexts	Establish regional agencies and institutions that regionally represent and implement national strategies and monitor regional STI developments and specializations (see also B))	Monitor international developments in STI fields and identify relative national strengths or weaknesses; source ideas on STI policies and promotion from other countries
Invest in the improvement of public research and (higher) education infrastructure, taking care of regional specificities and international contexts	Allocate and agglomerate research and education infrastructure to suitable places in the country offering corresponding sectoral structures	Organize and promote international exchanges of scientists, teachers and students; provide good access to international know-how and research documents
Set up financial programmes promoting industrial start-ups and collaborative innovation (also academia–industry) taking care of regional sectoral and know-how specificities and international contexts	Provide regional authorities with human capital, decision and financial powers to allow them to organize the support of start-ups and collaborative innovation in regional sectors and fields of specialization (see also B))	Identify sectors and fields of know-how in which the country is or may become internationally competitive according to regional sectoral and know-how-based specializations and which should be supported accordingly
Implement incentives and regulations for selectively attracting foreign direct investors in technology sectors according to regional sectoral and know-how specificities in the country	Care for the agglomeration of foreign technology investments in structurally sufficiently strong regions; enable local authorities to cater to the needs of investors (see also B))	Promote the attraction of foreign investments in technology sectors from abroad by explicitly using regional sectoral and know-how specificities in the country as a marketing argument
Implement incentives to attract re-migrant and foreign qualified labour	Care for the agglomeration of qualified re-migrants and immigrants in structurally	Establish communication links to communities of emigrated qualified labour in other countries,

according to regional sectoral and know-how specificities in the country	strong regions; enable local authorities to provide incentives to re-migrant entrepreneurship (see also B))	continuously informing them about STI developments, achievements and return incentives in the home country relating to particularly successful and attractive regions

B) Tasks on the regional scale

Major functions and focus	Link to the national scale	Link to the international scale
Regionally represent the national STI strategy, adapt and implement its components according to regional sectoral and know-how specificities; continuously monitor regional STI developments and processes of sectoral specialization	Represent region-specific interests in STI promotion against central authorities and care for their inclusion in the national strategy; regularly report regional achievements in STI fields and industrial sectors to central authorities	Monitor international developments in STI fields relating to regional sectoral and know-how specializations and identify relative regional strengths or weaknesses that should be regarded in regional and national approaches to STI promotion
Actively integrate regional public research and (higher) education institutions into regional communication, building upon social contacts	Suggest to central authorities which additional research and education infrastructure should be allocated to the region according to upcoming specializations and bottlenecks, and care for its realization	Contribute to a good institutional and social embedding of foreign scientists, teachers and students visiting the region via international exchanges; initiate programmes of international scientific interaction according to regional know-how specificities
Regionally organize and implement national support programmes for start-ups and collaborative innovation (also academia–industry) in regional sectors and fields of specialization, building upon social contacts	Suggest to central authorities which additional support programmes for start-ups and collaborative innovation should be designed according to upcoming specializations and bottlenecks, and care for their realization	Monitor international developments in start-up and collaboration support relating to regional sectoral and know-how specializations and identify options for new regional and national approaches; initiate border-crossing programmes of collaborative innovation

Table 4.1 *(continued)*

B) Tasks on the regional scale		
Major functions and focus	Link to the national scale	Link to the international scale
Engage in catering to the regional needs of foreign direct investors regarding 'paperwork', employment, local networking including social contacts (to firms, R&D, education)	Suggest to central authorities which additional incentives should be implemented nationally to attract foreign investors relating to regional sectoral and know-how specificities	Actively take part in international activities that aim at attracting foreign direct investors, representing and presenting the region as a site of location according to its sectoral and know-how specificities
Welcome and monitor inflows of return and foreign qualified migration; provide incentives to re-migrant entrepreneurship and help in local networking to sectoral and R&D partners	Report developments regarding regional inflows and entrepreneurship of qualified migrants to central authorities; suggest incentives taking account of regional sectoral specificities	Actively advertise for flows of return and foreign qualified migrants from other countries into the region according to its sectoral and know-how specificities

Source: Depiction by the author.

SPATIAL SYSTEMS OF INNOVATION PROMOTION IN INDIA AND INDONESIA

The recommendations concerning an NSSI strategy are now exemplified, or so to speak 'tested', by looking at the cases of India and Indonesia. These Asian NICs have recently experienced remarkable growth dynamics in some technology industries (India especially in information technology, Indonesia in electronics production). This raises the question of to what extent success is accompanied and supported by 'supersystemic' features, which could also offer good prospects for continued upgrading in the future. The presented insights partly draw on empirical research in both countries in the late 1990s.[3] Field work has produced valuable insights into specificities of knowledge-intensive collaboration in two technology-oriented regional economies, Bangalore (India) and Bandung (Indonesia), as well as into the – more or less successful – cofunctioning of different spatial scales promoting industrial innovativeness in these countries (Fromhold-Eisebith, 1999; 2001; Fromhold-Eisebith and Eisebith, 2002). The following subsections focus on scale-bridging and scale-combining aspects of innovation systems in India and Indonesia, followed by a section discussing the findings against the theoretical background from a comparative angle.

India

In terms of an initial overview, outstanding features of innovation systems in India regarding spatial issues are:

1. The paradigm shift from the exclusion of major international influences before 1991 to a selective, partially effective admittance of international agents and contacts thereafter;
2. The allocation of major R&D centres, academic organizations and technology-oriented public enterprises to certain regions where they have contributed to sectoral agglomerations;
3. A federal system that leaves some competences regarding the development of innovative industries to (regional) state authorities, which, however, act with a dominant logic of promoting regionally balanced growth and the industrialization of backward areas;
4. The implementation of central strategies of the internationally oriented promotion of certain technology industries on regional levels and via regional agencies.

The following paragraphs elucidate these aspects against the backdrop of the NSSI idea.

Regarding the first point, STI strategies in India had for many decades been dominated by the paradigm of self-reliance and import substitution, before the New Economic Policy of 1991 opened up the economy to more international inflows of investment and trade (Desai and Bhalerao, 1996; Aggarwal, 2001). Accordingly, the NSI and RSI of India have for a long period evolved from attempts to build up indigenous technological capabilities with just selective international influences on industrial development, excluding potentially beneficial ISI impulses. Foreign TNCs have been restricted with respect to investment, location, import and other allowances, which continues to have its impact on a complex system of regulation. India remained highly dependent on foreign technology, though. And the national system of STI and industry promotion can still be blamed for not displaying a really strategic appearance, composed of an irritatingly large number of various institutions and programmes that seem to lack coordination (www.mst.nic.in). In its overall institutional appearance the Indian STI approach hardly complies with the NSSI features described in Table 4.1.

Yet, the overall regime has changed especially regarding one industry: information technology (IT), and particularly the provision of IT services to international customers by domestic and foreign firms. This sector gains a growing stake in technology production and political support on various spatial levels (Heeks, 1996; www.nasscom.org), which provides evidence of a productive integration of international, national and regional processes. There is, however, a danger of growing imbalances between one highly competitive and increasingly innovative sector, and a lagging behind of the other politically less favoured ones. But the continuing success of IT is remarkable in an NSSI framework as it strongly relies on the combination of inward TNC investments and the use of regional human capital for the increasing integration of the Indian economy into international relationships, anchored in certain regions. The labour migration issue mentioned in Table 4.1 is particularly important here: first experiencing a growing international 'brain drain' of Indian IT experts to other countries, the national government has then implemented an incentive structure for returning 'Non Resident Indians' and tries to keep in touch with Indian scientists abroad. This helps to direct know-how from outside to industrial uses in the country and especially to attractive and absorptive RSIs such as Bangalore (Fromhold-Eisebith, 2002a).

Second, according to the logic of self-reliance India has put a lot of effort into building up the country's own scientific and technological capabilities, including international exchanges. These have brought some impressive, albeit isolated results regarding qualities of Indian scientists. However, the country experienced difficulty in raising its entire research and higher education system to international standards regarding industrial application, due to insufficient finance and imbalanced support (Prakashi and Phondke, 1995; Aggarwal, 2001). There has been a strong bias towards the development of indigenous defence technologies and products, which have occurred in well-functioning systems of interacting public R&D laboratories and state enterprises, but hardly induced innovation impulses to the wider industrial community.

Relating to the NSSI framework, central authorities, however, have often created favourable conditions for the emergence of regional technology-oriented collaboration by co-locating laboratories, outstanding academic organizations (such as Indian Institutes of Technology or the Indian Institute of Sciences) and state companies in one metropolitan area, in some cases creating well-networked RISs with certain sector foci (Bangalore again provides a good example; Fromhold-Eisebith, 1999; 2001). In some places this has provided a good seedbed for the local growth of agglomerating private indigenous and foreign technology firms which have especially been attracted to these centres of knowledge, connecting those regions to the ISI. The recent shift to IT has also led to an upgrading of regional academic capabilities in this field. While this helps profitably to integrate international agents into that particular RSI, education and research relating to other sectors are, however, not modernized.

Third, the federal system of India has enabled its states and their separately elected (regional) governments to establish their own regulative and incentive structures regarding issues of higher education or the attraction and location of private industries, also in technology sectors. Jumping onto the moving train and competing against each other as investment destinations, many states have implemented their own 'IT policies' (all also published via the Internet). They aim at attracting international firms which want to utilize specific regional assets that were often established by national authorities, linking the three system scales. According to a traditionally strong role of 'balanced regional growth' philosophies in India, states have tried to direct new investments to lagging areas within their territory, which, however, do not offer good regional conditions for collaboration and collective innovation. As a matter of fact, international IT investors in particular are not inclined to follow these directives but show a high preference for locating in and close to metropolitan RSIs, just as has generally been highlighted above. State agencies in India are flexible enough to allow for that combination of ISI and RSI potential, which pays off for many IT locations. As per their IT policies, some states have established additional sector-specific organizations of higher education, locating them close to industrial agglomerations (for instance Indian Institutes of IT) and thus enriching the RSI beyond NSI influences. Due to the triggered dynamics, objectives to vitalize secondary centres may get fulfilled over time by cost-driven spatial spillover effects (Fromhold-Eisebith, 2006).

Fourth, a major impulse that supports the growth of agglomerating export-driven indigenous and foreign IT companies in many Indian regions in the last decade has come from a central government initiative; hence the NSI has, just as recommended above, taken care of anchoring ISI elements in suitable RSIs. The Software Technology Parks of India scheme (STPI) provides a certain status and incentives to export-oriented industrial units. It is enacted nationally but is implemented via agencies located in close to 30 urban regions all over the country (www.stpi.soft.net). Many locations offer a particularly attractive institutional context for technology industries in terms of an RSI. The STPI offices care for certain infrastructural amenities (satellite communication) and cater to the needs of regional client firms. In many places this has laid the foundations for a respectable proliferation of entrepreneurship and employment in IT, leading to mutually beneficial and increasingly innovative interaction between various private and public, local and foreign technology agents (Fromhold-Eisebith, 2002b). The STPI programme therefore features as an effective way to trigger the development of successful RSIs by nationally integrating them internationally. The only caveat is that apart from the focus on IT, other sectors have not been able to profit from similar support so far.

In sum, India shows some elements of a functional NSSI, but they predominantly relate to one sector only, IT services. Here the country has managed to integrate ISI impulses – mainly through TNC investments and remigrating 'brains' as outlined in the theoretical section – into the development of several prospering RSIs. They together contribute to positive upgrading dynamics in IT industries on the national scale, nurturing the respective sectoral innovation system across scales. In line with suggestions in Table 4.1, respective regional assets have been created and enabled by a supportive NSI that has agglomerated important infrastructure elements in certain areas and allocated competencies to care for industrial promotion to regional agencies, backed up by tailored national initiatives. As a consequence, the entire national economy and government system profits from rising export income and technological standards. Insufficiencies of the Indian NSSI, however, prevail with respect to other possibly innovative sectors, such as engineering or biotechnology industries. There surely are nascent RSIs in these fields, but they lack both upgrading influences from the ISI and focused strategic support from national authorities. The latter should pay more attention to, and activate, underutilized regional innovative potential in non-IT sectors, opening them to international links where an attractive RSI offers a good seedbed of absorptive capacities.

Indonesia

An overview of innovation systems in Indonesia regarding expedient combinations of scales needs to highlight the following characteristics:

1. The early opening to, and strong dependence on, international influences regarding the development of technology-oriented industries, concentrated in a few regions;
2. The longstanding systemic and pronounced strategic orientation of national STI policies regarding the internationally supported development of R&D, education and technology-driven public enterprises in certain sectors and regions;
3. A centralist government system relating to STI and other sector policies that hardly devolves decision power and financial competences for technology and industrial promotion to lower level agencies;
4. National strategies that attract foreign technology firms to certain industrial areas promising fast and easy profit, but overlook RSI specificities and opportunities elsewhere.

When evaluated against the ideal of an NSSI, the first aspect speaks in favour of Indonesia's openness towards the international acquisition of technology, know-how and respective direct investments, as the NSI has tried to make use of ISI assets for a long time. Using international assistance had been a necessity for the country in order to master the urgent restructuring from an economy highly dependent on oil exports to a major producer of manufactured goods (Hill, 1996; Booth, 1998). Although this strategy brought some success in the early 1990s, the trajectory could not successfully be sustained because of economic failures and the socio-political problems which accumulated towards the late 1990s. The country has fallen short of deriving major self-reinforcing benefits from its good international connections, despite attempts to establish ISI links to indigenous firms on a national scale, rather than on regional ones.

One reason lies in an insufficient development of the country's own technological capabilities and human capital, as know-how and technology goods could too easily be acquired by buying them from abroad (Hill and Thee, 1998). This indicates the importance of building up sufficient indigenous NSI and RSI potential, which is necessary to allow for a constructive integration of ISI impulses in an NSSI context. Another reason for a lack of success is that international technology collaboration used to focus on certain prestigious sectors that offered little scope for proliferating learning processes in wider parts of the economy (this aspect relates to the second point further below). Additionally, internationally connected technology activities have been highly concentrated in the capital region of Jakarta and, lately, in insular peripheral locations like Batam Island, neither of which offer the best RSI conditions in terms of socially embedded collaboration opportunities or absorptive structures (see point four below). Consequently, know-how has hardly diffused to technology users in other parts of the country, and the NSI has only marginally profited from international integration.

Regarding the second aspect, the low technological achievements of Indonesia are somehow surprising since the national STI policies had once enjoyed quite strategic and systematic planning in terms of a well-coordinated NSI, yet with a fairly imbalanced focus (Lall, 1998b). For twenty years, until 1998, national research and technology minister B.J. Habibie had monocratically directed an entire system of various complementary and interacting technology support agencies, R&D institutes and state companies. He pursued a pronounced strategy to build up the country's technological competences from imitation to conducting fundamental research (Rice, 1998). This structure, which forms a major part of Indonesia's NSI, appears to be well designed and not marked by too many redundancies. Still, it has not been effective for several reasons.

Above all, there is the widespread tradition of poorly paid public sector and academic employees in Indonesia spending major parts of their work time earning extra money in various 'moonlighting' jobs (Fromhold-Eisebith and Eisebith, 2002), which fundamentally obstructs the functioning of the NSI and any RSIs in the country. And the national institutional system has not induced innovation because it over-ambitiously fostered capabilities in prestigious, but isolated technology fields such as aircraft building; even here it did not bear the promised success (as indicated by the failure of the national aircraft industry IPTN; Cohen, 2000). The costly efforts had their good sides, though, as several national organizations of research and production were co-located in regions that offer good human capital from universities, constituting a favourable base for emergent RSI in some industries, such as telecommunication electronics (Bandung and Jakarta provide good examples; Fromhold-Eisebith, 2001; Fromhold-Eisebith and Eisebith, 2002; Gammeltoft, 2004). These agglomerations have been connected to international R&D partners, yet hardly to technology-driven foreign investors, which leaves the ISI connection incomplete and ineffective. After 1998 the Indonesian STI system lost its one-sided strategic focus and shifted to programmes that rather emphasize the alleviation of poverty in wider parts of the country (www.bppt.go.id). This may strengthen some lower rank RIS, but with the ambitions go the good chances of Indonesia approaching international standards in technology production and generation.

Third, the strong dominance of the NSI over RSI in Indonesia has its roots in the overall government system. Despite efforts to decentralize, it highly concentrates decision and financial powers in the central ministries in Jakarta, with provincial and lower level authorities acting as executing and controlling units for central initiatives (Beier and Ferrazzi, 1998). This strongly contradicts basic NSSI requirements (Table 4.1). The imbalance applies in particular to STI measures and related industrial promotion which are all centrally designed and implemented from Jakarta, even when actually addressing other regions. As a consequence, provincial and, one level below, district authorities do not only lack the power to engage in regional innovation support but are also not even aware of region-specific assets that could be used for location promotion. This leads to the (quite incomprehensible) situation that even outstanding regional collaboration systems of technology agents – a major asset of respective RSIs – are hardly noticed by regional governments and, thus, receive no official local back-up, nor adequate NSI support. Innovation-oriented interaction operates essentially on a private basis (including 'moonlighting' relationships) and is marked by strong social embedding in the respective locality (Fromhold-Eisebith and Eisebith, 2002).

As promising RSI structures appear to exist outside the 'awareness space' of regional and national governments, fourth, Indonesia also misses out on adequately using its regional potential in nationally governed activities to attract and embed foreign direct investors in technology sectors, hence on using RSIs for productively connecting NSI and ISI. Over the past decades, the country had been quite successful in gaining investments from global players, for instance, in electronics production (Hill, 1996; Hill and Thee, 1998). National authorities, eager to reap highest short-term benefits, have mainly directed these investors to new industrial parks established at attractive and cheap, although infrastructurally rather bare, locations such as Batam Island near Singapore. Accordingly, the investments follow the usual pattern of exploiting low-cost low qualified labour for carrying out routine jobs (Dicken, 1998). There has been some success in the export production of electronics, but the industrial parks maintain an enclave character. The nation sacrifices options of a more sustainable integration of foreign technology firms, of ISI benefits for its NSI, which could be reaped from locating TNCs in well-established indigenous RSIs that offer relatively good absorptive capacities, such as Bandung (Fromhold-Eisebith, 2002b; Fromhold-Eisebith and Eisebith, 2002).

Summing up over the Indonesian case, the country shows particular deficits in utilizing its RIS potential in ways that would be necessary for a functional NSSI. On the one hand, this has negative implications for the national level as existing RSIs remain too weak for jointly supporting a move ahead of the entire NSI, which could also reduce the country's high dependency on international technology imports. On the other hand, the neglect and disregard of RSI opportunities also causes an insufficient embedding of international agents into collaborative structures in the country and the loss of chances for innovative impulses on domestic industries on a wider national scale. Generally, the country could have profited more from the previous systemic and focused national STI approach if it had been more balanced and less ambitious in the selection of thrust areas. By raising awareness of existing RSIs, by allowing them to more autonomously regulate, develop and market their assets, and by strategically integrating that potential into international strategies, as suggested in the NSSI framework (Table 4.1), the Indonesian innovation system could probably gain a great deal.

Comparing NSSI Features of India and Indonesia: Factors of Success and Failure

The depicted spatial innovation system structures of India and Indonesia are evaluated in more analytical terms against the backdrop of ideal type NSSI structures. This reveals the following commonalities and differences in the two Asian approaches.

As is suggested for a functional NSSI, in both India and Indonesia the national state and respective agencies have occupied a leading role in STI promotion. This takes account of the strong competencies and powers of national authorities to decide on overall political strategies, to develop an adequate legislative framework and build up infrastructure, institutions and incentives (referred to in Table 4.1), supported by the fact that central agencies usually pool the financial assets devoted to STI promotion (Lundvall, 1992b; Freeman, 2002; Bathelt and Depner, 2003). In India and Indonesia the state assumes a dominant role over markets: important prerequisites for the emergence of innovative nations and regions, notably infrastructure and integrated R&D promotion policies, are the responsibility of the public sector because private initiative has been rather weak so far. The ideal NSSI, however, requires a state which is far from centralist, authoritarian and parasitic, but combines strength, justice and fair balance in allocating sufficient power and finance to other levels. This seems to apply, to some extent, to the Indian example, but not at all to the Indonesian one.

Another commonality of the depicted two cases is that national level authorities, due to a good overview of international trends and strategic objectives, have identified lead technology sectors which offer particularly good and profitable development opportunities. Accordingly, both have agglomerated relevant R&D and education infrastructure as well as corresponding public sector enterprises in suitable regions, in line with suggested measures (Table 4.1). India, however, has been more successful with its choice (more sensitively reacting to real market signals) than Indonesia where individual prestige outweighs collective usefulness. In contrast to an ideal type NSSI, however, central agencies in India and Indonesia hardly monitor innovative activities in various parts of the country from a comparative perspective. Hence, they miss out on identifying promising localities, sectors and actor groups based on reports from the regions, and do not integrate this information into the design of further national strategic instruments.

Getting to major differences of the Indian and Indonesian models, only the former partly fulfils the NSSI requirement that national authorities, when conceptualizing and implementing STI policies, should especially take account of region-specific potentials and support developments there. Only here have federal decision-making structures allowed for a combination of top-down and bottom-up processes emerging from regional initiatives and competences in terms of an RSI. This pays off for India, as shown by enabled regional dynamics of interactive learning and collaboration between public or private agents, which also productively integrate international agents like TNCs. In this case regional authorities manage to exploit the fact that they are much closer – in a physical as well as social sense – to the actors to be taken care of than are national ministries. Hence, regional public support can more directly affect collaboration behaviour, also based on pre-existing social relationships between private and public actors that support the proliferation of innovation-related interaction in an effective RSI.

The Indonesian government, quite the opposite, has not enabled promising technology regions to fulfil such tasks and thus has not adequately supported functional RSI in the country. There has been insufficient allocation of strategic STI competences and financial powers to regional agencies, and no effort has explicitly directed industrial activities and foreign investors in technology sectors to the most absorptive regions for effectively linking them to the ISI; all this contradicts the major NSSI features (Table 4.1). It appears that regional developments in innovation related fields, although present in many regions, are hardly even noticed on the central level, even less taken into account in integrated STI, investment and industrial development strategies. No wonder that the national economy and NSI do not profit from wider benefits of well-running RSIs that enrich national capacities in an international setting.

Taken as a whole, only the Indian strategic approach has succeeded in using national regulation to integrate international agents into a scale crossing innovation system based on attractive and absorptive RSIs. This case is much closer to the recommended NSSI qualities than the Indonesian one. Indonesia ignores the insight that, regarding their complementary roles in innovation, the national and regional systemic support levels should each take over the functions they can, by their nature, provide best and mutually contribute to the functionality of the other ones, including links to the ISI.

CONCLUSIONS

'Systems of innovation are increasingly complex and intertwined, with regional, national, and international levels of integration of innovating activities' (Hotz-Hart, 2000, p. 444). Regarding the issue of how the system scales of RSI, NSI and ISI could effectively be linked in policy strategies for Asian NICs, this chapter suggests the scale-bridging approach of a 'supersystemic' NSSI. Realizations that the three major levels do not function independently from each other, but mutually rely on each other's strengths and specific system qualities in order to interact productively, should be regarded in innovation and technology related development strategies. While collaboration enhancing RSIs tend to emerge only from a good base of nationally established infrastructural and institutional foundations, it is necessary for there to be successful, socially embedded and networked RSIs for constituting a well-functioning NSI. In order to sustainably profit from an integration in ISI and exogenous inflows of know-how, it is essential to address suitably absorptive RSIs in the course of respective initiatives.

Against this backdrop the examples of STI promotion strategies of India and Indonesia have been evaluated, revealing some superiority of the Indian model compared to the Indonesian one with respect to functional NSSI features. India – more implicitly than explicitly – has substantially profited from scale-crossing system constellations in particular relating to the IT service sector. Due to the country's federal, decentralized and democratic policy setting, various partly competing RSIs have been endorsed to take over their beneficial role as nationally supported platforms for the creative combination of indigenous and foreign (TNC) technology capabilities. Indonesia, with its prevailing centralist approach, in contrast, has not realized or adequately utilized the powers of interacting scales of innovation systems, although possessing some good RSI and ISI potential that lack proper connection. In sum, however, neither country fully follows an NSSI strategy and systematically regards the virtues of the RSI level in national STI strategies, including ISI dimensions. Hence, neither of the two cases represents the 'ideal way' of adequately incorporating the regional, national and international scales.

Which general recommendations can be derived for STI policies in Asia? Insights developed here may lead to a better coordinated functional division of tasks between innovation system scales as suggested by the NSSI framework. *National* authorities should concentrate on aspects of innovation support that need to be actively governed, regulated, coordinated and supervised based on strategic, yet balanced policy guidelines and decisions. These integrate STI, investment and industrial policies in a common logic. Yet, to achieve 'supersystemic' qualities national agencies need to act as 'masters of scales' in conducting their tasks. This means that they need to actively distribute sufficient power to, and use the potential of, regional levels or RSIs with an international perspective, and to expediently monitor and combine regional and international assets in innovation support. Accordingly, national STI and industrial policies in Asian NICs ought to become more sensitive and reflective regarding spatial scales, conceiving their regions as the major base where technology-related organizations and institutions should be agglomerated and decisive processes of regional and international innovative interaction must be enabled. In this context it is especially important to direct foreign investors in technology sectors to the most absorptive RSI in the country.

At the same time a functional NSSI is characterized by sets of strong RSIs. Based on favourable infrastructural and institutional conditions created by the national system, the region is the level where self-dynamic processes of innovative cooperation should be stimulated. As these usually defy their forced production through policy measures but rather rely on socially embedded contacts, a major task of *regional* authorities is to create opportunities and increase chances for such contacts to materialize between the right technology agents. Including foreign or re-migrating domestic scientists and investors and improving their informal links to local communities is especially important for Asian latecomer economies because this supports the inflow, diffusion and application of internationally competitive know-how. The promotion of innovation-based entrepreneurship is also best delegated to regional authorities.

This chapter has surely left open many questions concerning concrete ways to organize an effective NSSI. Nations should decide by themselves on details of their systemic STI strategies since these will work best when coordinated according to nation-specific institutional structures, political cultures and ways of conduct, which allows for an evolutionary development (as marks every NSI; Lundvall, 1992b; Edquist, 1997). How to institutionalize 'supersystemic' links between scales, or whether national or regional tasks are better conducted by a 'single-window' agency or a network of specialized authorities, cannot be recommended here. Anyway, when countries orient their innovation-enhancing national policies towards the direction highlighted above, this should not be taken as a guarantee for success. But employing an NSSI strategy and thoughtfully combining scales in the creation of an innovation system is definitely more in line with factors of success identified by research than centralist NSI policies that ignore the potential of localities. It therefore is probably more likely to achieve systemic effectiveness.

ACKNOWLEDGEMENTS

The author thanks Jan Vang and an anonymous referee for their valuable comments and suggestions relating to an earlier version of the chapter.

NOTES

1. This is why country-specific analyses and policy recommendations drawing on the NSI concept tend to miss out addressing issues of regionalization (see, for example, Altenburg et al., 2004).
2. Empirical work in Europe has provided evidence that especially technology-intensive small and medium sized firms are highly oriented towards regional collaboration in innovation (Koschatzky and Sternberg, 2000). The advantages possibly induced by regional concentrations of actors and activities are captured by a whole range of 'territorial innovation models', such as industrial district, creative milieu, innovative cluster or learning region (for overviews see Malecki, 1997; Moulaert and Sekia, 2003).
3. Altogether the research team (special thanks goes to my team partner Günter Eisebith) carried out interviews with over 200 executives of technology-oriented companies, staff of universities and R&D organizations, and representatives of public and private industrial support agencies.

REFERENCES

Aggarwal, A. (2001), 'Technology policies and acquisition of technological capabilities in the industrial sector: a comparative analysis of the Indian and Korean experiences', *Science Technology and Society*, **6** (2), 255–304.

Altenburg, T. et al. (2004), 'Strengthening knowledge-based competitive advantages in Thailand', Reports and Working Papers 1/2004, Bonn: German Development Institute.

Archibugi, D. and J. Michie (1997), 'Technological globalisation and national systems of innovation: an introduction', in D. Archibugi and J. Michie (eds), *Technology, Globalisation and Economic Performance*, Cambridge, UK: Cambridge University Press, pp. 1–23.

Asheim, B.T. (2000), 'Industrial districts: the contributions of Marshall and beyond', in G.L. Clark et al. (eds), *The Oxford Handbook of Economic Geography*, Oxford: Oxford University Press, pp. 413–31.

Asheim, B.T. and P. Cooke (1999), 'Local learning and interactive innovation networks in a global economy', in E.J. Malecki and P. Oinas (eds), *Making Connections. Technological Learning and Regional Economic Change*, Aldershot, UK: Ashgate, pp. 145–78.

Bathelt, H. and H. Depner (2003), 'Innovation, Institution und Region. Zur Diskussion über nationale und regionale Innovationssysteme', *Erdkunde*, **57** (2), 126–43.

Beier, C. and G. Ferrazzi (1998), 'Fiscal decentralization in Indonesia: a comment on smoke and Lewis', *World Development*, **26** (12), 2201–11.

Booth, A. (1998), *The Indonesian Economy in the Nineteenth and Twentieth Centuries: A History of Missed Opportunities*, London: Macmillan.

BPPT/STAID (1993), *Science and Technology Indicators of Indonesia 1993*, Jakarta: BPPT.

Braczyk, H.-J., P. Cooke and M. Heidenreich (eds) (1998), *Regional Innovations Systems: The Role of Governances in a Globalized World*, London: UCL Press.

Bunnell, T. G. and N.M. Coe (2001), 'Spaces and scales of innovation', *Progress in Human Geography*, **25**, 569–89.

Chesnais, F. (1992), 'National systems of innovation, foreign direct investment and the operations of multinational enterprises', in B.-A. Lundvall (ed.), *National Systems of Innovation*, London: Pinter, pp. 265–95.

Chowdhuri, A. and I. Islam (1993), *The Newly Industrialising Economies of East Asia*, London: Routledge.

Cohen, M. (2000), 'New flight plan. Indonesia's heavily indebted state-run aircraft maker seeks new ways to keep the country's hi-tech dream alive', *Far Eastern Economic Review*, 2 March, p. 45.

Cooke, P. (1992), 'Regional innovation systems: competitive regulation in the new Europe', *Geoforum*, **23**, 365–82.

Cooke, P. (2001), 'Regional innovation systems, clusters, and the knowledge economy', *Industrial and Corporate Change*, **10** (4), 945–74.

Cooke, P., M.G. Uranga and G. Etxebarria (1997), 'Regional innovation systems: institutional and organisational dimensions', *Research Policy*, **26**, 475–91.

Cooke, P., M.G. Uranga and G. Etxebarria (1998), 'Regional systems of innovation: an evolutionary perspective', *Environment and Planning A*, **30**, 1563–84.

Desai, A.V. and N. Bhalerao (1996), *Industrial Economy of India*, 2nd edn, Mumbay, Delhi and Nagpur: Himalaya Publ. House.

Dicken, P. (1998), *Global Shift. Transforming the World Economy*, 3rd edn, London: Paul Chapman.

Edquist, C. (ed.) (1997), *Systems of Innovation. Technologies, Institutions and Organizations*, London: Pinter.

Enos, J.L., S. Lall and Y. Mikyung (1997), 'Transfer of technology: an update', *Asian-Pacific Economic Literature*, **1** (1), 56–66.

Freeman, C. (1987), *Technology Policy and Economic Performance*, London: Pinter.

Freeman, C. (1995), 'The "National System of Innovation" in historical perspective', *Cambridge Journal of Economics*, **19**, 5–24.

Freeman, C. (2002), 'Continental, national and sub-national innovation systems – complementarity and economic growth', *Research Policy*, **31**, 191–211.

Fritsch, M. (2002), 'Measuring the quality of regional innovation systems: a

knowledge production function approach', *International Regional Science Review*, **25** (1), 86–101.
Fromhold-Eisebith, M. (1999), 'Bangalore: a network model for innovation-oriented regional development in NICs?', in E.J. Malecki and P. Oinas (eds), *Making Connections: Technological Learning and Regional Economic Change*, Aldershot, UK: Ashgate, pp. 231–60.
Fromhold-Eisebith, M. (2001), *Technologieregionen in Asiens Newly Industrialized Countries. Strukturen und Beziehungssysteme am Beispiel von Bangalore, Indien, und Bandung, Indonesien*, Series Wirtschaftsgeographie, vol. 18, Münster and Hamburg: LIT-Verlag.
Fromhold-Eisebith, M. (2002a), 'Qualified labour migration and regional knowledge economies', in R. Hayter and R. LeHeron (eds), *Knowledge, Industry and Environment: Institutions and Innovation in Territorial Perspective*, Aldershot, UK: Ashgate, pp. 125–43.
Fromhold-Eisebith, M. (2002b), 'Regional cycles of learning. Foreign multinationals as agents of technological upgrading in less developed countries', *Environment and Planning A*, **34** (12), 2155–73.
Fromhold-Eisebith, M. (2006), 'Infotech industries and regional disparities in India', in S. Corbridge, S. Kumar and S. Raju (eds), *Colonial and Postcolonial Geographies of India*, New Delhi: Sage.
Fromhold-Eisebith, M. and G. Eisebith (2002), 'The Indonesian technology region of Bandung: high potential, low profile', *International Development Planning Review*, **24** (1), 41–57.
Gammeltoft, P. (2004), 'Development of firm level technological capabilities. The case of the Indonesian electronics industry', *Journal of the Asia Pacific Economy*, **9** (1), 49–69.
Heeks, R. (1996), *India's Software Industry: State Policy, Liberalisation and Industrial Development*, New Delhi: Sage Publications.
Héraud, J.-A. (2003), 'Regional innovation systems and European research policy: convergence or misunderstanding?', *European Planning Studies*, **11** (1), 41–56.
Hill, H. (1996), *The Indonesian Economy since 1966: Southeast Asia's Emerging Giant*, Cambridge: Cambridge University Press.
Hill, H. and K.W. Thee (eds) (1998), *Indonesia's Technological Challenge*, Singapore: Institute of Southeast Asian Studies.
Hotz-Hart, B. (2000), 'Innovation networks, regions, and globalization', in G.L. Clark et al. (eds), *The Oxford Handbook of Economic Geography*, Oxford: Oxford University Press, pp. 432–50.
Howells, J. (1999), 'Regional systems of innovation?', in D. Archibugi et al. (eds), *Innovation Policy in a Global Economy*, Cambridge: Cambridge University Press, pp. 67–93.
Koschatzky, K. and R. Sternberg (2000), 'R and D cooperation in innovation systems – some lessons from the European Regional Innovation Survey (ERIS)', *European Planning Studies*, **8** (4), 487–501.
Lall, S. (1998a), 'Technological capabilities in emerging Asia', *Oxford Development Studies*, **26** (2), 213–43.
Lall, S. (1998b), 'Technology policies in Indonesia', in H. Hill and K.W. Thee (eds), *Indonesia's Technological Challenge*, Singapore: Institute of Southeast Asian Studies, pp. 136–68.
Lundvall, B.-A. (1992a), 'Introduction', in B.-A. Lundvall (ed.), N*ational Systems of Innovation*, London and New York: Pinter, pp. 1–19.

Lundvall, B.-A. (ed.) (1992b), *National Systems of Innovation*, London and New York: Pinter.
Lundvall, B.-A. (2003), 'National innovation systems: history and theory', paper presented at the NSTDA-JICA seminar on innovations systems in Asian economies, Bangkok, 4–5 September, Aalborg.
Lundvall, B.-A. and P. Maskell (2000), 'Nation states and economic development: from national systems of production to national systems of knowledge creation and learning', in G.L. Clark et al. (eds), *The Oxford Handbook of Economic Geography*, Oxford: Oxford Universiy Press, pp. 353–72.
Malecki, E.J. (1997), *Technology and Economic Development. The Dynamics of Local, Regional and National Competitiveness*, 2nd edn, London: Longman.
Malerba, F. (2002), 'Sectoral systems of innovation and production', *Research Policy*, **31**, 247–64.
Marton, K. (1986), *Multinationals, Technology, and Industrialization: Implications and Impact in Third World Countries*, Lexington and Toronto: Lexington Books.
Maskell, P. and A. Malmberg (1999), 'Localised learning and industrial competitiveness', *Cambridge Journal of Economics*, **23** (2), 167–86.
Moulaert, F. and F. Sekia (2003), 'Territorial innovation models: a critical survey', *Regional Studies*, **37**, 289–303.
Mowery, D.-C. and J.-E. Oxley (1995), 'Inward technology transfer and competitiveness: the role of national innovation systems', *Cambridge Journal of Economics*, **19** (1), 67–93.
Nelson, R.R. (ed.) (1993), *National Innovation Systems. A Comparative Analysis*, New York and Oxford: Oxford University Press.
Nelson, R.R. and N. Rosenberg (1993), 'Technical innovation and national systems', in R.R. Nelson (ed.), *National Innovation Systems. A Comparative Analysis*, New York, Oxford: Oxford University Press, pp. 3–21.
Niosi, J. and B. Bellon (1994), 'The global interdependence of national innovation systems: evidence, limits, and implications', *Technology in Society*, **16** (2), 173–97.
Oinas, P. and E.J. Malecki (1999), 'Spatial innovation systems', in E.J. Malecki and P. Oinas (eds), *Making Connections: Technological Learning and Regional Economic Change*, Aldershot, UK: Ashgate, pp. 7–34.
Prakashi, S.C. and G.P. Phondke (eds) (1995), *Science Technology and Industrial Development in India*, New Delhi: Wiley Eastern.
Rice, R.C. (1998), 'The Habibie approach to science, technology and national development', in H. Hill and K.W. Thee (eds), *Indonesia's Technological Challenge*, Singapore: Institute of Southeast Asian Studies, pp. 185–98.

5. Thailand's national innovation system in transition

Patarapong Intarakumnerd

1. INTRODUCTION

While the study on the NIS concept as a whole is still at an early stage, the study on NISs in developing countries is at an even more primitive stage. Most studies on developing countries were on countries such as Korea, Taiwan and Singapore, which have more aggressive policies and 'intensive technological learning', hence, to a certain extent, successfully catching up with developed countries (see Kim, 1993; Hou and Gee, 1993; Wong, 1995; Wong, 1999). Little is known about the innovation, entrepreneurships, dynamics and changes in a less successful developing country where its innovation system can be characterized as weak and fragmented. It is also interesting to examine whether and how it can be transformed to a stronger and coherent one.

This chapter tries to supplement these studies of the NIS by exploring Thailand as a case study. It will especially focus on the transitional process of its NIS from being weak and fragmented to becoming stronger and more coherent.

The economic performance of Thailand during the past 40 years has been rather impressive. During this period, the growth rate of GDP of Thailand was around 7 per cent, more or less similar to those of the Newly Industrialized Economies (Korea, Taiwan, Singapore and Hong Kong). However, the policy regime adopted by Thailand was different, in that Thailand appeared primarily to take inspiration from the NIEs' export orientation and less from their state activism (this will be discussed in detail in section 3.1). Like NIEs, the Thai economic structure has also changed from an agriculture-based economy to an economy in which the industrial (manufacturing in particular) sector has gained distinctive significance. The share of the agriculture sector in GDP has been reduced remarkably from almost 40 per cent in the 1960s to approximately 10 per cent in the late 1990s, while that of the industry sector has experienced exactly the reverse situation. Interestingly, there was a change in the composition of Thai

Table 5.1 Distribution of manufactured export by technological categories (%)

Sector	Korea			Singapore			Taiwan			Thailand		
	80	90	99	80	90	99	80	90	99	80	90	99
Resource-based	9.0	6.8	11.6	44.4	26.9	13.2	9.8	8.2	9.2	21.7	13.8	10.7
Labour-intensive	49.2	40.8	23.2	10.6	10.3	7.6	54.3	41.2	31.0	47.0	45.5	35.8
Scale-intensive	23.6	19.3	21.0	9.3	5.9	5.5	9.1	10.3	10.6	7.8	6.3	7.7
Differentiated	11.3	15.6	18.7	20.5	22.3	21.2	12.4	20.6	20.4	22.2	14.1	19.5
Science-based	6.9	17.4	25.5	15.1	34.6	52.5	14.5	19.8	28.9	1.2	20.2	26.4

Source: Calculated from UN Comtrade data base.

exports along the line of NIEs. The share of once-dominating resource-based and labour-intensive exports has gone down while that of science-based and differentiated exports has gone up especially in the 1990s (see Table 5.1). Nonetheless, one cannot argue that Thai exports have become more technology-intensive, as the categories do not reflect the sophistication of technological activities required to produce goods; for example, what is categorized as science-based exports might only be assembled locally, while their technologically sophisticated and high-value-added components are imported. Nonetheless, this trend suggests a general change in the structure of the Thai economy.

In 1997, Thailand had the worst economic crisis in 40 years. Apart from other causes like untimely and under-regulated financial liberalization and the burst of the bubble economy, diminishing international competitiveness resulting from the weak and fragmented NIS was one of the main reasons (see details in Intarakumnerd et al., 2002). In return, the crisis also affected the actors of the Thai NIS. This will be discussed in detail later.

2. METHODOLOGY AND SCOPE OF WORK

Adapted from Arnold et al. (2000), we attempt to analyse the national innovation system of Thailand in a broad perspective by examining the evolution of roles, capabilities, and the linkages of the following actors:

- Government
- University
- Private firms

- Private bridging organizations (industry, trade, and professional organizations)
- Financial intermediaries/markets
- Institutional context (see Figure 5.1).

Taking into account idiosyncratic differences in terms of historical, institutional, cultural and international contexts, the study, to some extent, analyses the evolution of the Thai NIS in comparison with leading East Asian countries (namely, Japan, Korea and Taiwan) in order to shed light on its strengths and weaknesses.

As already mentioned, together with existing literature on Thailand's NIS and those of Japan, Korea and Taiwan, the main source of information for our study is the R&D/Innovation Survey, carried out twice in 2000 and 2002. The survey in 2000 was the first of its kind in Thailand and it covered both R&D and other technological innovation activities only in the manufacturing sector. The second survey, for the year 2002, included the service sector in order to gain a better understanding of the nature and differences of R&D and innovation activities in both manufacturing and services.[1] The survey adopted definitions and methodologies used by the OECD's Frascati Manual and the Oslo Manual, and was in line with those used in several countries in Asia (such as Singapore and Malaysia).

3. EVOLUTION OF THAILAND'S NATIONAL INNOVATION SYSTEM: MAIN ACTORS AND THEIR LINKAGES

3.1 Government: From a Passive Regulator to a Pro-active Competitiveness Promoter

After serious attempts at industrialization in the early 1960s when the country had its first National Economic and Social Development Plan, the government policy regime regarding science, technology and innovation can be divided into two phases: before the Thaksin government and after the Thaksin government.

From the 1960s up to the present government of Prime Minister Thaksin Shinawatra (starting in January 2001), the scope of the S&T policy in Thailand was rather narrow. It covered only four conventional functions, namely, research and development, human resource development, technology transfer, and S&T infrastructure development. This narrow scope of S&T was very much based on the perception that private firms were 'users' of S&T knowledge mainly produced by government agencies and

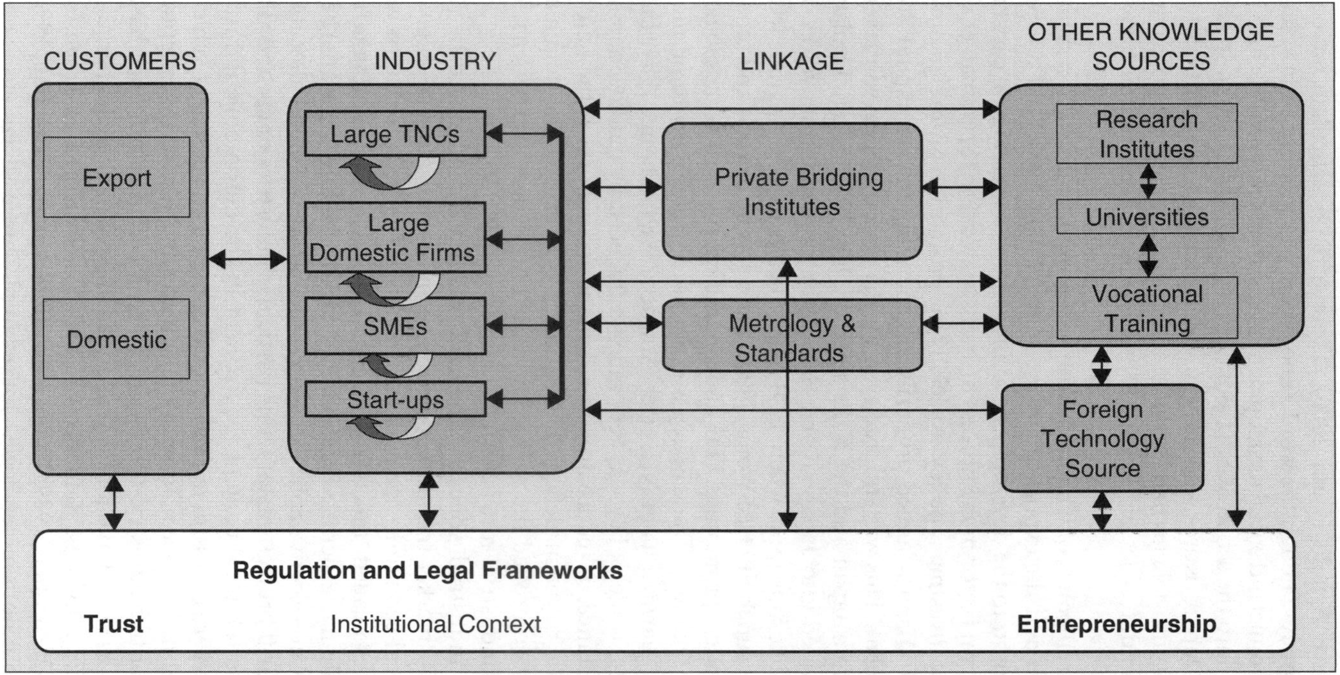

Source: Adapted from Arnold et al. (2000).

Figure 5.1 Framework for analysing Thailand's national innovation system study

universities (see Arnold et al., 2000). There was no articulate national innovation policy. Though the word 'innovation' was mentioned in several national economic and social development plans, it was not wholeheartedly incorporated into the scope of S&T policies (see Lauridsen, 2002). In addition, unlike Japan, Korea and Taiwan, S&T elements were not part of broader economic policies namely, industrial policy, investment policy and trade policy and, to a lesser extent, education policies (see Intarakumnerd et al., 2002).

The industrial policy of Thailand, both in the import substitution regime (1960s–1970s) and the export oriented regime (1980s–1990s) did not pay enough attention to the development of indigenous technological capability as an integral factor in the process of industrialization (Sripaipan et al., 1999: 37). The investment policy, especially the promotion of foreign direct investment (FDI), was aimed primarily at generating inward capital flow and employment. This was still true even for the period of heavy inflow of investment from Japan and the NIEs since the late 1980s when Thailand was incorporated into Japanese and East Asian production networks. In Thailand, unlike Singapore where FDI is specifically used to upgrade local technological capability (see Wong, 1999), there was no explicit and proactive link between promoting FDI and upgrading of local technological capability in Thailand. Trade policy was not used strategically to promote technological learning as in NIEs (see Amsden, 1989; Chang, 1994; Lall, 1996). Instead, trade policy was very much influenced by macroeconomic policy, for instance, to reduce domestic demand for imports at the time of the balance of payment deficit. The Ministry of Finance, the dominant agency which controlled the policy, had little knowledge or experience of industry and industrial restructuring (Lauridsen, 2000: 16–20).

Moreover, industrial policy in Thailand was heavily influenced by the World Bank's 'market-friendly' approach to industrialization and, given the neoclassical economics inclination of leading Thai technocrats, it was limited to the so-called 'functional' intervention such as promoting infrastructure building, general education, and export push in general. There were virtually no selective policy measures, such as special credit allocation and special tariff protection, targeting particular industries or clusters. The exception was the local content requirement in the automobile industry, which was rather successful in raising local contents of passenger vehicles to 54 per cent in 1986 (see Doner, 1992). Interestingly, with the exception of the automotive industry, there were no reciprocal performance-based criteria (such as export and local value-added and technological upgrading targets) set for providing state incentives such as in Korea or Japan (see Johnson, 1982; Amsden, 1989; Evans, 1989; 1998; Chang, 1994; Lall, 1996). Investment promotion privileges, for example, were given away once approved. The intention to attract foreign direct investment and promote export overshadowed the need to develop local initiatives and indigenous technological capabilities. As a result, linkages between multinational corporations and local firms were also weak. Unlike Taiwan, the governmental protection and promotion, without strengthening absorptive capabilities of Thai suppliers, left a profound impact on the weak technology and suppliers' network of industries (Vongpivat, 2003). Later, in the late 1980s and 1990s, the Thai government subscribed to the so-called 'Washington Consensus' as serious attempts were made to promote deregulation, privatization and liberalization (both in trade and financial systems). By pursuing this, an implicit assumption was that this kind of policy would lead to an increase in technological capability. The loss in the country's competitiveness evident by much slower export growth in the mid-1990s, leading to the economic crisis in 1997, proved that the Washington Consensus policies did not work as expected (see details in Intarakumnerd et al., 2002).

The major shift in policy regime came recently under the present Thaksin government. The media and academics in Thailand and Southeast Asia labelled this government's distinctive policy as 'Thaksinomics' (Thaksin's economics). Dual track policy is the main thrust of Thaksinomics. The government is trying to enhance the nation's international competitiveness by strengthening the 'external' side of the Thai economy, namely, export, foreign direct investment and tourism. At the same time, it is attempting to increase the capabilities of domestic and grass-roots economies by implementing projects like the Village Fund (one million Baht to increase the local capabilities of each village); a three-year debt moratorium on farmers' debt; the One Tambon, One Product project (supporting each Tambon[2] to have one outstanding product); and the People's Bank (giving loans to underprivileged people without asking for collateral).

This government, unlike its predecessors who concentrated on macroeconomic stability, focuses more on enhancing meso- and micro-level foundations for international competitiveness. The high priority of the 'competitiveness' issue on the government's agenda is illustrated by the establishment of the National Competitiveness Committee chaired by the Prime Minister. It is the first time that the Thai government has had serious 'selective' policies addressing specific sectors and clusters. The government has declared five strategic clusters which Thailand should pursue: automotive, food, tourism, fashion and software. Clear visions have been given to these five clusters, namely: the Kitchen of the World (food cluster), the Detroit of Asia (automotive cluster), Asia Tropical Fashion, World Graphic Design and Animation Centre (software cluster), and the Asia Tourism Capital. Building the innovative capabilities of the nation is highly regarded as a very important factor increasing and sustaining Thailand's international competitiveness. An 'innovative nation with wisdom and learning base' is one of seven Thailand Dreams projected by this government. To make this dream come true, several strategies have been devised. These include continuous investment in R&D and technology, an environment for attracting and stimulating innovation, high accessibility to knowledge and information across the nation, fluent English as a second language, a strong learning basis such as a passion for reading, better accessibility to cheap but good books, thinking school developed together with an innovation movement (see Phasukavanich, 2003).

The new ten-year Science and Technology Action Plan (2003–2013) places the concept of the national innovation system and industrial cluster at its heart. The scope of the plan is much broader than the aforementioned four functional areas. Measures to stimulate innovation and to strengthen the national innovation system and industrial clusters are explicitly highlighted. Equally important, the Board of Investment (BOI) has substantially changed its policy by paying more attention to issues underlying the long-term competitiveness of the country, including development of indigenous technological capability and human resources. A special investment package promoting 'Skill, Technology and Innovation or STI' has been initiated. Firms can enjoy one or two years' extra tax incentives if they perform the following activities in the first three years: spending on R&D or design at the level of at least 1–2 per cent of their sales; employing scientists or engineers with at least a bachelor's degree to constitute at least 5 per cent of their workforce, spending at least 1 per cent of their total payroll on training of their employees; and spending at least 1 per cent of total payroll on training personnel of their local suppliers.

To carry out these changes, the government has introduced the private sector's management style to improve efficiency and effectiveness in the public sector. Chief Executive Officer (CEO) style is now being implemented both at central and local government levels in order to integrate related government policies under clear leadership. Also, a Performance-Based Management (PBM) has been put in place, which clearly illustrates a contractual relationship and delegation of authority in the bureaucratic lines of governance.

3.2 Private Firms

Several studies of Thai firms conducted since the 1980s state that most firms have grown without deepening their technological capabilities in the long run, and their technological learning has been very slow and passive (see Bell and Scott-Kemmis, 1985; Chantramonklasri, 1985; Thailand

Development Research Institute, 1989; Dahlman and Brimble, 1990; Tiralap, 1990; Mukdapitak, 1994; Lall, 1998). The recent World Bank's study (see Arnold et al., 2000) confirms this long-standing feature of Thai firms. Only a small minority of large subsidiaries of transnational corporations (TNCs), large domestic firms and small–medium enterprises (SMEs) have a capability in R&D, while the majority are still struggling with increasing their design and engineering capability. For a very large number of SMEs, the key issue is much more concerned with building up more basic operational capabilities, together with craft and technician capabilities for efficient acquisition, assimilation and incremental upgrading of fairly standard technology. The slow development of technological capability of Thai firms is quite different from what characterized firms in Japan, Korea and Taiwan. Firms in these countries moved rather rapidly from mere imitators to innovators. As early as the 1960s, Japanese firms became more innovative, invested heavily in R&D and relied less on importation of foreign technologies (Odagiri and Goto, 1993). In general, firms in Korea and Taiwan, where industrialization (beginning with import substitution) started more or less in the same period as in Thailand, were more successful in increasing absorptive capacity (of foreign technology) and deepening indigenous technological capabilities in several industries (see for example, Amsden, 1989; Kim, 1993; Lall, 1996; Hobday, 1995; Kim, 1997). In the electronics industry, for instance, Korean and Taiwan firms were able to climb technological ladders (from simple assembly to own design and R&D) by exploiting institutional mechanisms such as OEM and ODM to help latecomer firms in those countries to acquire advanced technology, and access demanding foreign markets (see Hobday, 1995).

Comparison between the Thailand and Korea Innovation Surveys both conducted in 2002 illustrates the differences in terms of innovative capabilities of these two countries. Table 5.2 clearly shows that companies in Thailand lag far behind companies in Korea with respect to innovation. More than 40 per cent of Korean firms carried out innovations against just above 10 per cent in Thailand. It is striking that a much higher share of companies in Korea carry out product innovations. This could be an indication that Thai companies are at the stage where they would rather use their resources to improve the production process than the product itself, which in turn could hint at a rather OEM-oriented economy. At the same time very few companies in Thailand do both product and process innovations, which is very common in Korea. This reflects the more advanced innovation behaviour of Korean companies.

However, more intense competition in the global market and the economic crisis in 1997 has, to some degree, led to a change in behaviour

Table 5.2 Share of innovating companies in Thailand and Korea (%)

	Thailand	Korea
Innovating	11.2	42.8
Product and process innovation	2.9	21.0
Only product innovation	4.1	17.0
Only process innovation	4.3	4.0

Source: Thailand R&D/Innovation Survey 2002 and Korean Innovation Survey 2002.

among Thai firms. The first innovation survey indicates that more than 80 per cent of R&D performing firms express strong interest in increasing their spending in the following three years. This finding is supported by recent studies of Thai firms after the 1997 economic crisis (see, for example, Thailand Development Research Institute, 1998; Arnold et.al., 2000). It shows a few interesting phenomena:

a. Several large conglomerates recently increased their R&D activities.
b. After the crisis they changed their long-standing attitude of relying on off-the-shelf foreign technologies . . .
c. toward developing in-house R&D capabilities.
d. A number of smaller companies recently increased their technological efforts by collaborating with university R&D groups in order to stay ahead in the market or to enter a more profitable market section.
e. Recently, several subcontracting suppliers in the automobile and electronics industries were forced by their TNC customers/partners to strengthen their efforts to modify product design and improve efficiency and were able to absorb the design and know-how from foreign experts.
f. There were emerging new start-up firms (less than 50 employees) relying on their own design, engineering or development activities. These companies were managed by entrepreneurs who had a strong R&D background from studying or working abroad. Many of them were 'fabless' (no fabrication activities) companies.

A more recent study by NSTDA's researchers also indicates the positive change in Thai firms. Several locally-owned OEM manufacturers experiencing external pressure, especially from foreign customers that adopted global sourcing strategies, started to develop products through their own designs and brand names (see Intarakumnerd and Virasa, 2002).

3.3 Universities and Government Research Technology Organizations (RTOs)

From the Thailand R&D/Innovation Survey 2002 and Korean Innovation Survey 2002, it can be seen that universities and research institutes were regarded as much more important sources of information by Korean firms than by Thai firms (see Table 5.3).

The technological activities of public Research Technology Organizations (RTOs)[3] mainly focus on R&D and providing technical services such as testing and calibrating. They are not particularly involved in assisting firms in building up their 'internal' technological capabilities, especially 'non-R&D' capabilities such as technology assimilation, adaptation, designing and engineering, which are the technological thresholds typically faced by most Thai firms. In this respect, Thai RTOs behave differently from those of Japan and East Asian NIEs, when their levels of development were more or less at the current level of Thailand. A recent study done by the College of Management of Mahidol University summarizes the gaps in industry–academia collaboration. It demonstrates the weaknesses on both sides, which obstruct meaningful collaboration (see Table 5.4).

Table 5.3 Importance of external information sources

Thailand		Korea	
Clients	77.4	Customers	77.7
Internet	63.0	Competitors	69.3
Parent/associate company	61.2	Exhibition	65.5
Locally-owned suppliers	59.9	Internet	64.9
Specialist literature	56.6	Component suppliers	61.7
Professional conference & meetings	55.2	Patents	59.8
Foreign-owned suppliers	54.8	Equipment suppliers	57.7
Fairs and exhibitions	53.1	**Universities**	53.6
Competitors	42.1	Enterprises within the group	52.9
Technical service providers	40.2	**Public Research Institutes**	52.6
Universities or other higher education institutes	35.8	New personnel	51.9
Business service providers	33.1	Trade Associations	44.2
Patent disclosures	32.0		
Government or private non-profit research institutes	29.5		

Source: Thailand R&D/Innovation Survey 2002 and Korean Innovation Survey 2002.

Table 5.4 Gaps in industry–academia collaboration

Industries	Gaps	Academia
• Passive actors in initiating cooperative projects • No tangible/substantial activities that might lead to collaboration with education institutes	• Lacking continuous cooperative projects or activities and motivation for collaboration • Clear goals and objectives of the collaboration are missing • Lacking mediators who can understand both sides, coach, and foster the relationship • Lacking analysis of problems from the industry's perspective	• Major activities are not two-way cooperation. Education institutes usually initiate and dominate the relationship • Linkages are more in terms of asking for help than achieving the project together for maximum benefit of both parties • No substantial linkages in term of R&D projects

Source: College of Management, Mahidol University (2003).

Nonetheless, public RTOs and universities are under pressure from the present Thaksin government and the Budget Bureau to increase their revenue, hence reducing their reliance on the national budget. They are being forced to become more relevant to the needs of industry in order to earn extra income. In the next few years, the Thai public universities will become autonomous, and so will several public RTOs. The idea is to take them out of the red-tape bureaucratic system and let them enjoy more freedom financially. Most of their budget will still consist of government subsidies but they are expected to generate relatively more income from other sources, especially from the private sector. Therefore, they have to conduct research and other activities that are more relevant to industry. Recently, universities have increasingly tried to increase industry sponsorships and to forge links with industry through collaborative R&D and training activities (College of Management, 2003).

3.4 Private Bridging Organizations

This section analyses the roles and capabilities of non-profit organizations like trade and industrial associations, professional associations, and private networking/bridging institutes in supporting technological capability development and innovation activities of firms. Concerning the innovation support, there are just a small number of these organizations disseminating knowledge and promoting the innovation capability of firms.

This is quite different from Japanese industry associations which play significant roles in the diffusion of knowledge and new technologies among member firms. For instance, industry associations played a major role in establishing and running cooperative research in the camera industry and automobile parts industry in the 1960s (see Goto, 1997). Although Taiwan's trade and industry associations tend to be government-sponsored rather than voluntary gathering of private enterprises (see East Asia Analytical Unit, 1995), some are rather successful in helping members enhance their capabilities. The largest and most influential manufacturers association is TEEMA, the Taiwan Electrical and Electronic Manufacturers' Association. The association has been actively assisting its members to upgrade the manufacturing technologies, expand international marketing ability and develop operation management. Besides, TEEMA also serves as a bridge in communication between the industry and government.

In Thailand, the Federation of Thai Industry (FTI) and Thai Chamber of Commerce (TCC) are the most powerful private-sector organizations. Their influence on the government's economic policies is strong. They can pressure government to induce policy changes. Most of their activities, however, aim at protection of their short-term interests and gaining leverage in negotiation with government (Laothamatas, 1992; Phongpaichit and Baker, 1997a: 150), such as export quotas, import levies and tax regimes. They are not very active in promoting the innovation capability of Thai firms. History does matter as well. Their members come from the commercial capital, rather than the industrial capital (Samudavanija, 1990: 275). Therefore, they pay more attention to short-term commercial gains rather than long-term capability development.

The FTI and TCC voiced their needs and concerns in the Joint Public–Private Consultative Committee (JPPCC), in an attempt to be given investment privileges and commercial advantages (Phongpaichit and Baker, 1997b). The role of this committee was very prominent in the mid-1980s when the idea of 'Thailand Inc.', promoted by the government during that period, was popular. Since then, both the FTI and TCC have represented the interest of the private sector in several national-level committees. The importance of the JPPCC, nonetheless, has since substantially declined.

The roles of the business associations and JPPCC have experienced another turning point under the current government. The government thought that the JPPCC was rather passive and, finally, changed the style of its operation from large assemblies which took place sporadically to less formal meetings every Friday. This new form of informal meetings between the Prime Minister and the private sector led to clearer national strategic goals, with more up-to-date concepts such as supply chain management and industrial clustering being introduced.

There is a small range of activities of the FTI and TCC, which aim to encourage diffusion of technological knowledge among their members. Examples are management consulting services, promotion of ISO certification and clean technology, and training programmes in energy saving, sanitary standards, entrepreneurial management, design and technological skills upgrading ('Federation of Thai Industries', n.d.). These activities are more active in the 'strategic' sectors designated by the government. Firms in these industries are more open to change. Some sectoral groups within the FTI are more enthusiastic about change than others, especially those having explicit concerns over the loss of national competitiveness in comparison with other latecomer countries.

Concerning trust-building among members, which is a kind of social infrastructure of knowledge diffusion and innovation, the roles of the FTI and TCC are not very impressive. They could create a certain level of trust among members by getting them together, exchanges of ideas and opinions, and sharing information among members. Mostly trust emerges gradually from joint activities such as marketing campaigns and trade fairs. However, internal organizations of the FTI and TCC are politically divided. For example, since the TCC proliferated with a growing number of provincial members, the organization has been more divided and fragmented because of regional power politics (*Phuchatkan*, 14–15 November, 1992). The provincial chambers criticized the fact that Bangkok-based business groups manipulated the TCC (Chotiya, 1997: 258–9).

There are a few private-sector organizations that specialize as bridging institutions, diffusing knowledge within the national innovation system. The Technology Promotion Association (Thailand–Japan) or TPA and the Kenan Institute Asia (KI Asia) are the most prominent. They diffuse knowledge and educate Thai firms and Thai society on the importance of innovation. KI Asia, in particular, accomplishes its vision of 'knowledge partner' by positioning itself as an agent bridging knowledge, expertise, and information between government agencies, universities and private firms. The TPA plays a similar role by means of technology transfer. Most of its activities are in education, training and technical services. Comparing the works of the two organizations, the TPA, with a 30-year long history, has more impact. It has achieved success in supporting firms in various technical aspects, such as instrumental calibration, productivity improvement, IT and automation, and manufacturing management. Both the TPA and KI Asia are actively involved in the Thai government's policy of strengthening SMEs' capability and entrepreneurship. For instance, they took part in the large programme called Invigorating Thai Business (ITB) initiated by the Ministry of Industry to revitalize Thai SMEs heavily affected by the economic crisis in 1997 by enhancing their capabilities in management, marketing, technology management, finance and so on ('Invigorating Thai Business', n.d.).

According to the first innovation survey in year 2000, around 34.4 per cent of sample firms used services provided by the TPA; while around 25.9 per cent of firms used services from other industry/business associations. This is higher than the percentage of firms using services from government agencies.

3.5 Financial Intermediaries/Markets

Like Japan and Korea, Thailand's financial system supporting industrial development is bank-based. The commercial banks are the ones that finance most of the private-sector investment in Thailand. Many entrepreneurs in Thailand develop innovative projects that require external financing but face the problems of not knowing where they should go for funds. According to the survey carried out by the Ministry of Industry, the major problem that limits the entrepreneurial firms in taking up innovative activities is the lack of capital funding (see Advance Research, 1997). Moreover, while the entrepreneurs want to tap the money from financial institutions, they face difficulties in obtaining a bank loan. Banks are conservative in granting loans and generally would not support risky businesses. Most start-up firms face many years of negative earnings and are unable to make the interest and principal payments that would be required on a bank loan.

Industrial/technological development banks/funds

Similar to Japan and Korea, several industrial development banks were set up to provide long- and medium-term finance. Activities of the four most important ones are the Industrial Finance Corporation of Thailand (IFCT), SME Bank, Small Industry Credit Guarantee Corporation (SICGC), and Innovation Development Fund (IDF).

Some of these financial institutions are not well known to private firms and they are not operating efficiently because of chronic bureaucratic red tape. While the maximum loan limit under the programmes is rather low, the interest rates are claimed to be similar to those charged by commercial banks. The application processes are complicated and time-consuming. The loan processing activity, in most cases, takes several months. This discourages firms, especially SMEs, from seeking institutional loans and forces them to take loans from informal sources where they can obtain credit more quickly (*The Nation* newspaper, various issues; *The Bangkok Post*, various issues). Some institutions like the IDF achieved rather modest performance in providing necessary funds to firms' innovative projects, since the project evaluators, mostly university professors, lack understanding of firms' innovation processes and capability in business terms (Turpin et al., 2002: 73). Nonetheless, the Fund has performed very well as the pioneer in 'non-financial issue' that is, encouraging an 'innovation culture' in the private sector and the Thai society at large through innovation awards, public relations and training programmes (Altenburg et al, 2004: 161).

The capital market

In Thailand, there is no stock market especially established to promote such high-tech start-ups as those in Japan (JASDAG, NASDAQ-Japan, and MOTHERS), Korea (KOSDAQ), and Taiwan (TAIDAQ and TIGER). The Market for Alternative Investment (MAI) is a business unit of the main market, the Stock Exchange of Thailand (SET). It was set up in 1999 as a new secondary market for trading SME shares. MAI's requirements for initial public offering have been adjusted to allow SMEs flexibility in entering the capital markets. However, this market is not specifically aimed at promoting knowledge-intensive start-ups. In practice, the MAI gets little interest from SMEs because the founding shareholders are reluctant to enact common stock rights issues that would effectively dilute their stakes in the listed companies. Since most SMEs are family-controlled, this reduces the willingness to enact equity issues (for fear of diluting levels of ownership and control). Therefore, the capacity of the MAI as a conduit for small businesses is constrained. Moreover, many SMEs see that the MAI requirements tend to disqualify most small- and medium-sized enterprises for being below the minimum capitalization level. As a result, there are too few outstanding shares to trade adequately on the market (Freeman, 2000).

Venture capital financing

In 1994, the Thai Venture Capital Association (TVCA) was set up to provide firms access to finance. At present, approximately half of the Thai Venture Capital Association (TVCA)'s members, the ordinary members, are Thai and international Venture Capital and/or Private Equity Fund Management firms. The other half, the extraordinary members, operate businesses related to VC or Private Equity, such as financial advisory firms, accounting firms, legal firms, securities firms and finance companies. Understanding the policies to supply the venture capital finance, the Thai government has supported several VC funds, such as the SME Venture Capital Fund (THB 1 billion), the Thailand Equity Fund (US$50 million) and the Thailand Recovery Fund (US$250 million). The government is also considering giving tax incentives to promote more VC investment in Thailand.

Compared with Taiwan, Thailand's venture capital is lagging both in terms of growth of venture capital industry itself and the impacts on financing innovation and emergence of knowledge-intensive start-ups. The venture capital investment in Thailand tends to finance firms at an expansion or mezzanine stage, not at an early start-up phase as in Taiwan (see 'Taiwan Venture Capital Association', n.d.).

3.6 Institutional Context

The aforementioned actors operate in the specific Thai institutional context. Here we examine entrepreneurship, attitude to failure, and trust, as they influence the innovativeness of firms in Thailand.

With the exception of Indonesia, the Thai economy is rather unique in South-East Asia because no class of indigenous big business entrepreneurs exists. Even smaller businesses in Bangkok, especially in retailing, are mostly owned and operated by Sino-Thais (East Asian Analytical Unit, 1995: 78). The dominance of family-owned enterprises established by immigrant Chinese entrepreneurs in Thailand has long been rooted in the Thai business norms and cultures. Therefore, historically and culturally, entrepreneurship in Thailand does not differ greatly from Chinese-dominated countries like Taiwan.

In terms of trust, Chinese-owned businesses tend to be built as family-affiliated corporations that allow ownership- and kinship-led rather than skill-based management. This 'family-ownership-control-type business' (Suehiro, 1992: 392), characterized by low stock ownership diffusion and more family-related CEOs, has led to business and joint investment cooperation among different companies within the same family affiliates *but only few* cooperations among various enterprises of different families (Suehiro, 1992: 390 and East Asian Analytical Unit, 1995: 78). Although many Chinese-run firms have grown into big conglomerates covering many business areas, the founding family still keeps the ultimate rein. Subsequently, firms under the same family umbrella overlap and compete, leading to intra-family conflicts. In sum, cooperation is less likely in inter-family businesses, and in the intra-family enterprises, cooperation often draws family complexity and contention.

An effect of Chinese–Thai entrepreneurship on the attitude towards acceptance of failure can be translated into two contrasting views. While the first view sees Sino-Thai influence as a threat towards innovation, owing to their low acceptance of failure and a lack of merit-based management, the second view sees the Chinese–Thai business culture as a positive condition, which tolerates risky ventures, needed for long-term planning and investment.

First of all, due to the fact that Chinese-run enterprises expand their businesses for the main purpose of their 'total fortune of the family'

(Suehiro, 1992: 403), they advance into areas such as finance and real estates. This evidence shows their risk-averse characteristics in doing businesses. The up-front profit from the trading and property business is far more attractive than the expensive technology-intensive manufacturing that will only earn longer-term gains. As a result, technological deepening or long-term sustainability is not much of a concern. Political capability in terms of gaining access to lucrative oligopolistic sectors seems more important than technological capability in this case.

The structural and political context also affects behaviour of Sino-Thai firms. Most of the domestic expansion and diversification rationale comes from the fact that Sino-Thai firms take advantage of the government's industrial promotion and other tax incentives while diversifying into foreign ventures for reasons of scale and scope, given the limited domestic market and intensive local competition (Suehiro, 1992: 400). Therefore, liberalization and high industrial growth of the 1980s, together with many outside favourable conditions unrelated to the fundamental capability of the Thai industries, lured the Thai conglomerates into new diversification, technologically unrelated to their original businesses. In order to do that, the underlying capability these firms accumulated is the ability to establish and maintain political connections with government authorities, rather than technological and innovation capabilities (see Intarakumnerd et al., 2002).

The second view, however, sees the Chinese–Thai entrepreneurship positively. The fact that 'Sino-Thai families traditionally were reluctant to relinquish ownership and management of their companies' (East Asian Analytical Unit, 1995: 80) allows them to create a long-term vision for their very own family businesses. While some list their assets on the stock market, many still prefer to raise capital conservatively through loans and offshore bonds giving the opportunity to benefit from different international interest rates. The continuous vision passed down from fathers to children protects them from the short-term concerns of stock prices or threats of acquisition. The deep-rooted corporate culture and tacit learning of the family members create a qualified decision base for risky projects (Intarakumnerd, 2000: 16). Therefore, they are capable of embarking on risky ventures on the expectation of future success without being distracted by their stockholders.

Enterprepreneurship in Thailand is experiencing interesting changes. The changes in attitude and behaviour towards entrepreneurship in Thailand come from exposure to modernism, innovative culture and new technologies of the West, which have infiltrated through the overseas education of the newer generation. This factor is where the two contrasting views of the Sino-Thai business culture finally merge. The combination of the fast decision traits and the long-term plan assets will create a condition that allows Thai business to grow both horizontally and vertically. It is likely to create a business structure, though still of the family-run type, that becomes increasingly innovative and adaptive to the changing environment. The attitude that favours kinship rather than managerial skills has also started to change. Professionalism of management is growing despite the tight family control (see Intarakumnerd, 2000), allowing a better prospect for competency building and technology development.

The innovation surveys show a higher acceptance of failure, from 10.5 per cent in the year 2000 to 19.5 per cent in 2002. As high as 63.5 per cent of surveyed firms in 2002 considered *establishing long-term strategic partnership with other firms* to be rather important or important. There was also a positive change in other attitude indicators, such as openness of customers to innovation. This indicates a better innovative environment in Thailand.

Furthermore, the Thaksin government is trying very hard to make Thai society more entrepreneurial. It is encouraging Thai people to change their attitude from being employees of the government or big corporations to being self-made entrepreneurs. The Ministry of Industry has a firm intention to produce 5000 new entrepreneurs per year. As a result, financial incentives, technical support and training courses have been provided by government agencies and education institutes to individuals and start-up businesses.

4. CONCLUSION

In conclusion, the experience of Thailand illustrates that a national innovation system of a latecomer country can transform it from one that has long been weak, fragmented and slow-learning to one that can become stronger, more coherent and faster-learning, if there is a significant change in the behaviur of a key actor that can cause positive repercussions among other actors. External factors that have cross-cutting effects on all actors in the system, in different degrees, may also bring change.

Thailand's national innovation system is in transition. Passive and slow technological learning of firms, ineffective and incoherent government policies, isolated education and training institutes, technologically unsupportive and risk-averse financial institutions, incompetent trade/industry associations and an unfavourable institutional context have been circumstances that have prevailed for the past 50 years of Thailand's industrialization. These have begun to change. One of the main actors of the NIS, the government, is spearheading the change. A major shift in government policies and practices is encouraging and pressurizing other actors of the system to change as well. The present Thaksin government has initiated policies that are making Thai firms move faster in developing their own technological capabilities. It has pressured universities to conduct more research and to become more relevant to the industry. It has worked more closely with private-sector linked organizations. It is stimulating entrepreneurship in Thai society. Change in the external environment led by the crisis in 1997 has also had significant impact on most actors. It has forced some actors of the NIS to innovate or die.

Nonetheless, the transformation is slow and difficult. Above all, it is difficult to change the mindsets and routines of some actors. The rates and depth of transformation, therefore, vary from one actor to another. While emerging signs of change in the government and significant numbers of private firms are quite noticeable, the rest, especially universities, seem to adjust more slowly. A longer timeframe is needed for serious examination regarding whether the extent of these changes is large enough to make significant impacts on Thailand's innovation capabilities and long-term competitiveness.

NOTES

1. For the first survey, a total of 2166 firms were selected, using stratified random sampling based on the firm's size and industry, from the top 13 450 companies by revenues in 1999. It basically included all firms with total revenues over 12 million Baht as reported to the Commercial Registration Department. Of the 2166 firms sampled, a total of 1019 completed questionnaires were received, being a 47 per cent response rate. Of these, 223 firms carried out innovation activities. For the second survey, the sampling frame of firms was drawn up based on the first survey plus firms with annual revenue of at least 12 million Baht. The sampling frame in the manufacturing sector consisted of 14 870 firms (including 1019 manufacturing firms which responded to the first survey). For the service sector, a total of 6082 firms were selected from the 26 162 firms. The overall response rate received for both sectors was 37 per cent with 2246 completed questionnaires returned. Of these, 261 carried out innovation activities.
2. The Tambon is a unit of local government administration. One Tambon comprises several villages.
3. These include the National Science and Technology Development Agency, Thailand Institute of Scientific and Technological Research, Synchrotron National Research Laboratory, National Institute of Metrology, and Geo-Informatics and Space Technology Development Agency.

REFERENCES

About Kenan Institute Asia (n.d.), 'Your knowledge partner', retrieved 24 July, 2003, from http://www.kiasia.org/aboutus/index.html.

About TPA (2003), retrieved 24 July, 2003, from http://www.tpa.or.th/newtpa/about/main_about.html.

Advance Research (1997), 'The study of SMEs in Thailand', report submitted to the Department of Industrial Promotion, the Ministry of Industry, Thailand: Advance Research.

Altenburg, T. et al. (2004), 'Strengthening knowledge-based competitive advantages in Thailand', GDI Working Paper 1/2004, German Development Institute, Bonn, Germany.

Amsden, A. (1989), *Asia's Next Giant: South Korea and Late Industrialisation*, New York: Oxford University Press.

Amsden, A. and T. Hikino (1993), 'Borrowing technology or innovating: an exploration of the two paths to industrial development', in R. Thomson (ed.), *Learning and Technological Change*, New York: St. Martin's Press.

Arnold, E. et al. (2000), *Enhancing Policy and Institutional Support for Industrial Technology Development in Thailand: The Overall Policy Framwork and the Development of the Industrial Innovation System*, Washington, DC: World Bank.

Bell, M. (1984), 'Learning and accumulation of technological capacity of developing countries', in M. Fransman and K. King (eds), *Technological Capability in the Third World*, London: Macmillan.

Bell, M. (2002), 'Knowledge-capabilities, innovation and competitiveness in Thailand: transforming the policy process', A report for National Science and Technology Development Agency (Thailand), Washington, DC: World Bank.

Bell, M. and D. Scott-Kemmis (1985), 'Technological capacity and technical change', draft working paper No. 1, 2, 4 and 6, Report on Technology Transfer in Manufacturing Industry in Thailand, Science Policy Research Unit, University of Sussex, UK.

Brooker Group Public Company Limited (2001), 'Technology innovation of industrial enterprises in Thailand', final report submitted to National Science and Technology Development Agency, September, Thailand.

Brooker Group Public Company Limited (2003), 'Thailand's 2nd R&D/innovation survey in manufacturing and service sectors and database development', Final Report submitted to National Science and Technology Development Agency, May, Thailand.

Chang, H.-J. (1994), *The Political Economy of Industrial Policy*, London: Macmillan.

Chang, H.-J. (1997), 'Institutional structure and economic performance: some theoretical and policy lessons from the experience of the Republic of Korea', *Asia Pacific Development Journal*, **4** (1), 39–56.

Chantramonklasri, N. (1985), 'Technological responses to rising energy prices: a study of technological capability and technological change efforts in energy-intensive manufacturing industries in Thailand', Unpublished D.Phil. Thesis, Science Policy Research Unit, University of Sussex, Brighton, UK.

Chotiya, P. (1997), 'Changing role of provincial business', in K. Hewison (ed.), *Political Change in Thailand*, London: Routledge, pp. 258–9.

College of Management (2003), 'S&T needs and production of manpower in the manufacturing sector', June, draft final report submitted to National Science and Technology Development Agency, Thailand (in Thai).

Dahlman, C. and P. Brimble (1990), 'Technology strategy and policy for industrial competitiveness: a case study of Thailand', paper prepared for the World Bank, April, Washington, DC: World Bank.

Dahlman, C. et al. (1991), 'Technology strategy and policy for industrial competitiveness: a case study of Thailand', in *Decision and Change in Thailand: Three Studies in Support of the Seventh Plan*, World Bank's Report,Washington, DC: World Bank.

Doner, R. (1992), 'Politics and the growth of local capital in Southeast Asia: auto industries in the Philippines and Thailand', in R. McVey (ed.), *Southeast Asian Capitalists*, Southeast Asia Program (SEAP), New York: Cornell University Press.

East Asia Analytical Unit (1995), *Overseas Chinese Business Networks in Asia*, Department of Foreign Affairs and Trade, Australia.

Evans, P. (1989), 'The future of the developmental state', *The Korean Journal of Policy Studies*, **4** (2), 129–46.

Evans, P. (1998), 'Transferable lessons? Re-examining the institutional prerequisites of East Asian economic policies', *Development Studies*, **34** (6), 66–86.

Federation of Thai Industries (n.d.), retrieved 24 July, 2003, from http://www.fti.or.th/nfti/org/index.html.

Freeman, N. (2000), *Constraints on Thailand's Equity Market as an Allocator of Foreign Investment Capital: Some Implications on Post-Crisis Southeast Asia*, Institute of South East Asian Studies, Singapore.

Goto, A. (1997), 'Cooperative research in Japanese manufacturing industries', in A. Goto and H. Odagiri, *Innovation In Japan*, Oxford: Oxford University Press.

Government Information Office (2003), 'Taiwan's educational development and present situation', retrieved 24 June, 2003, from http://www.gio.gov.tw/taiwan-website/5-gp/yearbook/chpt17.htm.

Hobday, M. (1995), *Innovation in East Asia: the Challenge to Japan*, Aldershot, UK and Brookfield, US: Edward Elgar.

Hou, C. and S. Gee, (1993), 'National systems supporting technical advance in industry: the case of Taiwan', in R. Nelson (ed.), *National Innovation Systems: A Comparative Analysis*, Oxford: Oxford University Press.

Intarakumnerd, P. (2000), 'Thai telecommunication business groups: an analysis of the factors shaping the direction of their growth paths', unpublished D.Phil. thesis, Science Policy Research Unit, University of Sussex, Brighton, UK.

Intarakumnerd, P. and T. Virasa (2002), 'Taxonomy of government policies and measures in supporting technological capability development of latecomer firms', a paper presented at the 6th International Conference on Technology Policy and Innovation, August, Kansai, Japan.

Intarakumnerd, P. et al. (2002), 'National innovation systems in less successful developing countries: the case of Thailand', *Research Policy*, **31** (8–9), 1445–57.

Intellectual Property Institute (2003), 'Science and technology policy in Thailand', a report to Ministry of Science and Technology, Bangkok, Thailand.

Invigorating Thai Business-ITB (n.d.), retrieved 24 July, 2003, from http://hypeshop.tripod.com/itbproject.html.

Johnson, C. (1982), *MITI and the Japanese Miracle: The Growth of Industrial Policy, 1925–1975*, Stanford, CA: Stanford University Press.

Kim, L. (1993), 'National systems of industrial innovation: dynamics of capability building in Korea', in R. Nelson (ed.), *National Innovation Systems: A Comparative Analysis*, Oxford: Oxford University Press.

Kim, L. (1997), *Imitation to Innovation: The Dynamics of Korea's Technological Learning*, Cambridge: Harvard Business School Press.

Kim, L. and R. Nelson (eds) (2000), *Technology, Learning, and Innovation: Experiences of Newly Industrializing Economies*, Cambridge: Cambridge University Press.
Lall, S. (1996), *Learning From the Asian Tigers: Studies in Technology and Industrial Policy*, London: Macmillan Press.
Lall, S. (1998), 'Thailand's manufacturing competitiveness: a preliminary overview', unpublished paper for conference on Thailand's Dynamic Economic Recovery and Competitiveness, Papers for Session 4, 20–21 May, Bangkok.
Laothamatas, A. (1992), *Business Associations and the New Political Economy of Thailand: From Bureaucratic Polity to Liberal Corporatism*, Boulder, US: Westview Press.
Lauridsen, L. (1999), 'Policies and institutions of industrial deepening and upgrading in Taiwan III-technological upgrading', International Development Studies Working Paper No. 13, Roskilde University, Roskilde, Denmark.
Lauridsen, L. (2000), 'Industrial policies, political institutions and industrial development in Thailand 1959–1991', International Development Studies Working Paper No. 21, Roskilde University, Roskilde, Denmark.
Lauridsen, L. (2002), 'Coping with the triple challenge of globalization, liberalization and crisis: the role of industrial technology policies and technology institutions in Thailand', *European Journal of Development Research*, **14** (1) (June), 101–25.
Ministry of Science and Technology: Republic of Korea (MOST) (2003), 'Science and technology policy in Korea: vision and strategies for the 21st century', retrieved February 2003 from http://unpan1.un.org/intradoc/groups/public/documents/apcity/unpan008041.pdf.
Mukdapitak, Y. (1994), 'The technology strategies of Thai firms', unpublished D.Phil. thesis, Science Policy Research Unit, University of Sussex, Brighton, UK.
Odagiri, H. and A. Goto (1993), 'The Japanese system of innovation: past, present and future', in R. Nelson (ed.), *National Innovation System: A Comparative Analysis*, Oxford: Oxford University Press.
Phasukavanich, C. (2003), 'The pace of Thailand through the year 2020', power point presentation, Bangkok, Thailand, 20 May.
Phongpaichit, P. and C. Baker (1997a), *Thailand: Economy and Politics*, Singapore: Oxford University Press.
Phongpaichit, P. and C. Baker (1997b), 'Power in transition: Thailand in the 1990s', in K. Hewison (ed.), *Political Change in Thailand*, London: Routledge.
Phuchatkan (Thai Daily Economic Newspaper) (1992), 14–15 November, p. 4.
Samvdavanija, C. (1990), *State and Society: Triple Characteristics of Thai State in a Plural Society*, Bangkok: Cholalongkorn University (in Thai).
Science and Technology Policy Institute: STEPI (1995), 'Review of science and technology policy for industrial competitiveness in Korea', a report prepared by Science and Technology Policy Institute, Seoul, South Korea.
Science and Technology Policy Institute (2002), 'Korean innovation survey: manufacturing sector', a report prepared by the Science and Technology Policy Institute, Seoul, South Korea.
Sripaipan, C., S. Vanichseni and Y. Mukdapitak, (eds) (1999), *Technological Innovation Policy of Thailand*, Bangkok, Thailand: National Science and Technology Development Agency (Thai version).
Suehiro, A. (1992), 'Capitalist development in post-war Thailand: commercial

bankers, industrial elite, and agribusiness groups', in R. McVey (ed.), *Southeast Asian Capitalists*, New York: Cornell University Press.
Suehiro, A. (1993), 'Family business reassessed: corporate structure and late-starting industrialisation in Thailand', *Developing Economies*, **41** (4), 378–407.
Taiwan Venture Capital Association (TVCA) (2003), retrieved 24 July, 2003, from http://www.tvca.org.tw/.
Technology Promotion Association (2003), retrieved 24 July, 2003, from http://www.tpa.or.th/newtpa/services/main_services.html.
Thailand Development Research Institute (1989), 'The development of Thailand's technology capability in industry', reports Volumes 2–5, Bangkok, Thailand.
Thailand Development Research Institute (1998), 'Effective mechanisms for supporting private sector technology development and needs for establishing technology development financing corporations', a report submitted to the National Science and Technology Development Agency, Bangkok, Thailand.
Tiralap, A. (1990), 'The economics of the process of technological change of the firm: the case of the electronics industry in Thailand', unpublished D.Phil. thesis, Science Policy Research Unit, University of Sussex, Brighton, UK.
Turpin, T. et al. (2002), 'Improving the system of financial incentives for enhancing Thailand's industrial technological capabilities', a report prepared for the World Bank, June, Washington, DC.
Vongpivat, P. (2003), 'A national innovation system model: industrial development in Thailand', unpublished Ph.D. thesis, The Fletcher School of Law and Diplomacy, Tufts University, Cambridge, US.
Wong, P. (1995), *National Systems of Innovation: The Case of Singapore*, Seoul, Korea: Science and Technology Policy Institute.
Wong, P. (1999), 'National innovation systems for rapid technological catch-up: an analytical framework and a comparative analysts of Korea, Taiwan, and Singapore', paper presented at the DRUID's summer conference, Rebild, Denmark.

6. Hong Kong's innovation system in transition: challenges of regional integration and promotion of high technology

Erik Baark and Naubahar Sharif

INTRODUCTION

In recent years Hong Kong has regained its traditional position as the key transit point for the exchange of goods and services between China and the international economy. Sophisticated and reliable intermediary services occupy a key role in maintaining this status, so Hong Kong's future seemingly depends on the capacity of its intermediary firms to maintain a considerable share of business in Asian markets and the global economy (Meyer, 2000, p. 247).

Hitherto, however, technological innovation in Hong Kong has been undervalued as an element in Hong Kong's developmental experience and the few studies to have addressed the issue have emphasized the laissez-faire policies that shaped the process of industrialization in Hong Kong (for example Hobday, 1995). Hong Kong's entrepreneurs have skillfully exploited technology available on the international market, but they have not generally carried out research and development for the purposes of creating proprietary technology (Davies, 1999). Technological innovation has therefore only recently begun to attract serious attention in Hong Kong, where the Government in 1998 launched a new strategy in pursuit of knowledge-intensive economic growth.

Our point of departure for this chapter is the proposition that a system of innovation has been emerging in Hong Kong for at least the past century, conditioned by major economic and political upheavals at the global level accompanied by gradual institutional change at the local level. This transition in innovation has accelerated lately, as influential economic and political forces have reasserted themselves with the return of Hong Kong to Chinese sovereignty and the Asian financial crisis of the late 1990s. Hong Kong has of course passed through several stages of development; the latest transition represents elements of continued growth and expansion as well as important breaks with the past. Simultaneously with important changes in the economy that we shall briefly describe below, a growing trend towards policy change has provided new frameworks for innovative activities in Hong Kong. Some of these policy initiatives have been designed to advance the pace and direction of technological change directly, while others have influenced the process indirectly. Finally, geopolitical events, especially Chinese political and economic developments, have significantly shaped the conditions for innovation in Hong Kong.

CONCEPTUAL AND METHODOLOGICAL ISSUES

We shall analyse the emergence and transition of Hong Kong's innovation system through two complementary perspectives: the historical and the spatial. If the analysis of innovation systems is based on evolutionary and history-friendly approaches to theorizing, as we think it should be, it is important to include dynamic concepts pertaining to tensions between continuity and change, as well as critical trends or events, in the conceptual toolbox.

Viewing Hong Kong's innovation system along the historical dimension, we seek to identify important stages in its development that are characterized by cultural, social or institutional landmarks. We note periods surrounding critical events that have challenged the established trajectory of a previous stage and brought about a transition to the next. The dynamism that we explore in this perspective feeds off the tension between, on the one hand, continuities and trajectories that are shaped by the accumulation of cultural or social institutions and, on the other hand, perturbing factors and events that would require adaptation and reform in the system – whether such factors are endogenous to the system or elements of the external environment.

The second dimension – the spatial – is based on our conviction that the impact reflected in Hong Kong's constantly shifting position in the global and regional political and economic landscape must be taken into account in the conceptualization of significant factors influencing its innovation system. These factors are situated in political environments that frequently reach beyond the purely 'national' scale: the financial power of global networks has often directly affected the regional level, accentuating the need for local/regional development. Although Hong Kong's government has been relatively autonomous during both colonial rule and since the return to China – exercising a level of authority much akin to that of a national government – the international context and indeed the ideology of the government itself (which has generally espoused a hands-off or laissez-faire economic policy) have left much of the regulation of business in Hong Kong to market forces. The result is that most aspects of the development of Hong Kong's innovation system are co-determined at three important spatial levels, namely 'national', regional and global. In other words, what we observe in our analysis of the innovation system reflects causal relationships that transcend the national scale. There is then a need to adopt a multiscalar analytical framework to capture the transitions that the Hong Kong innovation system has experienced during the post-war period.

In the presentation of Hong Kong's history and the evolution of its innovation system we shall therefore discuss the ramifications of various influences and consequences of economic, institutional or political changes on several scales, according to whether these appear most appropriately analysed at a national, regional or global level – or at all three of those levels simultaneously. For example the economic integration of Hong Kong with South China has been driven by 'open door' politics in China at the regional level, together with the intensification of global production chains and the structural transformation of businesses shifting to higher value-added activities in Hong Kong. To discuss the transformation of innovation processes in Hong Kong associated with the economic integration taking place in South China with reference only to its regional aspects would be to disregard vital international aspects of the process. Similarly, it would be dangerous to assume the direct and unilateral influence of global forces since local contexts are often intermeshed with the wider global system, as portrayed in the concept of 'glocalization' (Swyngedouw, 1997).

In Table 6.1, we have tried to illustrate the intersections of these two dimensions. On the one hand, we shall roughly distinguish four historical stages in the development of Hong Kong's innovation system. These stages are of various durations, and it is of course possible to suggest sub-stages for each of the four. We have then tried to situate a range of key trends and events within the framework of the three above-mentioned spatial dimensions.

In the following sections, we will use this historical/spatial framework to explore the twists and turns in the evolution of Hong Kong's innovation system. Within the limited space at our disposal, we shall try to illustrate the influence of most of the factors and characteristics outlined in the table, but we will be unable to do justice to some of the complexities and deep-seated features of Hong Kong's innovation system.

Ultimately it is our ambition to identify some strengths and weaknesses that characterize innovation processes in Hong Kong and to discuss the suitability of recent initiatives to accelerate the pace of innovation in the

Table 6.1 Factors shaping the development of Hong Kong's innovation system

Development stage	Spatial dimension		
	National	Regional	Global
1. Early 20th Century/Pre-World War II	• Trade hub • Small-scale industrialization • Expansion in South-East Asia	• Migration and links with South China provinces • Center of overseas Chinese capitalism	• Strong trading links to China and Europe/US
2. Cold War Period: 1950s to 1970s	• Migrants • Large-scale Shanghai industrialists	• Loss of the China market in East Asia • Expansion of OEM networks	• Boycott of China • Hong Kong strategic center
3. Opening of China: 1980s and 1990s	• Structural change: from manufacturing to services • Expansion of financial center	• Movement of production to Guangdong	• Production-chain linkages and management
4. Return to Chinese Sovereignty and Post-1997	• Government initiatives to promote knowledge-based economy	• Pearl River Delta integration efforts • Closer Economic Partnership Agreement	• Asian financial crisis • China joins WTO

territory. We shall outline a few arguments in that direction in our concluding section.

A BRIEF HISTORICAL BACKGROUND TO HONG KONG'S DEVELOPMENT

The Hong Kong story makes a fascinating tale of how what was a barren rock a little more than 150 years ago has emerged as a dynamic and vibrant international city.[1] In reality, the phenomenon of Hong Kong's economic growth has transpired over a shorter period covering the last four or five decades. Nevertheless, the foundation was laid over a longer period of time and it is worthwhile exploring the development of its innovation system over the past century.

Early Twentieth Century

Studies of Hong Kong's economic development in the early part of the twentieth century identify a variety of informal institutions and state initiatives that have supported industrialization, relying primarily on small-scale manufacturers linked in familial or ethnic networks and connected with expanding markets for relatively low technology products in China, South East Asia and Europe/US (Clayton, 2000).

Official British colonial history has tended to neglect the growth of such industries in the territory of Hong Kong, partly on account of the influence of the perspective of the Major British 'Hongs' – trading houses – which had little commercial interest in manufacturing and instead emphasized the promotion of the entrepôt trade (Loh, 2002).[2] It is nevertheless important to recognize the existence of this 'undergrowth' sector of small-scale industrial firms in Hong Kong for two reasons: first it provided opportunities for Chinese entrepreneurs to accumulate technical and managerial skills that could be successfully deployed in subsequent stages of development; second it gave Chinese firms opportunities to practice organizational modes that supported networking, subcontracting relationships, and the international search for markets. These opportunities in effect 'rehearsed' critical features of subsequent industrialization in Hong Kong that prevail to this day.

Cold War Period: 1950s to 1970s

The overthrow of the Kuomintang (KMT) regime of General Chiang Kai Shek in 1949 by the current government of the People's Republic of China remains one of the most significant events in Hong Kong's history. The overthrow caused an exodus of about one million mainland Chinese to Hong Kong. The people of Hong Kong, including its migrants, grew up and developed in a community that had Chinese roots but were under British administration. These migrants in turn made a huge contribution to Hong Kong's economic explosion by escalating the establishment and size of manufacturing industries, further expanding the role that Hong Kong had played hitherto as an entrepôt. In the face of the declining power of the KMT, Shanghai textile barons in particular transferred enormous amounts of capital and managerial expertise in textile manufacturing to the colony (Wong, 1988). The international blockade of China, declared by the United Nations in 1950 (which remained until 1953), combined with the US imposition of an embargo on all goods of Chinese origin, cut Hong Kong off from its natural hinterland in Southern China and sapped the entrepôt trade that sustained its economy.[3] Lacking the trading networks necessary for export purposes and because the mainland market was closed to them, these Shanghai industrialists turned to the British trading houses in Hong Kong, which were experienced in the entrepôt trade and had established links with the British and other international export markets (Tsui-Auch, 1998, p. 59). At the same time they set about building export-oriented, transnational production operations across Southeast Asia, further developing the management of the dispersed manufacturing capabilities that are distinctive of Hong Kong to this day (Enright et al., 1997, p. 5).

The late 1950s brought the beginning of industrial diversification into the manufacture of plastics commodities and electronics products. By 1959 the value of manufactured exports had surpassed that of the entrepôt trade; by the early 1960s Hong Kong was among the largest manufactured product suppliers in the developing world (Hong Kong, 1961). In the 1960s and 1970s an increasing number of foreign and overseas Chinese industrialists invested in industry, trade, finance and tourism in Hong Kong, which was a free trade port with no control over foreign exchange (Tsui-Auch, 1998, p. 60). Increasing global trade, particularly in the West, and liberal international trade relations benefited Hong Kong's export-oriented production. All the while continued turmoil in the region during this period triggered periodic surges of migration both into and out of Hong Kong. Today it is estimated that more than half of Hong Kong's more than 7 million citizens are descendants of post-1949 migrants.

Hong Kong's government can be characterized as embracing an economic policy of minimum intervention in the belief that free markets allocate resources most effectively. This tradition began to change with the establishment of important institutional frameworks and organizations in Hong Kong around the late 1960s, early 1970s and beyond. These institutions both *supported* innovative activity in Hong Kong and *intervened* in it to varying degrees. Primary among them was the Hong Kong Productivity Council (HKPC), initially conceived by the Working Committee on Productivity in 1963, which was created in 1967. Ferguson (2001, p. 2) states that the HKPC has 'concerned itself with technological upgrading through consultancy and vocational training with most of its efforts to diffuse off-the-shelf best practice technology rather than create new techniques'. Another move in this direction was the Advisory Committee on Diversification, authorized in 1977, which recommended a more active role for the government and greater support for the provision of technological infrastructure (Ferguson, 2001, p. 2).

This period also witnessed the development and operation of infrastructure services and facilities that contributed significantly to the growth-enabling environment, including monopolies in utilities that since the 1970s have been governed by schemes-of-control. Investment in infrastructure in Hong Kong during the post-war period has been very high compared with that in many other developing countries (Mody, 1997, pp. xii–xiv). Consequently the supply of reliable electric power, telecommunications and transport services in Hong Kong has occurred on a scale that has enabled rapid economic growth. The transition to a service economy in Hong Kong has further underscored the importance of government policies that can support competition and raise efficiency in sectors such as telecommunications and professional business services (Cheng and Wu, 1998).

Opening of China: 1980s and 1990s

Under the umbrella conditions described above, industries in Hong Kong such as electronics evolved into competent companies exporting around US$7.5 billion worth of electronics products in 1991, equivalent to 60 per cent of the total exported by Taiwan at the time. Many overseas electronics firms first selected Hong Kong as a manufacturing base in Asia, and the electronics industries thrived on original-equipment-manufacture (OEM) subcontracting arrangements (Henderson, 1989; Hobday, 1995). The migration of electronics manufacturing to the Chinese mainland in the 1980s did not necessarily impede export-led technological learning, but an emphasis on low-cost production among Hong Kong networks in South China made more likely the capture of industries in OEM patterns that had been successfully employed earlier (Chiu and Wong, 2001). In fact some of the most important industry clusters competing from a base in Hong Kong were created in the wave of expanding manufacturing and services businesses in the decades of the 1970s and 1980s. These clusters include light manufacturing, transportation, tourism, financial and business services, and communication and media industries (Enright et al., 1997).

As part of its industrial policy the government sought at the time to facilitate the growth of industrial manufacturing through investments in infrastructure and human capital, while simultaneously upgrading institutions that served to enhance industrial production and penetration of overseas markets. Investments in infrastructure centered on the creation of industrial estates that enabled firms to build or rent manufacturing facilities without having to commit high levels of capital to physical infrastructure in the context of high land prices. Vocational training became another area of active policy initiative, as the government sought to meet the increasing demand for skilled labor. In many ways these actions reflected the persistence of an approach to industrial policy that emphasized a reluctance to target support to specific sectors or firms.

Given Hong Kong's singular position as a British Crown Colony on the doorstep of the most populous country in the world, political matters naturally served to shape its innovation system significantly. In this respect there is little doubt that the two most significant events affecting Hong Kong's innovation system in this period were (1) the modernization program that the late Chinese leader Deng Xiaoping promulgated in 1978 and (2) discussions between the Chinese and British governments that opened in 1982 over Hong Kong's sovereignty. The latter negotiations ended in 1984 with the signing and ratification of the Sino-British Joint Declaration, which stated that Hong Kong (HK) would become a Special Administrative Region (SAR) of the People's Republic of China and that Hong Kong's capitalist system and 'way of life' would be preserved for 50 years. The 'one country–two systems' framework under which Hong Kong is presently governed was enshrined in the 'Basic Law', the present constitution for the HKSAR. Coupled with the opening-up process these two events catalysed the transformation of Hong Kong's innovation system. In many ways, the opening of China precipitated and accelerated the Hong Kong people's learning curve, which ultimately proved helpful not just economically but also in accepting that their fate and future lay with that of Mainland China.

The most striking change in Hong Kong's innovative landscape that was triggered by the opening of the mainland in 1979 was the decreasing role of manufacturing in Hong Kong and the simultaneous rise in the services sector (Table 6.2). At its peak in the mid-1980s, the manufacturing sector in Hong Kong employed 41.7 per cent of the active labor force but by 1995 it employed only 15.3 per cent (Berger and Lester, 1997, p. 9).

Table 6.2 Percentage contribution to GDP by economic activity

Economic Activity/Year	Agriculture, Fishing, Mining, Electricity, Gas and Water	Construction	Manufacturing	Services
1980	2.5	6.6	23.6	67.3
1985	3.5	5.0	22.0	69.5
1990	2.8	5.4	17.5	74.4
1995	2.6	5.3	8.3	83.7
2000	3.3	5.2	5.8	85.7
2002*	3.5	4.4	4.6	87.4

Note: * Figures are preliminary and subject to revision.

Source: Census and Statistics Department, Hong Kong SAR.

Just prior to 1997 important changes occurred whereby the Hong Kong Government participated more actively in the transformation of Hong Kong's innovation system, a move that began during the Cold War period. Although little was done in response to specific 1979 recommendations by the Advisory Committee on Diversification (which called for greater support for technological deepening), the late 1980s saw the government establishing industrial estates in Tai Po, Yuen Long and Tseung Kwan O for companies engaged in technology-intensive techniques. Furthermore Hong Kong's third major university devoted to scientific and technological research – The Hong Kong University of Science and Technology – was also founded in the late 1980s.

These policy changes were triggered because Hong Kong had hitherto been primarily a low-cost producer. In the prelude to its return to China in 1997, it found its competitive advantage being eroded by rising wage and land-rental rates as well as the emergence of other developing economies in the region.

From Crown Colony to Special Administrative Region: Run-up to 1997 and Beyond

The role that Hong Kong came to occupy in Asia has been fundamentally shaped by its geographical and political position. Hamilton (1999) argues however that it has also been the organizing center of Chinese-led capitalism: 'Because Chinese modes of capitalist acquisition are based on bottom-up individual and family-based strategies of seizing opportunities wherever they exist, rather than on top-down corporatist strategies of linking state administrative capabilities with elite economic opportunities, Chinese capitalism is integral to world capitalism itself' (Hamilton, 1999, p. 16). The organization of Chinese family businesses involved firms of many sizes, a paternalistic management model, and perhaps most importantly a network of social and economic relationships from the outset (Redding, 1990). These networks were crucial for the rapid expansion of subcontracting networks and long commodity supply links serving the US and European markets during the 1960s and 1970s. The management of vertically disintegrated supply chains and networks remains a notable strength of Hong Kong entrepreneurship.

These events meant that as Hong Kong approached its return to China in 1997, it was proudly boasting that no other society had more experience in investing and producing in China. Ever since the mid-1980s, Hong Kong has been the largest source of foreign direct investment in China. Although the exact figures are impossible to determine, various statistical sources estimate that Hong Kong's contribution to realized foreign investment in

China comprised about two-thirds of the total by 1994 (Berger and Lester, 1997, p. 5). It is on this basis that Enright et al. (1997, p. 7) accurately describe how Hong Kong's historical role as a city of departure from China has laid the foundation for a reverse flow of business investments during the 1990s not only back to Hong Kong, but to mainland China through Hong Kong. They claim that this has 'helped Hong Kong become the de facto capital of the 50 million or more overseas Chinese who today play such an important role in the economic modernization of the Asian region and in the reconstruction of China's market economy' (Enright et al., 1997, p. 7). The economic impact is considerable, since overseas Chinese investors – often Hong Kong companies or investors operating out of Hong Kong – now employ at least 14–15 million people in China. It is also crucial to understand that the migration of production facilities to the Pearl River Delta (the PRD) in many ways represented growth, rather than decline, of Hong Kong's engagement in manufacturing. For political reasons such growth was, however, categorized as outside the territory, even if it was, from a historical perspective, a reintegration into Chinese markets. The effects on service industries must also be seen as beneficial in the sense that most of the migration spurred further growth and increased sophistication of producer business services (Tao and Wong, 2002).

In establishing and upgrading these networks Hong Kong firms have exploited their traditional strategies of imitation and followership, while emphasizing the development of organizational know-how rather than formal research and development for new products.[4] The bulk of R&D expenditure by private firms in Hong Kong is devoted to redesigning and improving products as well as making them easier and cheaper to produce. In other words process innovation has often taken precedence over product innovation in Hong Kong industries, and the support of bridging institutions such as the Hong Kong Productivity Council has often served to underscore these efforts to improve production efficiency. While learning extensively from OEM contacts overseas, Hong Kong firms have been instrumental in setting up and improving production facilities in Mainland China – transferring innovative production technology and management organization rather than product innovations.

Soon after 1997, however, the Asian Financial Crisis hit the entire region particularly hard and contributed to a dramatic bursting of the asset bubble that had bolstered much of Hong Kong's economy. Simultaneously economic reforms on the mainland were increasing in pace. These events, combined with China's accession into the World Trade Organization, put Hong Kong in a situation in which it could no longer rely as heavily on its traditional methods for accumulating wealth. The magnitude of these changes cannot be overemphasized. Following a tradition of relatively high and stable levels of growth, Hong Kong society suddenly found itself facing quarter upon quarter of negative economic growth and rapidly mushrooming unemployment. This was the first time Hong Kong had ever experienced such deep-seated economic turmoil. At an individual level those who had invested in either the stock or property markets saw their asset values slashed, falling into a situation of 'negative equity' whereby the amount of outstanding loans payable on their assets exceeded the assets' market value. As such, society at large was palpably yearning for strong leadership and a new vision to pull Hong Kong out of its economic difficulties.

These sudden changes in the economic landscape forced the Hong Kong government to participate even more actively and rapidly in the transformation of Hong Kong's innovation system so that it could identify a new role for itself as well as a new engine for its continued economic growth. This search culminated in the appointment of a Commission on Innovation and Technology (CIT) in March 1998. The members of the Commission comprised mostly American-trained academics and professionals. In its first report (1998) the CIT noted in its vision statement that 'innovation and technology are vital to the future prosperity of Hong Kong', proposing such a vision in response to the challenges that Hong Kong was facing at the time (HKSAR, 1998, p. 13). The commission's second and final report made eight concrete recommendations (HKSAR, 1999, pp. 5–7) which identified measures to promote high technology innovation in order to haul Hong Kong out of its worst ever recession.

- Coordinate the government's policy functions
- Merge the Hong Kong Science Park, Hong Kong Industrial Estates Corporation and the Hong Kong Industrial Technology Center
- Invest in education
- Seek overseas talent
- Relax immigration restrictions on talent from the mainland
- Expand the government's incubator program
- Forge closer ties between academia and industry
- Explore the feasibility of a co-investment scheme providing government venture capital on a matching basis with private funds

We will analyze two cases to illustrate the new face of the Hong Kong government's official posture: its attempts to follow up and further promote the economic integration taking place in the PRD, and its policy initiatives aimed at promoting high technology innovation such as the Cyberport and the Science Park.

INTEGRATION WITH THE PEARL RIVER DELTA REGION

The Pearl River is considered one of China's three main waterways. Formed at Guangzhou, the Pearl River flows east and south to form a large estuary between Hong Kong and Macau. The river links Guangzhou to Hong Kong and the South China Sea and is one of China's most important waterways for trade. The PRD is found along the estuary of the Pearl River.

Although the territories of Hong Kong and Macau are geographically integrated parts of the PRD, the 'special' status of these two territories often sets them apart from the rest of the region; therefore in the literature and debate that has emerged in recent years the term 'Pearl River Delta' frequently serves as shorthand for the administrative zones, municipalities and districts of the PRD in mainland China (excluding both the Hong Kong and Macau SARs). Some reports use the term 'Greater PRD Economic Region' when they include Hong Kong and Macau (for example, Federation of Hong Kong Industries, 2003).

Guangdong's geographic position as a peripheral province on the southern coast meant that it was far from China's industrial heartland. Guangdong's scarce natural resources limited its ability to contribute to the development of heavy industries, which were the focal points of China's five-year plans in the 1950s, 1960s and 1970s. The lack of convenient transportation links with the rest of China also hindered Guangdong's development. As such, Guangdong was an economic laggard as compared with other Chinese provinces over the period 1949–1979. All of this, however, changed in 1979 when Guangdong Province was put at the forefront of China's reform program. It was chosen for special treatment due to its proximity to Hong Kong and Macau, its distance from the heartland of the Chinese mainland, and the fact that it was not advancing as quickly economically as the other coastal provinces.

As a consequence Guangdong Province was given greater political and economic autonomy than other jurisdictions in the Chinese mainland. The main areas over which it was granted greater autonomy were finance and fiscal matters, foreign trade and investment, commerce and distribution, allocation of materials and resources, the labor system, and prices. Guangdong was allowed to keep a larger share of its output and foreign exchange than other provinces but it was required to be self-sufficient in terms of capital investment. The province was given greater control over economic planning, approval of foreign investments and foreign trade. Guangdong also took control over several state-owned enterprises located in the province. These measures launched rapid economic development in

Guangdong Province, mostly in the Special Economic Zones established in the PRD.

It was at this point that deeper economic links began to emerge between Hong Kong and the PRD as Hong Kong's economy began to switch from manufacturing to services and manufacturing began to move from Hong Kong to the PRD region. The open door policy coupled with economic reforms not only provided an enormous production hinterland and market outlet for Hong Kong's manufacturers, but also generated abundant business opportunities for a wide range of its service activities. In particular these include freight transport, storage, telecommunications, banking, real estate development, and professional services such as legal services, insurance and accounting. This emphasis on services allowed Hong Kong businesses and their managers to build an unparalleled fund of knowledge about what it takes to operate production systems distributed across long distances and to turn out high-quality goods in a wide range of industries in China. For this reason Hong Kong's experience in the PRD region stands as a benchmark for working in China. While 'Made in Hong Kong' manufacturing therefore declined, 'Made by Hong Kong' manufacturing – manufacturing in Hong Kong-owned and managed plants in the PRD region – flourished (Berger and Lester, 1997, p. 5).

By shifting parts of their operations to China, Hong Kong industrialists vastly increased the scope of their enterprises. By 1997 Hong Kong manufacturing companies were estimated to employ some 5 million people in their plants in Hong Kong and China (Berger and Lester, 1997, p. 10) – more than five times the workforce they had employed in Hong Kong at the peak of the manufacturing era in 1984. Today the figure is estimated to be in the region of ten million (Cheng in SCMP, 2003, p. 26). Over the period from 1980 to 2001 the PRD region was the fastest-growing portion of the fastest-growing province in the fastest-growing large economy in the world (Enright et al., 2003, pp. 21–25). A study of economic interaction between Hong Kong and the PRD region sponsored by the Hong Kong-based '2022 Foundation' outlined several clusters of service-enhanced industrial development that involved a division of labor between international services located in Hong Kong and production located in the PRD (Enright et al., 2003). Another report sponsored by the Federation of Hong Kong Industries similarly underscored the close economic linkages in the region and the strengthening of the overall infrastructure to facilitate R&D activities among companies in Hong Kong and the PRD. It was hoped that such initiatives would take advantage of the various strengths in the region, such as the intellectual property rights protection framework in Hong Kong and the availability of affordable R&D staff in the PRD (Federation of Hong Kong Industries, 2003).

Given the close economic linkages that now exist in the region, a regional innovation system is clearly also in the making. If, however, one employs the definition of an innovation system as the set of institutions that jointly and individually contribute to the development and diffusion of new technologies, it becomes an important issue as to *how* such an integrated system is actually emerging. The fundamental conditions that brought about Hong Kong's extraordinary societal and economic flourishing in the past included institutions such as the rule of law, limited or 'small' government, honesty and transparency in administration, an effective civil service, free private enterprise, public commitments to relatively high levels of social goods and equity, and protection of individual freedoms of expression, association and belief. While it is possible to debate the degree to which these fundamental conditions still exist as intensely as they did previously – particularly given Hong Kong's recent experiences – there is universal agreement among all analysts that the conditions enjoyed by Hong Kong are not as readily present on the mainland, including in the PRD region. The conundrum this poses is whether there should be some degree of equivalence among basic, elemental conditions in Hong Kong and the PRD for there to be a truly regional innovation system. In particular, need there be correspondence in terms of the rule of law, promotion of private enterprise and so on? However these questions are answered, what cannot be mistaken are the closer ties that Hong Kong has been forging with the PRD region on a number of fronts.[5] This is after all the region whence many of Hong Kong's present day residents trace their roots. Despite these ties, especially at the industry, trade and firm levels between Hong Kong and the regions north of its border, there remain pointed policy disjunctions between Hong Kong and the PRD. These differences question even further the extent to which a regional innovation system is developing and make it more difficult to understand how Hong Kong's innovation system is being transformed in an attempt to embrace the PRD region more readily.

A recent survey of R&D in Hong Kong and the mainland indicates that many firms in Hong Kong were carrying out R&D in both Hong Kong and the PRD. Based on the information supplied by 229 firms (49 per cent of the sample of firms operating in both Hong Kong and the mainland), it was clear that both R&D outsourcing and investments in R&D beyond the borders of Hong Kong were very significant. Only 17 per cent of the total R&D staff of these firms were located in Hong Kong, while 53 per cent were located in Guangdong Province, 3 per cent in the Yangtze River Delta, 19 per cent in other mainland provinces and 8 per cent overseas (Federation of Hong Kong Industries, 2003, pp. 47–8). Figure 6.1 shows that many Hong Kong firms are contracting out R&D services on the Chinese mainland or overseas.

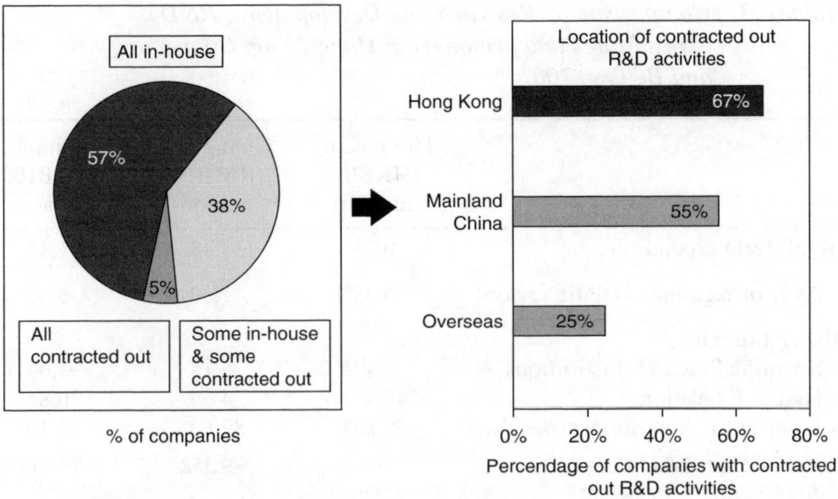

Source: Federation of Hong Kong Industries (2003), *Made in PRD: The Changing Face of HK Manufacturers*, p. 46.

Figure 6.1 Pattern of contracting out R&D by Hong Kong firms

The primary reason for locating R&D in the mainland was the supply of talent and research facilities; research costs ranked only third. The majority of firms with mainland operations surveyed (78 per cent) indicated that they planned to continue or expand their R&D efforts, and almost half (46 per cent) planned to recruit more R&D staff in Guangdong. Only 13 per cent had plans to recruit more R&D staff in Hong Kong. Table 6.3 indicates the comparative proportions of R&D expenditures and personnel in Hong Kong, Guangdong and Beijing. Given the substantial amount of R&D undertaken in Guangdong by Hong Kong firms, the figures for Hong Kong R&D expenditure probably understate the total R&D effort made by these firms; the table nevertheless illustrates that Guangdong has become a much more important site for innovation and that Hong Kong could benefit significantly from a closer association with this province.

Among the initiatives with the potential to promote Hong Kong's economic development is the Closer Economic Partnership Arrangement (CEPA). Under this arrangement, which came into effect on 1 January 2004, 273 Hong Kong products qualify for zero-tariff status under rules governing origin of manufacture.[6] It has been estimated that Hong Kong will save HK$750 million from zero-tariff exports (SCMP, 17 October 2003). Eighteen service sectors are allowed easier access to mainland markets,

Table 6.3 Comparison of Research and Development (R&D) expenditures and personnel in Hong Kong, Guangdong and Beijing, 2001

	Hong Kong (HK$100 million)	Guangdong (RMB100 million)	Beijing (RMB100 million)
Total R&D expenditure	70.76	137.43	171.17
– As % of regional GDP (in region)	0.55%	1.29%	6.02%
Expenditure by:			
– Scientific Research Institutions	1.47#	5.18	91.04
– Higher Education	48.47	4.65	20.86
– Large & Medium Enterprises	20.83*	89.60	21.10
Total FTE of R&D Personnel	7365	79052	96255
FTE of R&D Personnel in:			
– Scientific Research Institutions	280#	4209	43982
– Higher Education	3791	9949	18171
– Large & Medium Enterprises	3294	43279	12277

Notes: # Government; * Business.

Source: Based on Table 4.2 in Federation of Hong Kong Industries (2003), p. 54.

including telecoms, banking, accounting, logistics and tourism and there is 'enhanced cooperation' in various areas of trade and investment. In essence, CEPA is designed to allow Hong Kong firms to benefit early from the liberalization of the mainland's restricted sectors, which will open up to all foreign companies from 2005 as a result of its accession to the World Trade Organization. For this reason others have branded CEPA as more talk than action. To be sure CEPA still includes limitations on the operation of Hong Kong firms on the mainland market and all benefits accrue either to goods made in Hong Kong (of which there are increasingly fewer as manufacturing moves to the mainland) or to Hong Kong-based service firms providing a limited range of products in cooperation with mainland partners. CEPA will nevertheless promote economic integration with the whole of the mainland, including the PRD region (Hong Kong Trade Development Council, 2003).

A final noteworthy point concerns the co-evolution of systems of innovation. While this chapter focuses on Hong Kong's innovation system, it is important not to forget that China's innovation system is also undergoing massive changes with its move from a socialist to a market-oriented economy. While it is therefore analytically justifiable for the purposes of scope and concentration to examine only Hong Kong's innovation system, changes north of the border also merit independent assessment. We conjecture that changes to the PRD's innovation system continuously affect the development and transformation of Hong Kong's innovation system. While Hong Kong may thus endeavor to *pre-empt* its future direction and niches, there are many times when for all its best efforts it must, either by design or by default, *react* to developments in the PRD's innovation system. In this way one possible fruitful subject of analysis is the co-evolution of Hong Kong's and the PRD's systems of innovation: are the two systems converging?

POLICIES FOR DEVELOPMENT OF KNOWLEDGE-BASED INDUSTRIES IN HONG KONG

In March 1999, the Hong Kong Government announced plans to build Cyberport, a US$1.76 billion technology park in Pokfulam, in collaboration with the Pacific Century Cyber-Works (PCCW) company. Cyberport is designed to create a strategic cluster of leading IT and service companies in Hong Kong in the shortest possible time. It will concentrate on communication-oriented industries, calling for the building of telecommunications, network and wireless communications, optical electronics, and Internet appliances in Hong Kong. The giant project will purportedly serve as a multimedia and information technology hub, with state-of-the art wiring, room for 130 companies and adjacent housing. Designed as a pocket-size Silicon Valley, the futuristic office and residential park will occupy 64 acres on the southwest coast of Hong Kong island. The project is expected to generate more than 12 000 jobs in Hong Kong, while approximately 4000 jobs will be created in the construction industry to build Cyberport. With the first phase completed in 2002, Cyberport is expected to generate demand for support services such as accounting, legal and other back-office functions (Pun and Lee, 2002, p. 9).

When the Cyberport project was announced, it was seen as a vehicle with which to attract international high tech companies. Additionally, it was hoped that Cyberport would also be home to 100 smaller companies with the aim of incubating local business and retaining Hong Kong's indigenous technological talent. With the global economic slowdown and the technology bubble bursting, there have however been reports that Cyberport has had problems attracting new IT operations to Hong Kong. Sin Chung Kai, who speaks for the IT industry in the LegCo, thus comments: 'one problem is that the fall in the property market has taken away Cyberport's raison d'être'. Consequently Cyberport's ability to fill up the empty offices depends mostly on cannibalizing existing commercial space from tenants in Central district and Quarry Bay (Einhorn, 2002).

We can better understand our focus on Cyberport by looking at some of the initiatives that were undertaken before and shortly after the project was launched. The government had already created several industrial estates providing fully serviced industrial land through the government-funded Hong Kong Industrial Estates Corporation (HKIEC). In addition the Hong Kong Industrial Technology Centre Corporation (HKITCC) had provided, since being initiated in the mid-1990s, low-cost accommodation as well as marketing, financial and technical assistance through an incubation program. The center had hosted 80 new ventures in high technology during three-year incubation periods. Finally a range of consultancy studies had been conducted regarding the establishment of a science park in Hong Kong, and an ultimate decision had been taken to create a Provisional Hong Kong Science Park Company to undertake construction of new park facilities at reclaimed land near the Chinese University of Hong Kong. Both the HKITCC and the Science Park were designed to promote the creation of high technology industries and technological innovation in Hong Kong by supporting clusters of knowledge-intensive small and medium-sized firms. The HKIEC was also targeting more advanced industries and services, although the emphasis here was to supply land for manufacturing rather than services.

In May 2001 the three organizations were merged to form the Hong Kong Science and Technology Parks Corporation, which aims to provide a full range of services and infrastructure for high technology firms in Hong Kong. To a large extent, these facilities provide an environment that appears more appropriate for high technology industries and services than Cyberport, which can be seen as competing primarily on the basis of prestigious office space and price. The science park and the HKITCC are both located in close proximity to universities, while Cyberport is at a considerable distance from the University of Hong Kong (which has therefore rented office space at Cyberport to set up liaison contacts). Both Cyberport and the Science Park have been extremely eager to attract overseas tenants, conducting extensive marketing activities in advanced industrialized countries. In other words the two initiatives appear to be complementary in some respects while also overlapping rather unproductively in other respects.

Another of the important follow-up activities resulting from the work of the Chief Executive's Commission on Innovation and Technology was the creation of the Innovation and Technology Fund (ITF) in 1999, to replace the existing Industrial Support Fund and Services Support Fund (HKSAR, 1999). With a total endowment of HK$5 billion the ITF has supported 236 projects amounting to a total funding of HK$526 million during 2000–2002. A considerable proportion of the projects (approximately one-third) support R&D in the area of information technology (LegCo Panel, 2002). The majority of ITF funding (60 per cent) has, however, been earmarked for research projects to be undertaken by the Applied Science and Technology Research Institute, which will be located in a building in the Science Park. ASTRI has formulated initial areas of focus for its research and development programs that include photonics, wireless technologies, Internet content and applications, and integrated circuits design. Again we see the potential for overlap and duplication of development efforts. Cyberport's role in the commercialization of such technologies is unclear.

The general picture that emerges from an examination of Cyberport's role in the various policy initiatives related to innovation and high technology development in Hong Kong is one of dismal policy coordination. Most of the services that Cyberport intended to offer had already been planned for or directly provided by other organizations or programmes. Little evidence can be found of coordination between the Cyberport project and the broad range of schemes to support innovation and high technology development initiated by the SAR government in recent years.

Experience elsewhere in the world suggests that one of the key factors behind the rapid growth of high technology industries in a region is its ability to mobilize and attract skilled professionals and workers from all corners of the world (Saxenian, 1999). In addition, highly innovative clusters of high technology industries and services have generally emerged in places where advanced research organizations or educational establishments such as universities provide both a regular supply of highly skilled engineers and strong communicative links with new entrepreneurial ventures. Hong Kong has improved and expanded higher education during the 1990s, and has intensified the training of IT professionals as part of this effort. Unfortunately this expansion of educational opportunities has not yet created a sizeable local body of skilled specialists in IT and communication technologies; demand for these types of people still exceeds supply.

For this reason the SAR government promulgated an Admission of Mainland Professionals Scheme in 2001 for the purpose of encouraging the immigration of mainland professionals in finance and information technology to Hong Kong. The policy has not, however, been very popular among firms in Hong Kong or among the mainland Chinese IT professionals, primarily because of restrictions that require the potential employer to document whether a local professional could fill the post. Other restrictions could be cited as well, such as not allowing professionals to bring their families with them to Hong Kong or requiring mainland students who graduate from Hong Kong universities to return to the mainland before they can qualify for an application. Belatedly recognizing the weaknesses of the scheme, the SAR government has recently adopted an 'Admission Scheme for Mainland Talents and Professionals' that seeks to improve the conditions for this special class of immigrants from the mainland, including letting them bring their families to Hong Kong.

An important component of the growth of high technology industries in advanced industrialized countries – and in particular in Silicon Valley, the precedent upon which Cyberport is allegedly modeled – is the availability of venture capital (Florida and Kenney, 1988). In Hong Kong various venture capital funds have been set up during recent years, but these appear to have had little impact on the local high tech industry. The SAR government initiated a Venture Capital scheme in the late 1990s, which it left to a major bank to operate – with little success, as the scheme was discontinued due to difficulties in finding a project that would meet the bank administrator's criteria for investments (Kwong, 1997). On the other hand around 200 international venture capital firms have set up regional headquarters in Hong Kong, but these firms – with a portfolio of around US$10 billion in 2000 – primarily target opportunities for investment in the Chinese mainland. Overseas capital funds contributed 90 per cent of the funds available to the industry, and most of these are invested overseas in the regional market, to exploit beneficial tax treatment (Lowtax.net, 2005). Thus Hong Kong firms receive only 10 per cent of the disbursements. Although Hong Kong boasts one of the region's most advanced financial services sectors, the lack of a transparent institutional framework for high-risk financing and the preponderance of relational contracting has held back investments locally (Carney and Gedajlovic, 2000).

A new Growth Enterprise Market was set up in November 1999 to provide an alternative fundraising channel for emerging growth companies under a well-established market and regulatory infrastructure. At the end of August 2002 152 companies were listed on the GEM, with a total market capitalization of US$8 billion. The emerging high tech stock boom was hijacked, however, by property developers and short-term speculative interests, quickly eroding serious confidence in high technology investments and bursting the 'high tech bubble' following the NASDAQ landslide in 2000. Under these conditions it has become extremely difficult for genuinely innovative people and firms in Hong Kong to raise the capital necessary to commercialize their technology.

The Cyberport project, the Science Park, and the Applied Science and Technology Research Institute, like most of the other high tech innovation policy initiatives in Hong Kong, reveal the persistence of the 'linear paradigm' of high technology innovation, a model that has been increasingly questioned in recent decades. Technological innovation is interpreted, according to this paradigm, as originating in the results of scientific research, where these scientific results are subsequently developed into commercial technologies and finally marketed by firms. Being envisaged as a linear process (like a river flowing to the sea), the knowledge available from 'upstream' research by universities or research institutes will be transferred to 'downstream' research and development (R&D) in enterprises. The construction of infrastructure for such 'downstream' R&D in manufacturing and service industries consequently becomes a worthwhile project – not unlike the facilities constructed in Cyberport. If there is a potential demand for the ultimate 'innovation', the market will ensure that entrepreneurial people and capital will undertake such commercial 'downstream' R&D.

The problem is that innovation processes seldom follow such simple linear sequences. An innovation process typically provides opportunities for parallel processes involving the exploration of new knowledge and the exploitation of existing knowledge, as well as for interaction among many actors in an extended network that supports learning (Rosenberg et al., 1992; Van de Ven, 1999). Endogenous innovation processes are thus promoted in a milieu that fosters interactive relationships and creative learning. They depend to a considerable extent on the wider institutional context and the resources available to individual actors. Innovation policies formulated in the vein of the outmoded linear paradigm simply provide funding for research and perhaps infrastructure for commercialization by firms, but assume automatic transfer and diffusion in the market. In contrast, interactive innovation policies attempt to promote learning and diffusion of knowledge, often implemented through intermediary organizations providing technological services. Such policies are oriented towards the needs and demands of small and medium-sized firms and particularly of local productive systems (Vazquez-Barquero, 2002).

CONCLUSIONS

Our research on the transitions made by and challenges facing Hong Kong's innovation system during the past century offers two types of contributions to the literature. *Empirically* our research has highlighted key facets of the evolution of Hong Kong's innovation system as it traversed periods of industrialization and de-industrialization and extended its linkages to the global marketplace. *Theoretically* these empirical observations open multiple pathways to methodological and analytical advances in the field of innovation studies.

A commonly neglected feature of Hong Kong's innovation system, overlooked by analysts and scholars alike, is its early history. All too often histories of Hong Kong's industrial structure and economic development begin 'out-of-the-blue' during the 1950s, following the takeover of China by the Communists. This is supposed to have initiated a train of events leading to Hong Kong's becoming a prosperous entrepôt over the five ensuing decades. While the significance of the events of 1949 on Hong Kong's innovation system should not be downplayed, this chapter attempts to show that there were in fact antecedents to the innovation system that developed in the 1950s and evolved continuously until the handover of Hong Kong and beyond. In this way, we show that – and how – the transformation of Hong Kong's innovation system has been taking place over a far longer period than is commonly supposed. By tracing two important traits that characterize Hong Kong's people and thereby its firms – resourcefulness and entrepreneurial spirit – to a period before the magic milestone of 1949, we uncover the roots of Hong Kong's innovation system in the early 1900s. As a result, the empirical analysis that follows is better grounded in fact and therefore more insightful in helping one understand Hong Kong's present-day innovation system. The implications of this finding for innovation systems theory in general are twofold: first it suggests that innovation systems are in continuous evolution. When this evolution is studied within a broad time frame (of, say, a century) transformations between various periods can be identified, as we have done in the case of Hong Kong. Second it suggests that, in order to better understand present-day innovation systems, it is imperative that a country's commonly understood assumptions and trajectories are unpacked and 'opened-up' in order to determine the extent to which they hold true.

This leads to the second contribution suggested by our empirical analysis of the two main challenges facing Hong Kong – the need for a clear awareness of the opportunities and limitations attaching to a more active role played by the government in shaping the future of innovation in Hong Kong. Targeting closer integration with the PRD front and the development of knowledge-based industries in Hong Kong, we can identify a range of areas where private and public initiatives to develop innovative capabilities coexist. In some cases, these initiatives appear to be mutually supporting. In other cases public policies seem preoccupied with serving narrow business interests, sometimes to the point of contravening the professed ambitions of innovative industries.

Lying at China's doorstep, opportunities abound for Hong Kong's further integration with the PRD and thereby the extension and knowledge-based upgrading of its industrial clusters. This calls for a more systematic analysis of existing trends and resources with a reduced emphasis on meeting the demands of a handful of strong business interests in Hong Kong. This implies not that the Hong Kong government should start planning technological change in detail but rather that government policies should be based on extensive and continuous monitoring and analysis of Hong Kong's innovation system. In this endeavor the government should be guided by broader social and economic needs than those reflecting the interests of a limited group of tycoons. In other words the role of government should be grounded in the principles of comprehensive and coherent policy-making that have been informed by innovation systems research (see, for example, OECD, 1997; 1999 and 2002).

NOTES

1. When Hong Kong Island was ceded to the British in perpetuity, it was only a fishing community, inhabited by about 150 000 people, and dismissed by the then British Foreign Secretary, Lord Palmerston, as 'a barren rock'.
2. Loh (2002) ascribes this to the seizure of Hong Kong for trade purposes rather than for territorial aspirations – for Hong Kong to serve as a base for penetration of China and other Asian nations.
3. This blockade was imposed to penalize China for its support of North Korea during the Korean War.
4. Several surveys of electronics firms in Hong Kong, for example, have found that 60–70 per cent of these have copied or modified other products instead of initiating independent product design (Yu and Robertson, 2000).
5. The massive Severe Acute Respiratory Syndrome (SARS) outbreak in the first half of 2003, for example, drove home the fact that Hong Kong and its PRD neighbors live and die together, literally. Whatever affects one affects the other.
6. Certificate of Hong Kong origin is a requirement for eligibility of zero tariffs.

REFERENCES

Berger, Suzanne and Richard K. Lester (1997), *Made by Hong Kong*, Hong Kong: Oxford University Press.

Carney, Michael and Eric Gedajlovic (2000), 'East Asian financial systems and the transition from investment-driven to innovation-driven economic development', *International Journal of Innovation Management*, **4** (3), 253–76.

Cheng, Christopher (2002), 'Opening address to Conference "Forging a New Economic Force"', *South China Morning Post*, 5 July.

Cheng, Leonard K., Changqi Wu, Hong Kong Centre for Economic Research and Hong Kong Economic Policy Studies Forum (1998), *Competition Policy and the Regulation of Business, Hong Kong Economic Policy Studies Series*, Hong Kong: City University of Hong Kong Press.

Chiu, Stephen W.K. and Wong Ka-Chung (2001), 'Growth without catching up: organizational dynamics and restructuring of electronics industry in Hong Kong', in *Hong Kong Institute of Asia-Pacific Studies Occasional Paper Series*, **115**, Hong Kong.

Clayton, D.V. (2000), 'Industrialization and institutional change in Hong Kong', in A.J.H. Latham and H. Kawakatsu (eds), *Asia Pacific Dynamism, 1550–2000*, London, New York: Routledge.

Davies, Howard (1999), 'The future shape of Hong Kong's economy: why high-technology manufacturing will prove to be a myth', in P. Fosh et al. (eds), *Hong Kong Management and Labour: Continuity and Change*, New York: Routledge.

Einhorn, Bruce (2002), 'Why Hong Kong's developers are throwing bricks: state-backed Cyberport is helping push rents down', *Business Week*, 16 December, http://www.businessweek.com/magazine/content/02_50/b3812167.htm, accessed 5 March 2004).

Enright, M., E. Scott and D. Dodwell (1997), *The Hong Kong Advantage*, Hong Kong: Oxford University Press.

Enright, Michael J., Ka-mun Chang, Edith E. Scott and Wen-hui Zhu (2003), *Hong Kong and the Pearl River Delta: The Economic Interaction*, Hong Kong: 2022 Foundation.

Federation of Hong Kong Industries (2003), *Made in PRD: The Changing Face of HK Manufacturers*, Hong Kong, available at http://www.hku.hk/hkcer/ prd/ full_ report/English_Full_Report.pdf (accessed 6 June 2005).

Ferguson, Robert (2001), 'From ships to chips: research issues surrounding Hong Kong's technology policy', mimeo, Hong Kong: The Hong Kong University of Science and Technology.

Florida, Richard and Martin Kenney (1988), 'Venture capital and high technology entrepreneurship', *Journal of Business Venturing* (Fall) **3** (4), 301–19.

Hamilton, G. (1999), 'Hong Kong and the rise of capitalism in Asia', in Gary G. Hamilton, *Cosmopolitan Capitalists: Hong Kong and the Chinese Diaspora at the End of the 20th Century*, Seattle: University of Washington Press.

Henderson, J.W. (1989), *The Globalisation of High Technology Production: Society, Space, and Semiconductors in the Restructuring of the Modern World*, London, New York: Routledge.

HKSAR (1998), *Commission on Innovation and Technology First Report*, Hong Kong.

HKSAR (1999), *Commission on Innovation and Technology Second Report*, Hong Kong.

Hobday, Michael (1995), *Innovation in East Asia: The Challenge to Japan*, Aldershot, UK and Brookfield, US: Edward Elgar.

Hong Kong Government Information Services (1961), *Hong Kong Colonial Reports 1961*, Hong Kong.

Hong Kong Trade Development Council (2003), CEPA and Opportunities for Hong Kong, October.

Kwong, Kai-sun, Hong Kong Centre for Economic Research, Better Hong Kong Foundation, and Hong Kong Economic Policy Studies Forum (1997), *Technology and Industry, Hong Kong Economic Policy Studies Series*, Hong Kong: City University of Hong Kong Press.

LegCo Panel on Commerce and Industry (2002), 'Review of the innovation and technology fund', http://www.legco.gov.hk/yr 01-02/english/panels/ci/papers/ ci0311cb1-1232-3e.pdf (accessed 17 March 2004).

Loh, Christine (2002), *Hong Kong SMEs: Nimble and Nifty*, Hong Kong: CLSA, April.

Lowtax.net (2005), 'Hong Kong venture capital sector', http://www.lowtax.net/ lowtax/html/hongkong/jhkvent.html (accessed 5 June 2005).

Meyer, D.R. (2000), *Hong Kong as a Global Metropolis*, Cambridge: Cambridge University Press.
Mody, Ashoka (ed.) (1997), *Infrastructure Strategies in East Asia: The Untold Story*, Washington, DC: The World Bank.
Pun, Ngai and Kim Ming Lee (2002), 'Locating globalization: the changing role of the city-state in post-handover Hong Kong', *The China Review*, **2**, 1–28.
Organisation for Economic Co-operation and Development (1997), *Policy Evaluation in Innovation and Technology: Towards Best Practices*, Paris: Organisation for Economic Co-operation and Development.
Organisation for Economic Co-operation and Development (1999), *Managing National Innovation Systems*, Paris: Organisation for Economic Co-operation and Development.
Organisation for Economic Co-operation and Development (2002), *Dynamising National Innovation Systems*, Paris: Organisation for Economic Co-operation and Development.
Redding, S.G. (1990), *The Spirit of Chinese Capitalism*, Berlin, New York: W. de Gruyter.
Rosenberg, Nathan, Ralph Landau and David C. Mowery (1992), *Technology and the Wealth of Nations*, Stanford, CA: Stanford University Press.
Saxenian, AnnaLee (1999), *Silicon Valley's New Immigrant Entrepreneurs*, San Francisco: Public Policy Institute of California.
SCMP (2003), *South China Morning Post*, 17 October.
Swyngedouw, E.A. (1997), 'Neither global nor local: "Glocalization" and the politics of scale', in K.R. Cox (ed.), *Spaces of Globalization: Reasserting the Power of the Local, Perspectives on Economic Change*, New York: Guilford Press, pp. 137–66.
Tao, Zhigang and Y.C. Richard Wong (2002), 'Hong Kong: from an industrialized city to a centre of manufacturing-related services', in *Urban Studies, HKIEBS Working Papers*, **39** (12), 2345–58.
Tsui-Auch, Lai Si (1998), 'Has the Hong Kong model worked? Industrial policy in retrospect and prospect', *Development and Change*, **29**, 55–79.
Van de Ven, Andrew H. (1999), *The Innovation Journey*, New York: Oxford University Press.
Vazquez-Barquero, Antonio (2002), *Endogenous Development*, London and New York: Routledge.
Wong, Siu-lun (1988), *Emigrant Entrepreneurs: Shanghai Industrialists in Hong Kong*, Hong Kong, New York: Oxford University Press.
Yu, Tony F. and Paul L. Robertson (2000), 'Technological capabilities and the strategies of small manufacturing firms: the case of Hong Kong', in Nicolai J. Foss and Paul Robertson (eds), *Resources, Technology and Strategy*, London: Routledge.

7. The Indonesian innovation system at a crossroads

Peter Gammeltoft and Erman Aminullah

INTRODUCTION

One of the most conspicuous features of the Indonesian innovation system as it evolved under the 32-year rule of President Soeharto, was its segmentation: the co-existence of a set of divergent development orthodoxies worked to produce a corresponding set of relatively self-contained industrial subsystems, from government departments down through supporting institutions, to industries and supporting industries.

The end of Soeharto's rule in 1998 hurled the country into a series of economic, political and social reform processes. The further course of these reforms is as important for the future constitution and performance of the Indonesian innovation system as it is uncertain.

In this chapter we take stock of the development of the Indonesian innovation system. Indonesia is at a crossroads, having to come to grips with two major impetuses for reform: one is the multi-dimensional domestic reform process, the other is the one induced by what is commonly referred to as 'globalization'. Accordingly, in the first section we will look at how contemporary global economic dynamics, especially globalization and the changing economic role of knowledge and information, is redefining some of the core operating principles of the innovation system. Among the themes touched upon in this section are the necessity of avoiding excessive (borrowed) capital investment, strengthening technological capabilities, strengthening the SME sector, shifting focus from mere economic growth to techno-economic development, accommodating knowledge-based industrial transformation, strengthening absorptive capacity to better capture emerging opportunities offered by globalization, and strengthening national capacity to govern S&T policy.

In the second section we account for a set of underlying features of the Indonesian political economy, which have shaped the innovation system in the past and continue to do so today. The track record of the government to devise economic policies; the pervasive ethnic issue; the character of government–business interaction; and the presence and features of the aforementioned different policy orthodoxies are among the issues discussed here. The third section is devoted to a more factual outline of the Indonesian innovation system. We discuss the economic incentive system, technology policies, broader industrial policies, the training and education system, supporting institutions, R&D expenditure, and corporate structure. Finally, in the conclusion we propose a set of recommendations as to how the Indonesian innovation system may better meet contemporary challenges and become more integrated, efficient and sustainable in the future.

THE INDONESIAN INNOVATION SYSTEM AT A CROSSROADS

'Globalization' is a contested term that is rarely precisely defined. Globalization is popularly understood as the increasing reach of free market and democratic institutions, the irresistible and inevitable evolution of capitalism. Here, we simply take globalization to mean that an increasing number of people recognize change in their lives from non-local phenomena. Because of the many changes in science and technology, politics and economics, culture and social relations, ordinary life is increasingly affected by ideas from and interactions with constituencies outside the community and the country. Governments seem unable to oppose these trends, except to adapt to the variety of changes that might occur in the environment. Considering the case of Indonesia, in order to improve the Indonesian innovation system, we argue that three important issues should be dealt with: (i) knowledge-based industrial transformation; (ii) balanced technology absorption from R&D and learning, and; (iii) shifting the policy orientation from economic to techno-economic development. In the following, we will touch shortly upon those three issues.

Knowledge-based Industrial Transformation

There were three structural weaknesses in the Indonesian industrial system, which contributed to the economic crisis: excessive capacity, weak technological capability and a less-conducive environment for developing small and medium-scale industries (Aminullah, 2004). These factors were interrelated, with excess capacity created by aggressive investment based on hopes of high demand that would trigger future growth. In general, these investments were financed by loans intended to purchase technologies for the development of capital-intensive industries. Short-term loans, which were utilized for the funding of long-term investments, resulted in financial deficits causing the financial crisis, continuing into a crisis of confidence, shifting into the economic crisis and extending into a multidimensional crisis. Since small and medium-scale labour-intensive industries displayed low efficiency and productivity and therefore slow growth, a large increase in the rate of unemployment could not be avoided.

Balanced Technology Absorption from R&D and Learning

In Indonesia, public R&D institutions carry out most of the R&D activities. There are two primary problems encountered in most of these public R&D activities: (i) lack of funding leading to difficulties in hiring qualified researchers, and; (ii) lack of ties with the private sector, resulting in R&D programmes that are not responsive to the demands of industries. Many of the public R&D institutions designed the R&D projects without involving industries, as indicated by a very limited utilization by industries of public R&D output. Instead, Indonesia maintains the tradition of relying on foreign direct investment as the main source of building up industrial technological capability. However, the impact of FDI on Indonesia's industrial technological development indicates serious shortcomings in capability development (Thee, 2003; Okamoto and Sjöholm, 2001). Foreign investors' interest in technological transfer, for example to joint ventures, tends to be limited to good plant operation practices, whereas foreign-controlled firms mostly rely on capabilities developed by their parent companies in the domains of design and engineering.

Acquisition and absorption of technology from abroad have become increasingly important. Effective utilization of international technology spillovers is determined by the strength of industrial R&D and absorptive capacities more generally. However, focusing merely on industrial R&D will become increasingly inadequate in building technological capabilities, and more attention needs to be paid to technological learning from global technology spillovers. If the Indonesian industries succeed in striking a balance between local innovation and international learning, it will improve efficiency, raise productivity, and finally strengthen financial capability and stable business growth.

Accordingly, in Indonesia development of local industrial capability tends to originate from two sources outside local industries themselves, and both sources are associated with deficiencies: one outside source is the public R&D institutions which tend to fail due to weak linkage to the industrial sector; the other outside source is FDI which does not automatically generate technological capability due to weak industrial R&D infrastructure to absorb the technology.

Shifting the Policy Orientation from Economic to Techno-economic Development.

Economic actors have tended to focus on short-term profits and on trade and are not easily persuaded to give priority to long-term industrial upgrading. Combined with the speed and pervasiveness of technological innovation, the result has been that companies have tended to be dependent on producers of technology abroad. Dependence on foreign suppliers can result in fragile industries, as we were recently reminded by the jolts of the economic crisis of 1997, where industries dependent on foreign sources of technology were particularly seriously affected (Aminullah, 2003; Keller and Samuel, 2002).

Shifting from economic to techno-economic orientation refers to the process of changing the orientation of economic development from one stage to the next. Based on Indonesian experiences in coping with the economic crises, some innovative industries have moved from a capital-based towards an innovation-based and further into knowledge-based orientation. The economic crisis has forced industries to create new markets and demands through innovation processes (mostly through learning) to utilize over-investment. Some innovative enterprises and industries were able not only to overcome the crisis of over-investment, but also to carry out technological innovations, creating new knowledge and technological progress for the long-term survival of the business enterprise. However, although some innovative industries such as traditional drugs (jamu) and cosmetics, electric appliances industries and information and communication services have been successful in pursuing new roads to be competitive in the global market, most of the Indonesian industries are still highly dependent on foreign capital and technology, and therefore are not competitive globally.

POLITICAL ECONOMY

In this section we will discuss a number of features of the Indonesian political economy, which have shaped and influenced the innovation system in the past and continue to do so today. We consider the track record of the government to devise economic policies; the pervasive ethnic issue; government–business interaction; and the presence of contending economic development orthodoxies within the state.[1]

Government Capacity

The Indonesian state has traditionally been portrayed as being very centralistic and relatively weak in terms of capabilities, until the introduction of the regional autonomy law in 2001. We will elaborate on these two issues in the following. Due to its complex geographic, demographic, cultural and economic make-up, regional problems in general and relationships between regions and central government in particular have always been prominent issues in Indonesia.

Government centralism partially results from efforts to overcome the regional differentials of the colonial period. The highly uneven distribution of natural resources and the requirement that the revenue accrued to the nation as a whole rather than to the individual provinces induced the establishment of a centralized system to collect and redisburse income from the provinces. Subsequently, large inflows of foreign aid from the late 1960s and later still, huge oil revenues during the oil boom period in the 1970s, combined with weak regional administrative capacities and the need to initiate nationally coordinated and supervised physical infrastructure development projects, made a case for the adoption of a highly centralized mode of governance. A final reason for government centralism concerns the traditionally close ties between government and the Indonesian military, itself highly centralistic and wary of any developments which might threaten national unity. Therefore, ever since independence, national unity and the establishment and maintenance of central authority have been high on the government's agenda.

Today Indonesia has embarked on what is often characterized as the most ambitious and complex decentralization programme ever attempted, but widespread fear remains that decentralization could lead to unacceptable regional inequalities, the aggravation of ethnic and social conflicts, and possibly even to the demise of the Indonesian nation state. Yet, the economic case for decentralization is persuasive: central government no longer has the capacity to fund regional development to the extent it has done in the past; local, region-specific factors have become more important determinants of provincial performance; and successful efforts have already been made to upgrade the capabilities of local governments and planning agencies. Furthermore, there is ample evidence in the literature that development processes are frequently highly localized and contingent on the active involvement of local government, so increased devolution of powers may be required to release region-specific development dynamics.

Turning next to government capacity, when Soeharto assumed power in 1966, the civil service was, in the words of Mackie and MacIntyre (1994), 'notoriously unwieldy, inefficient, corrupt and torpid'. It has grown rapidly since 1975 at about twice the rate of growth of the labour force as a whole, and about 4 million people were employed in it in 1992 (Hill, 1996). Today, government bureaucracy is still renowned for its low efficiency, low productivity, low educational level of civil servants, low remuneration, and widespread corruption and rent-seeking practices.

Yet, significant progress has been made since 1966 towards a more capable, professional and disciplined bureaucracy in which at least key departments such as finance, home affairs and industry have acquired some capacity to execute policies of increasing technical complexity (Mackie and MacIntyre, 1994). Education levels in the civil service are rising quite rapidly, real wages have been increased somewhat to maintain competitiveness *vis à vis* the private sector and to curb 'moonlighting' and corruption (Hill, 1996). The inefficient and corruption-plagued trade and customs surveillance was privatized in 1985, banking reforms in the 1980s reduced the incidence of 'command loans' through which loans were granted to favoured individuals without proper commercial evaluation, and simplifications of government procedures have reduced some opportunities for corruption (Cribb and Brown, 1995: 154). The most significant step to curb corruption was the establishment of the powerful Commission for Eradication of Corruption (KPK) in 2003.

The Ethnic Issue

The issue of ethnicity has a pervasive and omnipresent impact on Indonesian everyday life, be it in business, in the political arena or elsewhere. Following independence, the new Indonesian state was faced with a situation in which the local business community was first of all puny and secondly overwhelmingly made up of Chinese rather than indigenous Indonesians. In the absence of a strong local bourgeoisie to undertake economic growth it was generally agreed among the political elite that the state would have to be actively involved in the economy (MacIntyre, 1994: 246).

Even though they constitute less than 5 per cent of the population, the Chinese Indonesians constitute by far the strongest capitalist group, but as is the case in other parts of Southeast Asia they are disliked and mistrusted by large sections of the indigenous community. Among the reasons are, historically, that they were perceived as being commercial middlemen for the Dutch during the colonial period, and today, owing to their economic success and, simply, their ethnic, religious and cultural difference. As a consequence, their access to government positions, to the army, and to universities has historically been restricted and they have kept a low political profile, relying instead on clientilist relations with the political and military elite. The domains of business and state have to a large extent been separated along ethnic lines.

Even though the cleavage remains, it has become less pronounced in recent years for various reasons: first, business and capitalism more generally is no longer regarded with the disdain it formerly was, and an increasing number of Indonesian business people and MBA graduates have found lucrative careers in the private sector (MacIntyre, 1994). Second, overlapping shareholdings and directorships between private Chinese Indonesian and politically well-connected *Pribumi* business groups are blurring ethnic and socio-economic boundaries. Many *Pribumi*, the Suharto family in particular, developed their businesses jointly with Chinese partners. For the Chinese, this is a way to share commercial risk and for the *Pribumi* it is a way to tap into Chinese expertise, capital and information networks. Finally, in response to the flight of Chinese capital and international pressure following the riots and the financial upheavals at the close of the 1990s, government has declared its intention to improve conditions for Chinese Indonesians.

The Chinese Indonesian dominance of the corporate sector does remain glaringly apparent, however. In 1993, Chinese Indonesians controlled 204 of the 300 top companies, comprising 80 per cent of total assets (Pangestu and Harianto, 1999), and among the ten largest conglomerates in a 1995 ranking by Pusat Data Bisnis Indonesia (PDBI), nine were controlled by Chinese Indonesians.[2]

How do these issues bear on the efficiency of the Indonesian innovation system? It would be significant if, say, the Chinese groups had an inclination to engage in particular activities. Some writers refer to the 'trading mentality of traditional Chinese businessmen', and any such mentality would be likely to be less conducive to manufacturing activities. During the colonial period, Chinese business activities tended to be mercantile in nature and family-based in structure (World Bank, 1994) but it is highly doubtful that this is ethnically grounded rather than the result of a particular, possibly ethnically specific, incentive structure. At any rate, such a mentality would be very much 'traditional' and have little contemporary relevance, as the many examples of Chinese engaging in industrially advanced activities and taking on more professional forms of organization demonstrate.

The importance is to be found elsewhere: even though the collaboration between large private business groups and political and governmental elites may, at least incidentally, accomplish some of the same business–government coordination as more formal business–government deliberation councils have accomplished in some East Asian countries, the main effect of the ethnic cleavage is a separation of activities conducted by the state from those of the private business groups. The state-owned enterprises and their well-funded supporting infrastructure, other state-controlled supporting institutions, and training and education institutions are less likely to forge linkages with the broader private business sector, and large Chinese conglomerates are less likely to cooperate with smaller *Pribumi*-owned companies. Accordingly, the potential for developing a well-integrated innovation system is constrained.

Political Co-optation and Control

The New Order regime strove vigorously and skilfully to organize societal groups, not with the purpose of channelling their aspirations into a democratic system of decision making but to co-opt and contain them. After coming to power in the mid-1960s, security planners in the New Order regime expanded and consolidated a state-controlled system of corporatist[3] representation put in place by the prior regime, with the intention of channelling political participation away from less controllable institutions, notably the political parties. The peak organization in this corporatist strategy, the state political party, *Golkar*, did not serve to aggregate and articulate political interests but rather to co-opt and contain interest groups and act as a buffer between societal interests and the state. In order to separate political and societal interests, all existing representative bodies covering labour, peasants, fishermen, youth and women were fused into five single and officially-designated organizations. Several other associations were established as part of an elaborate and far-reaching network of corporatist bodies, all subordinated to *Golkar*. Only *Golkar* was allowed to organize in the countryside, where the majority of the population is located.

These aspects have an important impact in the realm of business and economy. The Indonesian Chamber of Commerce and Industry (*Kadin*) is the peak corporatist body representing business and was established in 1968 (MacIntyre, 1991). It has three wings: the regional branches, a collection of 'aspiration groups', and the sectoral and subsectoral industry associations. The regional branches tended to be steered by *Golkar* officials who usually simultaneously held a political office and served particularly to inform the government about industrial issues and to brief industry on government policy intentions. The aspiration groups represent a few economically weak groups and are largely insignificant. The third wing, the industry associations, has widely been regarded as the only true representation of business interests (MacIntyre, 1991: 49), and while being formally subordinate to *Kadin* they have been able to operate relatively independently. *Kadin* itself is looked upon with disdain by broad sections of government and business alike, considered a corporatist tool of government, a rent-acquisition tool for privileged business leaders, and unprofessionally and ineffectively run.

There are several reasons why *Kadin* is marginalized (MacIntyre, 1991: 45): the credibility of *Kadin* as the peak business association is compromised by the dual facts that the economically powerful Chinese Indonesian are either altogether not or only symbolically involved and that the *Pribumi* leadership has relied on government patronage for their prosperity and thus may not be impartial in their undertakings. Further, government regards the *Kadin* leadership as seriously lacking in professionalism and seriousness, and is therefore reluctant to listen to its views on policy. Finally, and perhaps most importantly, an active involvement of *Kadin* in policy planning and implementation would have run counter to the corporatist approach of the New Order regime: the state-designated interest organizations are generally intended to control and co-opt societal groups rather than lend them a political voice.

Contending Economic Development Orthodoxies

Throughout most of the post-independence period, the policy stance of the Indonesian state has been markedly interventionist, both in economic and political affairs. But economic strategy has neither been uniform nor uncontested. Rather, different groups and different ideologies have been able to assert varying measures of influence at different points in time. Two groups are commonly recognized, 'the nationalists' and 'the technocrats'. In our context, it is appropriate to differentiate even further as we shall return to below.

One important contributing factor to state interventionism was the long colonial experience and the armed struggle against the Dutch (MacIntyre, 1994), which induced a profound scepticism towards liberal economic and capitalist ideas among the nationalist leaders, and to the present day there is a deep-seated distrust of market forces, economic liberalism and private (especially Chinese) ownership in large sections of Indonesian society (Hill, 1996). Accordingly, economic policy was ideologically informed by collectivist and socialist ideas about economic organization. Ideology aside, another contributing factor was that following independence, the indigenous business community was small and weak so that a need for the state to initiate, manage and guide industrial activities was perceived, particularly since reliance on market forces would leave the initiative to the decidedly stronger Chinese entrepreneurial community. Economic nationalism had two overarching concerns: increasing national control over the economy relative to foreign control, and increasing indigenous control relative to Chinese domination. Thus, nationalist policies have more or less explicitly had a *Pribumi* promotion component.

When Soeharto took control in 1966 of a country in economic ruin, a new group of economic policy-makers came to the fore, the so-called 'technocrats'. In order to renegotiate debt and attract badly-needed aid and foreign capital and technology, Soeharto installed a group of Western-educated economists in key economic bodies. They successfully argued for pursuit of a more outward-looking and orthodox liberal policy, spending on SOEs was cut, trade barriers loosened and investment laws reformed. The technocrats are generally committed to neo-classical growth economics, and primarily emphasize reliance on foreign capital and market forces, being cautious of state subsidy and protection of particular economic groups. The technocrats reside mostly in planning agencies, that is *Bappenas*, in the National Bank, the Ministry of Finance and the University of Indonesia.

Their influence soon declined, however. In the early 1970s, student protests were mounted against foreign influence, particularly Japanese, culminating in the 1974 Malari riots. At the same time, critics, most notably perhaps the Chinese academic and businessman Jusuf Panglaykim, associated with the Centre for Strategic and International Studies (CSIS), argued that without a coordinated national economic strategy, the penetration of foreign capital would be exploitative and destructive towards local industrial development (Robison, 1986). Finally and most importantly, the oil price hike in 1973 secured the state a revenue base, which allowed it to reinvigorate its interventionist policies. Protection and trade barriers increased, foreign entry was restricted, and programmes to establish local capital and intermediate goods industries and increase industrial deepening were undertaken, spearheaded by direct public investment.

With the plummeting of the oil prices in 1982 and 1986, the fall of other commodity prices, unfavourable currency realignments, and a general slowdown in the global economy, the technocrats came back into favour and economic policy swung back towards liberalization and private-sector investment in export industries. The financial sector was deregulated and numerous trade and investment barriers were lowered.

This later reform programme was, again, highly successful. High growth in non-oil exports and in manufacturing allowed for renewed growth of state-owned industries in the 1990s (Meyanathan, 1994) but this time with an explicit agenda of technological upgrading. A less frequently recognized group, the 'technologs', surrounding the former Minister of Research and Technology and later president, Habibie, argued for a strategy of technological leapfrogging, that is increase of domestic capabilities and value-added through the establishment of a number of high-tech 'strategic' industries along with supporting infrastructure. With its explicit technological agenda and its narrower focus on a limited range of state-owned industries, this group is meaningfully distinguished from 'the nationalists'. The technologs' influence increased from the early 1990s: an unprecedentedly large number of them were appointed to the cabinet in 1993, and Habibie's influence was growing (EIU, 1996). Following the financial crisis of 1997 and IMF reforms, funding was reduced or stopped, and restructuring of the strategic industries initiated. The IMF ordered an end to the subsidy of the aircraft manufacturer, IPTN (now Dirgantara Indonesia), in 1998.

In addition to the technologs, one might finally add a fourth group, 'the cronies', influential groups and individuals in the political and military elites that are less concerned with economic development than with the appropriation of state resources and state economic power, but that still significantly influence economic policy.

While the policies recommended by the technocrats have proved their worth with respect to macroeconomic policy, market-based policies are not likely by themselves to support the micro-level development of local technological capabilities. The main focus tends to be short-term allocative efficiency rather than long-term development. The technologs were explicitly concerned with indigenous technological upgrading, but the dirigisme and the extent of separation between the strategic industries and the bulk of private industry seriously question their sustainability and developmental impact. At the same time, they monopolize much of the technological infrastructure and budgets at the expense of private industry. It is also difficult to find successful examples of technological leapfrogging in history, and the literature generally discourages such attempts. At first glance, the ambitions of the economic nationalists are more likely to be conducive to the development of local capabilities, yet the strong *Pribumi* promotion component has worked against the pursuit of economically rational policies and excluded a large and important section of private business from policy focus. Futhermore, performance requirements were lacking, rent-seeking and patrimonialism abound, implementation was weak and supervision lacking or non-existent. If we identify 'cronyism' as a strategy, it is, as previously discussed, not concerned with long-term growth and upgrading but rather with short-term gain and acquisition of rents.

In addition to the restrictions that each of these ideologies place on the development of a strong and coherent innovation system, at a higher level, the struggle for political influence and economic wherewithal between the groups is likely to have caused policy and institutions to be less coherent and consistent than they might otherwise have been and to have led to interdepartmental conflicts.

THE INDONESIAN INNOVATION SYSTEM

In the following section we will outline the Indonesian innovation system by looking at the incentive system, technology policies, other industrial policies, the training and education system, supporting institutions, and at corporate structure. We find that the innovation system is generally weak and has insufficient linkages with private industry. One of the most conspicuous findings is that the innovation system is segmented and the segmentation is manifest at several different levels: different economic orthodoxies have prevailed at different times, resulting in distinct and historically varying trade, investment and competition regimes; different technology policies have been applied to the state-owned 'strategic industries' and to private industry; and supporting institutions are also separated along these lines.[4]

Incentives

Here we will consider the following elements of the incentive structure facing firms: macroeconomic policies, trade and investment regulations, and competition. The general incentive structure has varied considerably over the course of time depending on which development orthodoxy has been dominant. Three broad phases in the New Order government's (1966–98) approach to development should be distinguished: an adjustment phase (1966–73), an import substitution phase (1974–85), and an export orientation phase (1986–1996). As already mentioned, succeeding a period of severe economic stagnation and budgetary deficits under the Old Order (1948–65), import substitution policies were succeeded by adjustment and opening up to foreign investments and trade during the years 1966–73. From 1974 onwards, spurred by rising oil revenues and anti-foreign sentiments, increasing emphasis was placed on developing domestic industries and the policy regime became more restrictive. Induced *inter alia* by plummeting oil prices in the mid-1980s and the worldwide trend towards liberalization and deregulation, the Indonesian economy became more export oriented with fewer import restrictions and more openness towards foreign investments (EIU, 1996; Bhattacharya and Pangestu, 1993).

Throughout the New Order period, the government's macroeconomic policies were widely applauded: according to Hill (1995), Indonesia had the best performance among the OPEC countries since 1966, and a World Bank comparative study of 18 developing countries rated its macroeconomic management among the best performers, along with Korea and Thailand (Little et al., 1993). The importance of a stable and predictable macroeconomy as a prerequisite for firms' investment, trade, and financial transactions is uncontroversial. Even though the financial crisis from 1997 onwards brought some of these analyses into question, throughout the New Order period the macroeconomic groundwork for the development of a strong innovation system was advantageous.

With respect to the trade and investment regimes, during the pre-1973 adjustment phase, little promotion was given to industrial development because of budgetary constraints (Soesastro and Pangestu, 1998). The manufacturing sector was dominated by small and medium-sized companies in resource-based activities. The 1967 Foreign Investment Law was very liberal, allowing 100 per cent foreign ownership, and gave incentives such as tax holidays, accelerated depreciation, and duty-free imports of capital goods.

The years of import-substituting industrialization, 1974–1985, witnessed increasing protection, trade barriers and focus on capital-intensive industries of questionable efficiency (Soesastro and Pangestu, 1998). In the first half of the 1980s the government seemed ambivalent on whether to curb the current account deficit by reducing imports or by increasing exports, and a number of quantitative restrictions on imports were introduced (Booth, 1995), but eventually government turned decisively towards exports. The trade system was liberalized further in 1995 with substantial reductions of most tariffs and non-tariff barriers and the announcement of further reductions through to the year 2003.

With regard to investment, the regime was loosened from 1986 onwards: the application process was simplified and streamlined; licences were granted for a lifetime and expansion of production capacity up to 30 per cent was exempt from application for approval. Majority foreign ownership was allowed for export-oriented companies in 1986 and divestment requirements for foreign investments relaxed. In 1989, a long list of investment priority sectors was replaced with a shorter negative list of areas closed to foreign investment. In a major package in May 1994, most restrictions on foreign investments were removed: 100 per cent foreign ownership was allowed; the divestment requirement was virtually removed; the minimum investment amount was significantly reduced; and exports and domestic sales of goods produced by a foreign firm could now be undertaken by a joint venture. One of the intentions was to attract small and medium-sized investors in the components supply sector. This package had a substantial effect upon approved foreign investments. A June 1996 deregulation package introduced 'Super Highway' facilities for exporting companies, that is expedited services in agencies such as customs, tax and banks, and allowed foreign companies to sell products in the domestic market up to the wholesaler level.

Finally, we take a look at competition. The impact of competition is not straightforward. On the one hand, there is the infant industry argument that new entrants need some degree of protection or subsidy for some time to develop when faced with more competent competitors. On the other hand, competition is usually seen as bestowing a certain discipline and pressure to perform and innovate on firms, and studies of industrial districts around the world document the potentially beneficial impact of patterns of simultaneous cooperation and competition between firms. In Northeast Asia, the recipe applied was simultaneous competition in export markets and protection domestically. As we have seen, in Indonesia, the reforms since the mid-1980s have progressively reduced tariff barriers and quantitative import restrictions, and this has led to increased competition from imports. A provision that companies in export-processing zones can sell up to 25 per cent of their output on the domestic market, subject to duties and taxes, also exposes domestic manufacturers to more competition. Even though few restrictions remain on domestic competition, many domestic companies were established under the more protected regime.

What does this tell us about the Indonesian innovation system? Some policy initiatives in the past turned out to actually disfavour domestic producers, and other initiatives turned out to be too administratively complicated to be applied. Protection was not combined with performance requirements or with a strategy for purposeful industrial and technological upgrading. Companies were not induced to export, and increasing competition from imports and from foreign entrants has not been combined with support or incentives for exports, which does not square well with the positive Northeast Asian experiences with simultaneously limiting competition from abroad, and encouraging domestic competition and exports. Soesastro and Pangestu (1998) conclude that the structure of protection favoured production of final consumer durable goods and disfavoured production of capital and intermediate goods, had a heavy anti-export bias and that the result was the development of consumer goods assemblers with weak backward linkages. On the whole, the measures undertaken in the drive towards export orientation focused on measures to improve the 'high-cost business environment', to build a component industry, and to attract foreign investment. They did not, however, particularly address developing a high potency of domestic market, accelerating the emergence of a higher value-added industry, or improving company-level performance. Yet, given the further reductions in protection scheduled under various international agreements, the viability and performance of the domestic industry increasingly depends on initiatives in these latter areas.

Technology Policies

It has been claimed that Indonesia does not have a technology policy (Hill, 1995). Technically, this is not correct: both the five-year development plans (the *Repelitas*) and the 25-year development plans contain targets for science and technology (S&T) development. Policies for implementing the S&T strategies in these plans are issued by the Minister of State for Research and Technology through a publication, *Punas Ristek* (National Priority Program for Research and Technology) (Samadikun, 1998). The plan is based on proposals from the National Research Council (DRN), and detailed planning of the implementation is conducted by research institutes residing within the technical ministries (health, industry and trade, agriculture, mining and energy) and by national research institutes (Indonesian Institutes of Sciences, Atomic Energy Agency, Space Agency). Lall (1998) has stated that Indonesia does not have a 'technological strategy', perceived as 'a coherent set of policies'. Technically this does not seem entirely correct either. Issues of appropriateness and implementational success aside, Indonesia has had a strategy for scientific and technological development formulated by B.J. Habibie, who served as minister for research and technology from 1978 to 1998.

Habibie's strategy rested on a string of premises (Rice, 1990; 1998): Indonesia could not continue to depend on its rich resource base and low-cost labour for competitiveness due to deteriorating terms of trade of primary commodities, development of substitutes for simple labour-intensive commodities in developed countries and increasing competition among developing countries in the production of such commodities. Instead, the country should attempt to move into activities with higher value-added by developing human resources and technology. This would allow the country to maintain cultural and political integrity through economic development and participate in S&T-related decisions with important global consequences. Furthermore, high-tech products display a higher growth of world demand[5] and the presence of fewer competitors allows producers to set prices more favourably. Important reasons for the focus on human resources are that they are less internationally mobile, and that they appreciate through use, whereas physical resources depreciate. By developing human resources and the S&T infrastructure, Habibie foresaw that Indonesia would be able to attract more of the surplus capital accumulating in developed countries as profitable investment opportunities in those countries diminish.

These premises and principles led Habibie to advocate a strategy of 'picking winners', that is. industries particularly suitable as vehicles for technological and industrial development, and of state ownership of and control over those industries. Ten state-owned 'strategic industries' were established under the aegis of the 'Agency for Strategic Industries' (BPIS).[6] The primary mission of BPIS was to anticipate a shift from a resource- to a knowledge-based international business and a shift from comparative to competitive advantage.[7]

In the Repelita VII (1999–2004), the central priorities of the *Punas Ristek* are: (i) excellence and self-reliance in several key technologies. The strategic industries are put as a driving force behind this effort. (ii) International competitiveness. Through the establishment of new polytechnics, manufacturing laboratories at universities, and support in production techniques and training of operators and maintenance staff, manufacturing industries are to become more competitive in world markets. (iii) Development of technology-oriented small and medium-sized enterprises. (iv) Intellectual property rights. Efforts to import technology through licensing agreements and to increase production of domestic patents.

So, contrary to claims, Indonesia does have policies and strategies in the area of science and technology. Furthermore, the strategy formulated by Habibie concurs in many respects with the general recommendations of the literature on national innovations systems, for example the ambition to move into higher value-added activities, the emphasis on human resources and science and technology, the interdependence between routine manufacturing activities and innovation, and the need for support and coordination. Yet Hill (1996: 95) maintains that in spite of the various plans and associated agencies, there is surprisingly little detailed planning effort. If one requires a strategy to be 'coherent, stable, and well-implemented' to be a strategy proper, as Lall (1998) seems to do, the critique is more justified, as we shall see below. With the reliance on state-owned companies for technological upgrading, a strategy for using FDI for such upgrading, as has been present in other countries, is lacking. Finally, the strategy envisioned by Habibie may be characterized as one of 'technological leapfrogging', a vision which has generally been discredited both empirically and theoretically. Few outside observers have supported Habibie's strategy and the fact that the targeted industries were isolated from the bulk of private industry further reduces the prospects of a reasonable return on investment.

The most recent development in Indonesian technology policy was the 2002 bill on a 'National System of Research, Development and Application of S&T' (*Sisnasp3iptek*). The bill was intended to encourage the private sector to invest in research and innovation activities and, in terms of implementation, the government introduced various incentive instruments to stimulate indigenous R&D for private industries. The effectiveness of these instruments is yet to be seen and will obviously depend on the extent to which the agenda of strengthening knowledge-based activities is picked up by private industries.

Other Industrial Policies

In this section we will briefly review a few issues that were not covered in the previous section on more narrow technology policies.

A 1994 presidential decree from (*Keppres* 16/1994), applying to all government entities including SOEs, supports domestic producers through government procurement practices: if a domestic manufacturer is competitive in terms of cost and quality, the product must be bought locally. If there is no such manufacturer, the product must be bought through a local agent. If there is no local agent, the product can be bought from overseas suppliers.

Indonesia does not have an active technology licensing policy as Korea did. Apart from formal approval by BKPM of technology licences that are part of an investment project, technology licences are not regulated but left to individual firms. Given the level of administrative and technical capabilities, it is probably prudent to abstain from ambitious licence regulation programmes. Expenditure on R&D and training has been made tax deductible.

The Indonesian government is in the process of strengthening intellectual property rights (IPR) protection, and a new IPR bill was enacted in 2003. IPR protection is commonly taken to influence the willingness of foreign companies to transfer technology and use proprietary technology locally, the cost and ease with which foreign technology can be obtained by local firms, and the inclination of local firms to engage in R&D. During the earlier stages of development it is likely that weak IPR protection benefits local firms since it increases their ability to conduct reverse engineering or directly use proprietary technology without a legal agreement. However, there is a strong push for stepping up IPR protection under the WTO and the sanctions for violations may be severe, most likely with a greater negative impact than the potential benefits of weak protection. Yet if the issue of implementation is taken into account, IPR protection remains very weak in Indonesia, one of the reasons being the generally weak judicial system, particularly when it comes to settling individual business disputes.

The metrology, standards, testing and quality assurance (MSTQ) services system has been characterized as hampered by low quality awareness and the absence of a comprehensive set of industrial standards (Lall, 1998). The standards system supports industry in the following ways: formulation of standards and codes for products and production processes, implementation of calibration schemes for measuring instruments and equipment used in production, implementation of a laboratory accreditation scheme to regulate laboratories testing and certifying products. Product testing and certification serve to assure the quality of products and keep substandard and potentially dangerous products off the market. Obviously, export ability critically depends on these issues. Exports to the European Union must show evidence of compliance with ISO quality standards, for instance. Standards formulation and implementation are the responsibility of the individual ministries. For industry and trade the responsibility resides with MOIT, specifically the Centre for Industrial Standardisation (*Pustan*), which formulates Indonesian National Standards (SNI). Many standards have been below international levels and not recognized abroad, and measuring and testing laboratories have lacked international accreditation. The standards system has been very fragmented, with many different ministries involved and the private sector poorly represented, if at all. The national metrology institute cannot meet demand for its services and its calibration network is weak, and a regional calibration network operated by MOIT is not coordinated with the national institute. It provides very limited services and its staff are not adequately trained. Some of the testing laboratories under MOIT are poorly equipped and out of date (Lall, 1998).

Skills

In the context of technological capabilities, we will focus on the acquisition of more advanced skills rather than basic education and literacy. In this latter area, Indonesia has a very good track record, nearly having achieved universal primary education and almost obliterating illiteracy (Jones, 1994). According to the gross enrolment ratios in secondary and tertiary education, Indonesia is behind most of its Asian neighbours. This gives cause for concern in the light of the rapid structural transformation of the economy and the ambition to move into technologically more demanding activities, along with the fact that studies unanimously point to education as pivotal for the development of technological capabilities, the reasons being both the effect upon indigenous capabilities and that it encourages foreign investors to transfer technology since it is more easily absorbed.

The problem is further aggravated when not only the quantity but also the quality of education is considered: it is generally recognized that lecturers are insufficiently educated, some university professors still only hold a BA degree, and often have side jobs to supplement an inadequate income. If these jobs are closely related to the subject taught, this may enable them to relay relevant experiences but more generally it detracts from their teaching efforts. The curriculum is often inappropriate and too broad. Student/staff ratios are high and resources for teaching aids limited. Teaching style is authoritarian, emphasizes passive memorization, and does not encourage creativity or initiative. Vocational and technical education at secondary level is poorly developed (Jones, 1994). Tertiary education is skewed towards the social sciences and the humanities, with an insufficient supply of science and engineering graduates, partly because technical education is more expensive, and among the scientists and engineers who do graduate, the large majority seek employment in government rather than private industry.

Another widely used measure of the strength of the innovation system is R&D expenditure. R&D indicators are not easily compared, though, since terms are defined differently in different countries, and it varies between firms and governments as to which activities are categorized as R&D. Furthermore, as noted elsewhere, the absorption and diffusion of technology and incremental changes in the course of production may be more important than formal R&D, particularly for countries in the earlier stages of development.

If we look at actual R&D expenditure by both government and productive enterprises (see Table 7.1), Indonesia invested about 0.1 per cent of GDP in R&D in 1995, at the level of Thailand but lower than Malaysia and Korea. The bulk of expenditure is furnished by government (80 per cent in 1991), and the share of R&D activities carried out in the productive sector is modest. Considering that it is widely recognized, particularly in the literature on national innovation systems, that proximity between R&D, production and marketing activities is central to innovation and, further, that there is an inherent danger that supply-driven R&D undertaken by government is too remote from actual industry needs, both the expenditure levels, expenditure sources and agents of activities are likely to be disadvantageous to technological upgrading.

These problems were recognized by the government, and the target of R&D intensity in the second 25-year plan (1994–2019) to reach 2 per cent of GDP and 70 per cent private sector contribution by 2019 was ambitious (Samadikun, 1998). Simulation results (Aminullah, 1998) indicate that whatever policies will be adopted by the Indonesian government to accelerate domestic R&D, much time is required to achieve a significant level of industrial R&D intensity. In order to reach the level of industrial R&D intensity, amounting to 1 per cent of GDP, there is a need to increase the intensity by 15 times from the level of around 0.064 per cent of GDP (in 1994), despite the level afterwards tending to decrease to around 0.035 per cent of the GDP (1999) (see Figure 7.1).

The national innovation system literature points out, and it is commonly recognized, that many important competencies are not acquired through

Table 7.1 ***R&D expenditure, sources of expenditure and agents of activities, selected countries and years***

	R&D	Sources of R&D expenditure (%)			Agents of R&D activities (%)[1]		
	expenditure as % of GDP	Productive enterprises	Government	Foreign & other	Productive sector	Higher education	General service
Indonesia							
1986	0.3	–	–	–	–	–	–
1995	0.1	19[2]	80[2]	1[2]	33[2]	–	–
Korea							
1986	1.8	81	19	0	67	11	22
1994	2.8	84	16	0	73	8	19
Taiwan, 1991[4]	1.7	52	46	–	54	–	–
Malaysia, 1992[3]	0.4	43	53	4	45	9	46
Thailand							
1985	0.3	14	70	17	62[5]	30[5]	7[5]
1995	0.1	12	80	8	7	36	57

Notes:
[1] The 'productive sector' includes domestic and foreign industrial and trading establishments which product and distribute goods and services for sale; 'higher education' includes establishments of education at the third level as well as those research institutes, experimental stations, etc. serving them; 'general service' are various public or government establishments serving the community as a whole.
[2] 1991, source: Thee (1998).
[3] Data for the mid-1980s not available.
[4] Source: Thee (1998).
[5] 1987 source: UNESCO Statistical Yearbook, various years.

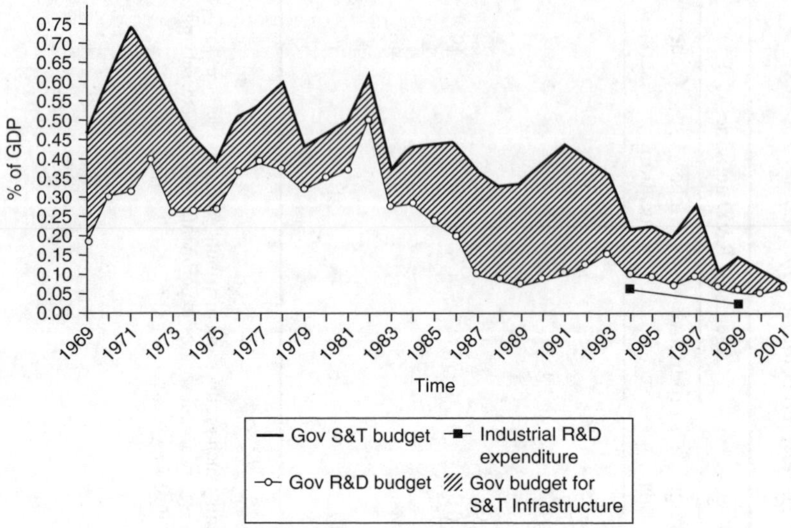

Notes: Government S&T budget is development budget for R&D activities and S&T infrastructures, Government R&D budget is development budget and routine budget (including maintenance and services of R&D facilities), Industrial R&D expenditure based on special survey conducted by the Central Bureau of Statistics (BPS) for Pappiptek-LIPI.

Sources: Data for the period of 1969–1994, quoted from Aminullah (1998). Data for the period of 1995–2001, for Pappipted-LIPI.

Figure 7.1 Trends of Indonesian science and technology (S&T) activities

formal schooling. Indonesia is thus faced with a classical chicken-and-egg scenario or in development studies parlance, a vicious circle: due to the technically less advanced state of Indonesian industry the opportunity to obtain relevant experience is limited. The lack of experienced personnel in turn impedes the development of domestic industries, and furthermore reduces the incentive for foreign companies to locate more complex activities in the country.

Supporting Institutions

Institutions are a central theme in the literature on national innovation systems: beyond the influence of specific institutions, institutions are at a more general level regarded as essential providers of the stability needed for innovative efforts to take place and to be successful and as repositories to create, store and transfer the knowledge, skills and artefacts feeding into innovative efforts or, put differently, mediators of externalities.

In this section we look at supporting institutions in the narrow sense of specific organizations supporting technological development in industry. In Indonesia such institutions have served two different purposes: one is to provide research, information and other support to industry in general; the other is to support the 'strategic industries'. Institutions and responsibility for policies are divided according to these lines. Consequently, there are two types of public sector research institutions: those under the Ministry of Industry and Trade (MOIT) and those under the Minister of State for Research and Technology (*Menristek*).[8] In addition to this, efforts towards long-term basic research are conducted by the Inter-University Research Centre (PAU) under the Ministry of National Education and independently by the scientific research establishment.

As far as the institutions under MOIT are concerned, through its Agency for Industrial Research and Development (BPPI), MOIT controls nine sectoral institutes, ten regional testing laboratories, and five industrial research and testing centres (Lall, 1998). They are primarily engaged in training and testing, product certification, standardization, and extension services. They undertake little technology development or R&D and tend to have poor linkages with industry. There are few incentives to sell technology and services to industry, and they have limited ability to market technologies. Salaries are low and staff insufficiently trained; laboratories have outdated equipment and are starved of resources, and management is weak. MOIT has also established semi-autonomous extension services, Technical Service Groups, to support textiles, engineering products and pulp/paper industries, with a 90 per cent subsidy to firms that use their services. Lall (1998) considers that these have done some valuable work but have limited geographical and industrial coverage, little firm-level reach and are dominated by expatriates.

Regarding the institutions under *Menristek*, most national expenditure on R&D is allocated to six non-departmental government institutes (LPNDs). They subcontract research from the ministries and are supposed to support private industry with research services and promote linkages with research at universities. They tend to be better staffed, funded and equipped than institutes under MOIT. But they too have established few linkages with private industry, are supply- rather than demand-driven, bureaucratic with respect to management and procedures and contribute little to the development of industrial technology (Lall, 1998).

The Agency for the Assessment and Application of Technology (BPPT) is in charge of formulating and implementing policies for industrial and other forms of technology, and was chaired by Habibie until he became president in 1998. Being one of the LPNDs under *Menristek*, the agency is relatively well endowed to provide technical information and support services to national industry. According to a personal interview with a BPPT director, the agency conducts four types of activities, with the first being the most important: (i) assisting in formulation of technology policy; (ii) coordination of assessment and application of technology, for example arranging discussion seminars with relevant parties before major decisions; (iii) services to government and private companies such as staff training; and (iv) in-house research. In practice, there are very few instances of service provision to private companies. They have attempted to reach private industry but have not been successful, according to the director because there is not much demand for such support, and because domestic companies tend to be linked to foreign principals and only want to use proven technology, rather than developing new technology. BPPT is well supplied with formally trained manpower but employees lack practical experience. There is practically no migration of experienced engineers into private companies. Even though activities of BPPT do not reach private industry, they can lower the cost of technology acquisition and make acquisitions more informed.

A science city, the National Centre for Science and Technology Research (*Puspiptek*) at Serpong, south of Jakarta, and established in the early 1980s, was initially proposed by *Menristek*. It has six BPPT laboratories and four LIPI (the Indonesian Institute of Sciences) laboratories and an industrial estate is being constructed to attract knowledge- and skill-intensive industries, which can draw upon the technological facilities there and contribute to commercializing technologies. The intention was that it should serve the strategic industries as well as complement and support other industry-related research done by MOIT, and it should have strong links with private industries. However, it has been subsumed under the somewhat self-contained system of *Menristek* and primarily caters for the strategic industries. Even though MOIT has laboratories nearby, they have little contact with the LPNDs. In Lall's (1998) assessment, it is too early to judge the longer-term benefits of *Puspiptek*, but currently it remains largely unlinked from the bulk of Indonesian industry.

In recognition of some of the problems mentioned above, the government launched a programme, Priority Partnership Research Program (RUK), in 1995, intended to encourage public S&T institutions to carry out activities more relevant to real industry needs. Under this programme, the government finances research, development and engineering activities, which are carried out in cooperation between public institutions and private or state-owned manufacturing enterprises, and aim at concrete challenges encountered in production. It may be premature to assess the initiative, but according to Thee (1998), most activities so far have been initiated by the public institutions and may consequently not correspond to industry needs.

Since 2001 several provincial governments have established a Regional Research Council (DRD) and a Regional Agency of R&D (*Balitbangda*) to act as a bridge between regional research activities and regional economic and industrial development. These are under the jurisdiction of the Ministry of Home Affairs (*Mendagri*).

In addition to the internal divisions between MOIT and *Menristek*, linkages between private companies and supporting institutions are generally very weak. The larger, more capable firms tend to regard their activities as irrelevant or of insufficient quality, while the smaller companies often do not know of their existence or services, or lack the motivation to seek them out. There are some links between companies and institutions under MOIT, particularly in the areas of management and quality assurance, but companies rarely engage with institutions under *Menristek*.

Corporate Structure

The dominant corporate players in the Indonesian economy during the Soeharto era were the state-owned enterprises (SOEs), Chinese Indonesian conglomerates, and an increasing number of *Pribumi* and joint *Pribumi*–Chinese Indonesian conglomerates. Foreign investors have been important in a limited number of sectors but relatively unimportant in the overall economy. Due to factors such as the post-independence lack of indigenous entrepreneurs; an aspiration to curb the economic power of the Chinese Indonesians; the nationalization of colonial property; and a desire to maintain strong central control of the economy, SOEs have had a strong presence in most sectors of the economy, albeit a declining one largely due to privatization of several SOEs in the reforms era.

With the inflow of oil revenues from the early 1970s to the early 1980s and the nationalist and protectionist turn of economic policy, a number of large private business empires were formed, primarily on the basis of clientele and patrimonial relationships between key government figures and company leaders. Consequently, the corporate sector became heavily dominated by a relatively small number of large conglomerates, predominantly owned by Chinese Indonesians.[9] With the end of the oil bonanza by the mid-1980s and the consequent structural reforms and push towards manufactured exports, the private sector became the principal engine of growth of a national economy previously dominated by the state.

However, as is often the case in developing countries, liberalization and deregulation had more impact on the formal regulatory framework than on the actual distribution of economic resources, and after the reforms in the mid-1980s, some existing business groups grew stronger and new ones entered the scene. The most prominent among these new entrants were then President Soeharto's various relatives and friends, and their relatives and friends in turn, who embarked upon lucrative business careers.

Economic concentration in conglomerates is not in itself adverse to economic development but may, rather, be a way to economize on scarce economic and entrepreneurial resources, as the case of Korea demonstrates. But in Korea, the attainment of rents was associated with stringent performance requirements, and non-compliance was penalized. If no such disciplinary measures are in force there is an obvious risk that large business groups focus on quick and easy acquisition of rents by way of licences, protection, monopolies, state credits and contracts, and so on. And indeed, Indonesian business groups are usually taken to be particularly dependent on rents (Amsden, 1995). Both the Chinese Indonesian and the *Pribumi* conglomerates depended heavily on political elite access and business privileges, but the *Pribumi* conglomerates were particularly renowned for their entry into government-related businesses, such as infrastructure (toll roads, telecommunications, electricity, ports) and other government-related contracts, and their businesses are largely confined to the non-tradable sectors.

According to Pangestu and Harianto (1999), the typical expansion pattern for Chinese conglomerates involves a transition from trading activities to raw material supplies, to manufacturers, to manufacturing and non-traded sectors. They describe the corporate structure of Chinese groups as follows:

> The internal structure of Indonesian Chinese family businesses, shaped by personal trust and suspicion toward outsiders, inherently limited the size of the organisation. A complex structure of a large bureaucracy was uncommon [. . .] and complex integration was avoided as much as possible. When the business expanded, the organisation was subdivided into a federation of small units of sub organisations. These various units were coordinated largely through the entrusted personnel assigned to head each unit. At the top, key positions were entrusted to family and small number of long-time employees and associates. Thus, strong overlaps among control, management and ownership prevailed, with the head of the family at the centre. (ibid: 9)

Pribumi business organizations are characterized by a similarly paternalistic structure. In addition, these latter, along with the SOEs, tend not to be export oriented or compete in international markets. The common view of the major business groups being predominantly 'crony' is not entirely undisputed, though: based on PDBI data, Wibisono (1991) estimates that 263 out of 300 conglomerates, generating 75 per cent of their total sales, have succeeded through 'hard work and productivity' rather than favouritism, but it is a fair bet that few other observers would agree with his optimistic outlook.

In addition to the direct and immediate risk that groups with special privileges shied away from activities requiring gradual and long-term technological capability acquisition, there are other less direct adverse consequences: enterprises which were not similarly privileged were faced with unfair competition and found the general business environment more difficult to negotiate. Competition is a basic incentive affecting capability development, and the corporate structure described is clearly distortive. Pangestu and Harianto are perhaps somewhat strong in their observations concerning the importance of particularistic relationships and personal trust in business transactions, suspicion towards outsiders, and disinclination towards vertical complexity worked against the formation of the innovative user–producer interaction and inter-firm linkages stressed in the literature on national innovation systems. Furthermore, crony groups can, qua their connections with government elites, pose a threat to the integrity or even survival of other enterprises: some authors report that successful entrepreneurs have been forced to sell businesses to predators with powerful political connections (MacIntyre, 1994; Fromhold-Eisebith, 1998).

In sum, it is likely that the Indonesian conglomerates, which emerged and thrived first and foremost on the basis of rent-seeking practices, have not been conducive to the development of an internationally competitive manufacturing sector. The corporate structure, without independent professional management, without a modern bureaucracy and with limited vertical complexity, all identified by Chandler (1990) as characteristics of successful modern industrial enterprises, would seem to place constraints on the technical, organizational and managerial sophistication of economic activities.

LOOKING TOWARDS THE FUTURE

In the previous sections we have analysed the Indonesian innovation system, described its underlying politico-economic dynamics, and identified important weaknesses. We identified and described a set of different factions within the state, the nationalists, the technocrats, the technologs, and the cronies, and argued that the different policy orthodoxies pursued by these factions had resulted in a visible segmentation of the wider Indonesian innovation system.

In the past, business performance had been hampered by fragmented innovation policies, as expressed by: (i) economic incentives were not successful to improve technology upgrading and export of higher added value products; (ii) technology policy was not able to create a fruitful interaction between strategic industries and private industries; (iii) industrial policy was not successful in stimulating indigenous R&D; (iv) HRD policy was not successful in developing more advanced skills, and; (v) policy on managing innovation infrastructure was not able to create synergy among supporting institutions.

Several steps could be taken to counter some of the threats and weaknesses identified: (i) increases in industrial capacity should be planned within the limits of realistic growth forecasts, taking into account fluctuations in economic demand; (ii) technological capabilities should be strengthened in large-scale industries, and; (iii) efficiency and productivity of small and medium-scale industries should be pushed both by strengthening technological capability and by integrating support from the large-scale industry initiatives.

A major policy challenge is to elevate the Indonesian economy from a path focusing narrowly on economic growth to one focusing on techno-economic development. The future challenge is to make industries realize fast that an increase in industrial productivity in the economy will need to be pushed by industrial technological capability. At the same time, technological innovations will also need to be pulled by demand from the economy. A strategy of economic development based on capital accumulation should be complemented with technology mastery and innovation through industrial R&D. This implies that future Indonesian economic development requires a shift in the mindset of industrial leaders from mere economic to techno-economic.

In the future, the Indonesian government might succeed in providing the better-developed technological infrastructure required to adopt, disseminate and upgrade technology. However, as long as the industries themselves have not learned from the adverse consequences of heavy dependence on foreign technology, innovations based on R&D will not assume an important role for long-term efficiency and productivity. Should this be the case, it can be anticipated that Indonesia will never reach the appropriate balance between trade-led industrialization and technology-based industrialization as the driver of long-term economic growth.

The need to shift the policy orientation from economic to techno-economic will require stepping up industrial R&D. The low industrial R&D in the past was due to an over-emphasis on trade with technological development being neglected. To further enhance technological capabilities, the following activities would appear particularly worthwhile on the part of the government: (i) quick restoration of the technological infrastructure, which was damaged during the prolonged economic crisis; (ii) strengthening the linkages between public research institutions and industry to allow more effective commercialization of innovations, and; (iii) better enforcement of fair industrial competition in the economy.

NOTES

1. For a more elaborate account of these issues, see Gammeltoft (2001).
2. The ten conglomerates were, in ranking order: Salim, Astra, Sinar Mas, Lippo, Gudang Garam, Bimantara, Bob Hasan, Gadjah Tunggal, Ongko and Djarum.
3. Corporatism has been defined as 'a pattern of state-society relations in which the state plays the leading role in structuring and regulating interest groups, organizing them along functional rather than class lines (in order to minimize collaboration and conflict), and typically granting official recognition to only one representative body in any given sector' (MacIntyre, 1994: 1).
4. We will not systematically document the reasons for this segmentation but point to the politico-economic issues discussed earlier, particularly the presence of different statal factions and the issue of ethnicity, see also (Gammeltoft, 2003; 2004). The four statal factions identified above have become associated with different industrial segments, with different affiliated institutions and policies: the technocrats with an export-oriented, foreign-dominated segment which grew rapidly during the 1990s; the technologist with the 'strategic industries'; the nationalists with the private domestic manufacturers which benefited from various forms of protection from the mid-1970s to the mid-1980s; and finally the 'cronies', most clearly represented by business ventures by former President Soeharto's family and friends.
5. And higher income elasticity, one might add.
6. Habibie's emphasis on the strategic industries represents a major shift relative to his predecessor, Professor Sumitro. Like Habibie, Sumitro envisioned transforming Indonesia into a technologically advanced nation, but instead of focusing on specific priority areas, Sumitro advocated carrying out research in a wide range of areas selected on the basis of its impact on employment creation, the use of domestic materials, and productivity. See (Rice, 1990; 1998) for a more thorough discussion and critique of Habibie's approach.
7. According to a personal interview with executives at one of the strategic industries (PT LEN), the purpose of BPIS was also to coordinate the activities of the different industries and avoid conflicts and undue competition between them.
8. Each line ministry has its own sector-specific research laboratories, but here we deal only with industry support.
9. Among the best-known conglomerates are the Salim group (Liem Sioe Liong), the Sinar Mas group (Eka Tjipta Widjaya), the Hasan group (Bob Hasan), the Barito Pacific group (Prajogo Pangestu), the Gunung Sewu group (Goh Sie Kie) and the Roda Mas group (Tan Siong Kee). The revenue of the Salim group alone was US$8–9 billion in 1990 (60 per cent generated from operations in Indonesia) (Schwarz, 1994), which is equivalent to about 8 per cent of Indonesia's 1990 GDP.

REFERENCES

Aminullah, Erman (1998), 'Pola dan Kecendrungan Riset dan Teknologi di Indonesia, Analisis Dinamika Administrasi Tekno-ekonomi, (Patterns and Tendencies of Research and Technology in Indonesia: Dynamic Analysis of Techno-economic Administration)', PhD dissertation, Program Pasca Sarjana, Universitas Indonesia.

Aminullah, Erman (2003), 'Managing technology policy for long-term economic development: comparative analysis between Indonesia and Malaysia', paper presented in *Seaf kaneka Public Forum on Science & Technology for Holistic Development,* Bangi, University of Kebangsaan Malaysia (UKM), 17 October.

Aminullah, Erman (2004), *Berpikir Sistemik untuk Pembuatan Kebijakan Publik, Bisnis dan Ekonomi (Systemic Thinking for Public, Business and Economic Policy Making)*, Jakarta: PPM Press.

Amsden, Alice H. (1995), 'Like the rest: South-east Asia's "late" industrialization', *Journal of International Development*, **7** (5), 791–9.

Bhattacharya, Amar and Mari Pangestu (1993), 'The lessons of East Asia: Indonesia', Development Transformation and Public Policy, Washington: The World Bank.

Booth, Anne (1995), 'The state and the economy in Indonesia in the nineteenth and twentieth centuries', in John Harris, Janet Hunter and Colin M. Lewis (eds), *The New Institutional Economics and Third World Development*, London: Routledge.

Chandler, Alfred D. (1990), *Scale and Scope: The Dynamics of Industrial Capitalism*, Cambridge, MA: Belknap Press.

Cribb, Robert and Colin Brown (1995), *Modern Indonesia: A History Since 1945*, London: Longman.

EIU (The Economist Intelligence Unit) (1996), *Indonesia Country Profile 1995–96*, London: EIU.

Fromhold-Eisebith, Martina (1998), 'Technology-oriented firms in the region of Bandung, Indonesia: supporting and repressing factors of their development', paper presented at the 6th Dutch-German Seminar on Economic Geography, West Terschelling, Netherlands, 7–10 October.

Gammeltoft, Peter (2001), 'Embedded flexible collaboration and development of technological capability: a case study of the Indonesian electronics industry', PhD dissertation, International Development Studies, Roskilde University.

Gammeltoft, Peter (2003), 'Embedded flexible collaboration and development of local capabilities: a case study of the Indonesian electronics industry', *International Journal of Technology Management*, **26** (7).

Gammeltoft, Peter (2004), 'Development of firm-level technological capabilities: the case of the Indonesian electronics industry', *Journal of the Asia Pacific Economy*, **9** (1).

Hill, Hal (1995), 'Indonesia: from "chronic dropout" to "miracle"?' *Journal of International Development*, **7** (5), 775–89.

Hill, Hal (1996), *The Indonesian Economy Since 1966: Southeast Asia's Emerging Giant,* Cambridge: Cambridge University Press.

Jones, Gawin W. (1994), 'Labour force and education', in Hal Hill (ed.), *Indonesia's New Order: The Dynamics of Socio-Economic Transformation*, Honolulu: University of Hawaii Press.

Keller, W.W. and R.J. Samuel (2002), 'Innovation and Crisis in Asia', MIT–Japan program (working paper 02.03).

Lall, Sanjaya (1998), 'Technology policies in Indonesia', in Hal Hill and Thee Kian Wie (eds), *Indonesia's Technological Challenge*, Singapore: Institute of Southeast Asian Studies.

Little, Ian, Richard Cooper and Sarath Rajapatirana (1993), *Boom, Crisis, and Adjustment: The Macroeconomic Experience of Developing Countries*, New York: Oxford University Press.

MacIntyre, Andrew (1991), *Business and Politics in Indonesia*, Australia: Allen & Unwin.

MacIntyre, Andrew (1994), 'Power, prosperity and patrimonialism: business and government in Indonesia', in Andrew MacIntyre (ed.), *Business and Government in Industrialising Asia*, Ithaca: Cornell University Press.

Mackie, Jamie and Andrew MacIntyre (1994), 'Politics', in Hal Hill (ed.), *Indonesia's New Order: The Dynamics of Socio-Economic Transformation*, Honolulu: University of Hawaii Press.

Meyanathan, Saha Dhevan (ed.) (1994), *Industrial Structures and the Development of Small and Medium Enterprise Linkages: Examples from East Asia*, Washington: The World Bank.

Okamoto, Y. and F. Sjoholm (2001), 'Technological Development in Indonesia', the European Institute of Japanese Studies (ELJS) (working paper No. 124).

Pangestu, Mari and Farid Harianto (1999), 'Corporate governance in Indonesia: prognosis and way ahead', paper submitted for the International Conference on Democracy, Market Economy and Development, Seoul, Korea, 26–27 February.

Rice, Robert (1990), 'Indonesian approaches to technology policy during the Soeharto era: Habibie, Sumitro and others', in Robert C. Rice (ed.), *Indonesian Economic Development: Approaches, Technology, Small-Scale Textiles, Urban Infrastructure, and NGOs*, Australia: Monash University.

Rice, Robert (1998), 'The Habibie approach to science, technology and national development', in Hal Hill and Thee Kian Wie (eds), *Indonesia's Technological Challenge*, Singapore: Institute of Southeast Asian Studies.

Robison, Richard (1986), *Indonesia: The Rise of Capital*, Australia: Allen & Unwin.

Samadikun, Samaun (1998), 'Indonesia's science and technology policies', in Hal Hill and Thee Kian Wie (eds), *Indonesia's Technological Challenge*, Singapore: Institute of Southeast Asian Studies.

Schwarz, Adam (1994), *A Nation in Waiting: Indonesia in the 1990s*, St Leonards: Allen & Unwin.

Soesastro, Hadi and Mari Pangestu (1998), 'Indonesia's electronics industry', in Christopher Findlay and Mari Pangestu (eds), *The Effects of Liberalisation in Asia's Textiles, Clothing and Electronics Industries*, Australian Pacific Economic Cooperation Committee.

Thee, Kian Wie (1998), 'Determinants of Indonesia's industrial technology development', in Hal Hill and Thee Kian Wie (eds), *Indonesia's Technological Challenge*, Singapore: Institute of Southeast Asian Studies.

Thee, Kian Wie (2003), 'The major channels of international technology transfer to Indonesia: an assessment', paper presented in Conference on Catch-up Growth And Technology Transfer, University of Groningen, 17–18 October.

Watanabe, Chihiro, Bing Zhu, B. Asgari and C. Griffy-Brwon (2001), 'Global technology spillover and its impact on industry's R&D strategies', *Technovation*, No. 21, 281–91.

Wibisono, Christianto (1991), 'Anatomi konglomerat Indonesia (the anatomy of Indonesian conglomerates)', in Kwik Kian Gie and B.N. Marbun (eds), *Konglomerat Indonesia: Permasalahan dan Sepak Terjangnya (the Indonesian Conglomerates: Problematic and Its manner)*, Jakarta: Pustaka Sinar Harapan.

World Bank (1994), *Indonesia's Small and Medium-Sized Exporters and Their Support System*, Washington: The World Bank.

8. Performance and sources of industrial innovation in Korea's innovation system

Kong-Rae Lee

INTRODUCTION

Nine years have passed since Korea experienced the economic crisis in 1997. Korea's industry still maintained competitive advantage in international markets in spite of economic adversity imposed by a crisis-ridden financial environment. Advantage has been possible due to the continuous innovations within firms introducing new products and services. Korean firms have striven to strengthen the technological capability that enables them to improve the quality of products and innovate new products and processes. Those firms that gained competitive advantage have showed advancements in their technological capabilities.

A study of firm-level innovation reported that Korean firms showed aggressive learning activities in order to catch up with advanced technologies and create new products and services in industries after the 1997 economic crisis (World Bank, 2002). It was also found that Korean firms currently put emphasis on manufacturing, process innovation and system integration. At the same time, they are eagerly attempting to move 'upstream' toward research and development and 'down-stream' towards distribution, marketing and brand-value development which usually require considerable investment.

This shift indicates that there has been a transition of the Korean economy from a simple manufacturing-led economy to a more knowledge-intensive and service-intensive economy. It also implies that the technological development of Korea's industry is evolving from an imitation stage to an innovation stage. In this transition period, creative learning of technological knowledge is essential for firms wishing to move further upstream from their base in manufacturing towards higher value-added. It is also closely associated with knowledge-intensive activities based on research, development and new product innovation.

This chapter aims to offer an understanding of the performance and sources of industrial innovation that is taking place in the transition period. The research questions addressed in this chapter are: how much did economic and R&D performance of the Korean economy change after the economic crisis in 1997? What are the sources of industrial innovation? What are the weaknesses and policy issues? This chapter attempts to discuss these questions from the viewpoint of innovation systems since the innovation process involves a complex system that has economic, social, cultural and institutional dimensions (Freeman, 1982; Pavitt, 1988; Lundvall, 1992; Nelson and Rosenberg, 1993).

The chapter presents the results in four sections. It presents a brief review of macroeconomic indicators, R&D indicators and business performance of three industries in the first section. In the second section, sources of industrial innovation at both the macro level and the intermediate level are presented. The next section points out weaknesses and policy issues of the Korea's innovation system, and the final section presents conclusions and lessons for other countries.

R&D AND BUSINESS PERFORMANCE OF KOREA'S INNOVATION SYSTEM

Economic Performance and R&D Expenditure

Economic performance of Korea's innovation system may be an indication of its effectiveness and efficiency. Economic performance of Korea's innovation system might be higher if its effectiveness and efficiency are strong enough to make fruitful business results. First, a look at the average annual growth rate of GDP from 1990 to 2002 shows it was 5.8 per cent, which is not bad considering the hardship that resulted from the economic crisis that took place in 1997. The long-term growth trend has been clearly downward as we see that the average growth rate of the 1970s was 7.1 per cent, but decreased to 4.1 per cent in 1980s, 3.0 per cent in 2001 and 6.3 per cent in 2002 as shown in Table 8.1.

Trade performance may be directly linked with the effectiveness and efficiency of the Korean innovation system since export sales accounted for 32.7 per cent of total GDP in 2002 as shown in Table 8.1. Korean exports amounted to US$162 billion, while imports reached US$152 billion, making the surplus of US$5.394 billion. Average growth rate of exports from 1990 to 2002 was 8.2 per cent, which is higher than that of imports, 7.3 per cent. We can see that the trade performance of Korea's innovation system is, by and large, not bad.

Table 8.1 Changes in economic indicators of the Korean economy

	1980	1990	2001	2002	AGR ('90–'02)
Population	38 124	42 869	47 343	47 638	0.9
GDP (US$, billion) (A)	62	253	422	497	5.8
Growth rate of GDP (%)	7.1*	4.1**	3.0	6.3	5.8
GDP per capita (US$)	1632	5900	8918	10 013	4.5
Trade balance (US$, million)	−4384	−2004	9341	5394	–
Exports (US$, million) (B)	17 214	63 124	150 439	162 471	8.2
Imports (US$, million) (C)	21 598	65 127	141 096	152 126	7.3
B/A (%)	27.8	25.0	35.6	32.7	–
C/A (%)	34.7	25.7	33.4	30.6	–

Notes:
* growth rate of GDP in 1980 is 10 years average growth rate of GDP from 1971 to 1980.
** growth rate of GDP is nine years average from 1981 to 1990.
AGR is average annual growth rate.

Source: Ministry of Finance and Economy.

What benefit is rendered to people from this relatively good performance of Korea's innovation system at the macro level? It is worth mentioning that good performance at the macro level indicators does not guarantee welfare of people and safety or soundness of the national economy. GDP per capita is an index to see more closely the welfare change of people that results from changes in macro indicators. As shown in Table 8.1, GDP per capita increased from US$5900 in 1990 to US$10 013 in 2002. It is an encouraging sign that the long-term economic performance of Korea's innovation system during the 1990s has been good in spite of the economic crisis that occurred in 1997, which may have improved the quality of life for people in Korea.

However, another economic problem that is not easily revealed in macroeconomic indicators emerged regardless of the per capita income increase at the beginning of the 2000s. That is the unemployment of young people, especially university graduates. The number of unemployed people was 825 000 and the unemployment rate was 3.6 per cent in December 2003, which does not appear to be too serious in terms of the number and the rate. However, it is yet serious because they are mostly young people, including university graduates, implying that young people lose opportunity to learn from work, and to weaken the national innovation capability in the long run.

Table 8.2 Evolution of R&D inputs in Korea's innovation system

	1970	1980	1990	2002	AGR ('90–'02)
GERD (US$, million)	33	428	4676	13 849	9.5
Government vs. Private	71 : 29	64 : 36	19 : 81	26 : 74	–
R&D / GDP	0.38*	0.77*	1.87	2.91	–
Researcher (persons)	5628	18 434	70 503	189 888	8.6

Notes:
* R&D/GNP.
AGR is average annual growth rate.

Source: Ministry of Science and Technology. Korea Institute of S&T Evaluation and Planning (2003), *Report on the Survey of R&D in Science and Technology*.

The Korean government is well aware that R&D is critical for strengthening the growth potential. The government has, therefore, consistently increased R&D expenditure since the Ministry of Science and Technology inaugurated national R&D programs for the first time in 1982. Total national R&D expenditures, including the private sector, reached US$13.9 billion in 2002, from US$4.676 billion in 1990 (see Table 8.2). The government accounted for 26 per cent of the total national R&D expenditure, and the remaining 74 per cent came from the private sector. Korea's NIS has a low ratio of government R&D expenditure compared to the 26.6 per cent in Japan (2001), 31.9 per cent in Germany (2001), 35.9 per cent in the UK (2001), 33.8 per cent in the USA (2002) and 40.3 per cent in France (2000).

Relatively, active R&D expenditure made by the private sector pushed up the R&D expenditure ratio out of GDP from 1.87 in 1990 to 2.91 in 2002. It surpassed those of most OECD member countries except for Japan (3.09), Finland (3.40), and Sweden (4.27) in 2001. However, generous R&D expenditure by the private sector has been made mostly in large firms, which accounted for 72 per cent in 2002. The five largest firms had a 37.5 per cent share of the private R&D expenditure, and the 20 largest firms 49.6 per cent in 2002. KOITA reported that the portion of large firms in private R&D expenditure tends to decline because small and medium-sized firms are increasing their R&D expenditure (KOITA, 2003).

The R&D performing structure of the Korean innovation system has been continuously evolving from a public-institutes centered one to a private-companies centered one. Public institutes accounted for the 49 per cent of the national R&D budget in 1980, 22 per cent in 1990 and recently 14.7 per cent in 2002. Particularly, the share of government-sponsored research institutes (GRIs) were radically reduced from 27 per cent in 1980 to 16 per cent in 1990, and finally to 10 per cent in 2002. This change reflects

Table 8.3 Evolution of R&D performing structure in the Korean innovation system (per cent)

	1970	1975	1980	1985	1990	2002
Public institutes	84.0	66.0	49.0	24.0	22.0	14.7
(GRIs)	(25.0)	(27.0)	(27.0)	(20.0)	(16.0)	(10.0)
Universities	4.0	5.0	12.0	10.0	7.0	10.4
Companies	13.0	29.0	38.0	65.0	71.0	74.9
Total	100.0	100.0	100.0	100.0	100.0	100.0

Source: Ministry of Science and Technology. Korea Institute of S&T Evaluation and Planning (2003).

the relative contraction of the role of government in national R&D activities. Universities that are not divided here into public and private, maintained the 10 per cent level of national R&D spending over the last decade.

Private companies appear to have been the most significant player in the Korean innovation system since the middle of the 1980s. They accounted for 74.9 per cent of the national R&D expenditure in 2002, which increased from 71 per cent in 1990 as revealed in Table 8.3. Private companies are obviously central innovating agents of the Korean innovation system and the leading actors of the Korean economy. Their innovation activities resulted in the creation of world-class Korean industries, such as semiconductors, automobiles, shipbuilding, steel and many others. It is worth investigating what factors were behind the success of their innovation in the Korean innovation system.

Business Performance of Major Industries

The business performance of Korean firms more directly indicates competitiveness of the Korean innovation system than other macro level indicators. In this section, we intend to briefly investigate business performance of major industries and leading firms that are based in and operating in Korea. It includes the semiconductor industry, the mobile telecom service sector and the automobile industry.

The semiconductor industry

The Korean semiconductor industry has been a world market leader since 1998. It captured 42 per cent of the world market share in 2001. The leading firm in the industry has been Samsung Electronics Co. (hereafter, Samsung) that has been not only the domestic market leader but also a world market leader over the last decade. Samsung alone captured 30.8 per cent of the world semiconductor market in 2002 (Choi, 2003) by achieving export sales of US$17 billion in 2001. Total sales of Samsung amounted to US$25 billion, making US$1.8 billion net profit in 2001 (Song et al., 2002).

Samsung annually spends around US$1.859 billion on R&D activities, 7.4 per cent of total sales, and employed 15 000 researchers at local R&D centers and 900 employees at the Samsung Advanced Institute of Technology (SAIT) in 2001 (Song et al., 2002). SAIT has been playing the role of central agent by collecting and disseminating strategic technological knowledge and information. Samsung submits applications for, and registers 200–300 patents per annum. Samsung has also invested in overseas R&D facilities in the UK, USA and Japan in order to access leading edge technologies. It has made agreements with such advanced multinational companies as Sony, Microsoft, Nortel, Yahoo, Thales, and many others in strategic alliances.

As a result of its aggressive R&D performance, Samsung has created many frontier technologies that relate to its core manufacturing business. Such products as DVDs, DVDRs, TFT-LCDs, cellular phones, displays and new generation DRAMs have emerged as a result of its in-house developments. These products have gone on to become Samsung's major new businesses and sources of profit making.

The mobile telecommunication service industry

Korea successfully developed CDMA (Code Division Multiple Access) technology for supplying mobile telecommunication services. Development of CDMA technology motivated local telecom service providers to improve their service quality. Encouraged by service quality, the number of mobile service users skyrocketed from 1.6 million in 1995 to 32.4 million in 2002 at the annual growth rate of 53.7 per cent for the period. The Korean mobile service industry has been providing world-class service to its users as it continuously updates its technological capability.

The leading mobile service provider has been SK Telecom, followed by KTF, a sister company of Korea Telecom (KT). SK Telecom began as a part of KT in 1985 and was sold to the SK Group in 1995. SK Telecom captured 53.3 per cent of the domestic mobile service market as its subscribers reached 17 million in 2002, while KTF maintained around 10 million subscribers. Total sales of SK Telecom amounted to US$6.689 billion in 2002, making a net profit of US$1.171 billion with a profit rate of 17.5 per cent (National Computerization Agency, 2003). SK Telecom employed around 20 people with PhDs and 150–200 people with a

Masters degree for R&D work. The World Bank (2002) reported that about half of the 400 employees in SK Telecom worked on mobile internet services.

SK Telecom scored very high in the innovation management evaluation by the World Bank. Management of innovation in SK Telecom was regarded as 'creative', which is the highest level among management types that have fully-developed sets of technological capabilities and undertakes a pro-active approach to exploiting technology for competitive advantage (World Bank, 2002). It has been pursuing an innovation strategy to move progressively downstream from systems and products to networking service solutions. In order to more quickly secure procurement of capital goods, SK Telecom switched equipment suppliers from foreign to local suppliers: Samsung, LG, and so on. Local equipment suppliers responded quickly and gave more attention to SK's particular needs.

The automobile industry

Korea's automobile industry served 5.2 per cent of global markets as exports amounted to US$13.7 billion in 2002. The industry produced 2.651 million passenger cars, of which 1.622 million (61.2 per cent) were sold in the domestic market and the remaining 38.8 per cent (1.029 million) were exported in 2002. With this production of passenger cars, the Korean automobile industry ranked 6th following the USA (1st), Japan (2nd), Germany (3rd) and France (4th) in 2002 (KOITA, 2003).

The industry's leading firm has been Hyundai-Kia Motor Co. (hereafter, Hyundai). Hyundai was established in 1967 with technical assistance from Mitsubishi, a Japanese car assembler. Hyundai produced 1.702 million passenger cars, accounting for 64.2 per cent of the Korean car production in 2002. In the global market, Hyundai ranked 13th, following GM, Ford, DC, Toyota, VW, Honda, Nissan, Fiat, PSA, Renault, BMW, and Mitsubishi. Hyundai has the ambition to become one of the top five global producers. It is, however, poorly supported by the local capital goods industry and relies heavily on foreign parts suppliers.

Hyundai has been judged to have well-considered technology strategies, impressive and growing technological competencies and well-honed search and acquisition capabilities (World Bank, 2002). It employed around 6000 R&D personnel and spent US$790 million on R&D activities in 2002.

SOURCES OF INDUSTRIAL INNOVATION IN THE KOREAN INNOVATION SYSTEM

Sources at the Macro Level

The above-mentioned performance of the Korean innovation system both at the macro level and the industry level seems to be impressive, although many innovation studies have pointed out its inefficiencies and underdevelopment. This chapter intends to find sources of innovation at both macro level and meso level, assuming that innovative performance of the Korean industry has been quite good in terms of the macro and the mid-level aspect.

Hardworking people and aggressive learning

The most general and significant source of industrial innovation is likely to be the hard-working attitude of people and resultant progressive learning. Hard-working does not guarantee productive innovations in industry. In Korea, the hard-working attitude of people is associated with learning activities that vary from simple imitation of foreign technology or production know-how to upgrading. Diligent learning behavior of people gave rise to the speedy mastery of foreign technology, sooner or later leading to the improvement of the mastered technology.

Why then, do the Korean people work hard and engage in learning? Kim (1997) argued that the perseverance of Korean people in turmoil and hardship inflicted by foreign invasions in the beginning of the 20th century, and the associated *han* psyche produced energy and established the hard work habit of the people. The hard work habit of Korean people seemed to be well harmonized with learning activities of the Confucian culture. The teaching of Confucianism has emphasized 'study', which means learning from respectable senior people or parents. This learning culture has been intensively applied to workers in hierarchical organizations such as companies.

Another cause of the hard-working habit may be associated with the dense population and severe cold weather in the winter. It may have forced Koreans to work competitively, and the memory of deprivation bred the hardworking trait into the country's workers. A hard-working habit strengthened an intensity of efforts, which led to aggressive learning in firms, ultimately quick mastery of imported technology and their successful innovation afterwards.

Export-oriented strategy: major stimulus

The export-oriented strategy of both the government and firms has been frequently referred to as a source of innovation and as a major stimulus of innovation on the demand side (Westphal, 1978; Westphal et al., 1984a, 1984b). At the absorption stage of foreign technology in the 1970s and the 1980s, Korean exporters made lump-sum investments for capacity in excess of local market size to achieve economies of scale. This forced local firms to accelerate technological learning to improve productivity and maximize capacity utilization.

Stimulus from foreign buyers is still an important source of innovations in the transition period of industrial development in the 1990s and the beginning of the 2000s. Surviving in the competitive international markets requires continuous local effort for Korean firms to improve product quality, which requires a heavy investment in technological learning. The more firms globalize, the more they acquire stimulus from international competitors, compelling continuous innovation of not only technology but also marketing skills and management.

Korean firms relied heavily on foreign buyers of original equipment manufacturing (OEM) for marketing their products internationally in the past (Hobday, 1995). They provided priceless help to Korean firms in acquiring necessary capability through interactive tutorial processes, allowing the firms to focus their efforts primarily on acquiring production capability. Now, at the beginning of the 21st century, competitive Korean firms are pursuing their own design manufacturing (ODM) and even extending to own brand manufacturing (OBM), which inevitably exposes them to a more demanding environment that brings in stimulus from the outside.

Sequential capability building

Industrial innovation requires sequential capability building by different stages of development. Korean firms successfully sequenced different learning phases in building technological capability. Firstly, they successfully performed reverse engineering through learning by using and learning by imitating. Secondly, they showed excellent performance in the assimilation of foreign technology through learning by designing. Finally, Korean firms succeeded in improving imported technology as they made independent designs and diversification of products through creative learning (Lee, 2000).

In the process of passing through this sequential capability building, a relatively well-educated workforce and the continued inflow of Korean–American scientists and engineers helped sustain the accumulation of tacit knowledge to match the rapidly advancing technological frontier (Kim, 1997). Their learning capability and high learning speed shortened the time to catch up to the technological level of advanced countries.

Heavy investment in R&D activities

Facing the need to shift to higher-value technology intensive products, Korean firms have rapidly raised their R&D investments in recent decades. Although the absolute amount of national R&D expenditure is small (US$13.8 billion in 2002), compared with advanced countries, its growth rates have been rapid in Korea. The average growth rate of national R&D spending in the private sector was 11.4 per cent from 1990 to 1997, and 11.2 per cent from 1998 to 2002 after the 1997 foreign exchange crisis. R&D spending of the private sector has been even more impressive. The average growth rate of R&D spending in the private sector was 20.7 per cent from 1990 to 1997, and 13 per cent from 1998 to 2002 (US$10.8 billion in 2002).

Such relatively heavy investment in R&D activities has strengthened the innovation capability of firms. R&D investment is necessary, but not a sufficient condition for firms to produce high quality products and services. Survey results reveal that most Korean firms are not employing sufficient high-caliber human resources with a Ph.D. degree, which may weaken their potential to innovate in the long run. The private sector consumed 78.3 per cent of the total national R&D expenditure, but only employed 14.7 per cent of R&D human resources with a Ph.D. degree in 2002. Moreover, government-produced R&D results are not well linked to the needs of private firms so as to underutilize the national R&D results (Lee, 1998).

Overcoming the crises through opportunistic learning

Prompted by overcoming such crises as the oil price increase, foreign currency crisis, radical policy changes, and economic slump, Korean firms made a great leap in technological capability. In particular, *chaebol* firms came out of a series of crises through technological learning by setting overly challenging goals in acquiring and assimilating foreign technologies (Kim, 1997). Top management often used crises as a major means to promote opportunistic learning for success in large investment with high risk.

Kim (1997) argues that crises were deliberately constructed to expedite technological learning within Korean firms. According to his argument, a firm facing a crisis has to exert 'a significant portion of staff energy to educating coalitions and organizational members to agree to crisis management, mitigating resistance to change, and unlearning past practice'. It is certain that effective learning and problem solving can be attained in the process of managing crises together within an organization. Rapid learning within Korean firms may, to some degree, have been generated from the management of various crises.

Government: an effective stimulator

As many studies assert, the Korean government played an important role in the technological development of industries. Since the Ministry of Science and Technology (MOST) implemented the national R&D program in 1982, various ministries have planned and pursued specific national programs that fit to their objectives. MOST alone has implemented such diversified programs as the Highly Advanced National Project (the HAN Project), Creative Research Initiative (CRI), National Research Laboratory (NRL), Biotechnology Development Program, Space and Aeronautics Program, and so on.

A survey (Cho and Lee, 1997) reported that firms participating in government R&D programs acquired portions of technological skills and knowledge from their partners. Government R&D programs must have been an effective stimulator for participating firms as they frequently contact academic professionals, research manpower of public R&D centers and partner companies. Government technocrats have identified winners and allocated necessary resources to them through national R&D programs to achieve ambitious goals for technology acquisition.

On the demand side, the government introduced various policy measures: antitrust and fair-trade legislation, trade liberalization, and financial liberalization to inject more market forces into the economy so as to stimulate firms to innovate. What is generally known as 'industrial policy' may also have positively affected the innovation activities of firms as they became oriented toward competition, liberalization and globalization, which caused firms to be further exposed to the outside world's abundantly more knowledgeable societies.

Sources at the Meso Level

The sources of industrial innovation mentioned at the macro level are specific to Korea's innovation system so that generalization may not be easy. Sources of industrial innovation at the meso level may be more clearly understood for general readers. I believe sources at the macro level are complementary with those at the meso level so that real sources of innovation are better identified, although this is subject to an in-depth empirical investigation. This section attempts to briefly identify sources of innovation at the meso level for the three industries: the semiconductor industry, the mobile telecommunication service industry, and the automobile industry.

The semiconductor industry

Studies on Korea's semiconductor industry have listed 'the economy of speed' (Kim, 1997), 'speedy decision marketing' (Choi, 2003), shortening learning time and development period as major sources of innovations of the industry. Kim and Choi argued that Korea's semiconductor manufacturers have realized the benefit of 'the economy of speed' so that they could be ultimately successful in consecutive innovations of DRAM as well as in marketing them (Kim, 1997). For example, Hyundai tried to set up its production process and acquire design technology concurrently in order to shorten the time from investment to marketing. 'Concurrent engineering' was once one of the most frequently used jargons due to the corporate culture emphasizing 'speed'.

Large R&D investment and effective R&D strategy were important sources of semiconductor innovations. Most Korean semiconductor manufacturers like Samsung, Hyundai and LG applied their massive R&D resources to technological learning and new product development, such as advanced memory chips and non-memory technologies. They not only conducted considerable in-house R&D work, but also made strategic R&D alliances with advanced firms in developed countries. The World Bank (2002) reported that Korea's semiconductor makers successfully built management capability to integrate internal R&D with outsourcing (World Bank, 2002).

Top management attention is also worth mentioning as one of the sources. There was strong top management of semiconductor manufacturers to provide a favorable environment for technological activities, that is total quality control and initiatives in developing new products by focusing on critical technologies (Song et al., 2002).

The mobile telecommunication service industry

Sources of innovation in the mobile telecom service industry may be different from other industries. The public sector, including the government and government-sponsored research institutes, played an important role in the innovation of the industry, unlike other industries. Public research institutes such as ETRI and KIST have accumulated the capability to develop a telecom system over a relatively long period. R&D activities to obtain technical solutions for the TDX project, which was an indigenous Korean technology project in the 1980s, were basically similar to those of the CDMA project (Chung and Lee, 1999). Therefore, know-how accumulated from the experience of the TDX project could be a basis in developing the CDMA project.

ETRI worked together with foreign technology sources like *Qualcomm* to utilize their resources at the pre-commercial stage of the CDMA telecommunication system. ETRI has accumulated know-how related to communication system technology over a relatively long period (Chung and Lee, 1999). It closely cooperated with private firms in a wide range of

R&D activities so that knowledge accumulated in ETRI smoothly spilled over to private firms. There were about 250 researchers involved in the ETRI project and about 200 researchers in each participating company. They worked hard and harmoniously to make the project come to fruition. ETRI has also experienced organization and management of large system-projects from the outset of communication technology development in Korea.

Government was an important source of industry innovation. First of all, it assisted in the creation of a market environment favorable for mobile telecom service providers to adopt the CDMA system by setting it as a national technical standard (Chung and Lee, 1999). The government eventually intervened in the whole process of development. In particular, the Ministry of Information and Communication provided administrative support for the project. MIC also gave financial support to ETRI by way of subsidizing the costs of the project.

The automobile industry

Aggressive learning of foreign technology by car assemblers might be the most important single source of the indutry's innovation. Passenger car assemblers have had a strong desire to enhance internal technological capabilities to establish their own brand at the outset of the industry's development. They have actively acquired foreign technologies from car assemblers in advanced countries, and rapidly caught up by the adoption of new technologies. For example, Hyundai adopted the multi-point injection engine one step earlier than foreign firms.

Development of production technology by in-house development of various capital goods seems to be one source of innovations in the industry (Lee, 2000). Hyundai Motor is a prominent example of firms making capital goods by in-house development. Hyundai Motor produced a variety of capital goods ranging from standardized machines such as machining centers, CNC copy milling machines, CNC turning centers, and CNC gear hobbing machines to special purpose machines such as transfer machines, rotary index type machines, flexible transfer machines, and various other items of industrial machinery. Hyundai has developed its learning activities for capital goods from simple learning by using, to more complex learning such as learning by designing, and finally creative learning as shown in Figure 8.1.

The role of top management should be included in the sources of the industry's innovation. Top management of Hyundai strongly supported R&D activities and provided a favorable R&D infrastructure, for example introducing a training program, purchasing R&D facilities, cooperating with the university. The technology policy of Hyundai's top management

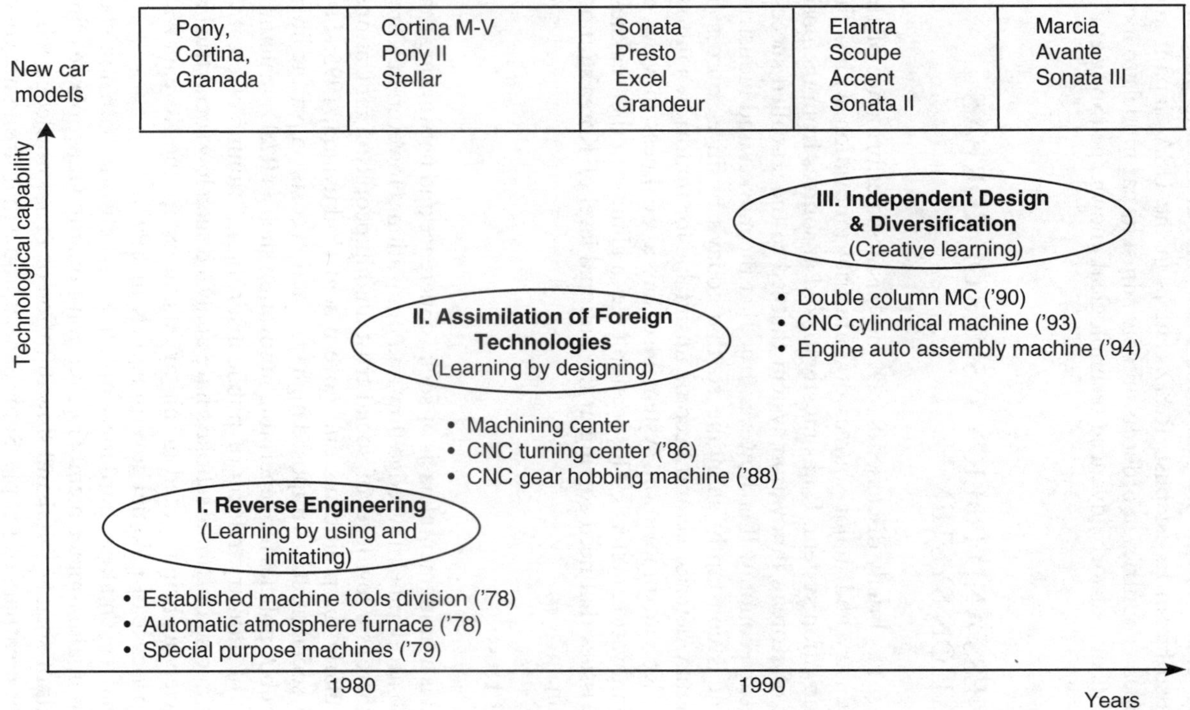

Source: Lee (2000).

Figure 8.1 Evolution of technological learning in Hyundai Motor Co.

has recently focused on the globalization of R&D. Hyundai started to cooperate with United Technologies Fuel Cells (UTCFC) for developing a new car model and joined the California Fuel Cell Partnership program. Hyundai also began to establish R&D centers in the USA, Japan, and some European countries in order to suitably adjust its car models to local environments and absorb advanced technological knowledge (Song et al., 2002).

WEAKNESS AND POLICY ISSUES OF KOREA'S INNOVATION SYSTEM

This chapter has briefly discussed R&D and business performance of the Korean industry and major sources of industrial innovation within the Korean innovation system. It has mostly looked at sources from the positive aspect of Korea's innovation system that led to the so-called success of the Korean economy. It is, however, true that the market and industrial environment is constantly changing so that today's positive sources of innovation may deteriorate and become unsuitable for tomorrow's innovation. The Korean innovation system also has a weakness in view of advanced innovation systems. This section takes a glimpse at the weakness and policy issues that need solving at the national level of Korea's innovation system.

Low Social Trust

Social trust is a fundamental basis of social coherence and networking that is needed to accelerate knowledge flow and innovation. However, in Korea generally there is a low level of social trust among people, organizations, and particularly for politicians and government (Fukuyama, 1995; Lee, 1998). Low social trust has imposed high transaction costs on private firms as well as public policies for generating innovation such as R&D, structural adjustment of industry, reform of public institutions, training of creative manpower, and so on. Some scholars have analyzed such low social trust in Korean society as being rooted in military dictatorship, the Korean War, and the experience of colonial government (Kim, 1997).

Low social trust makes it hard for people to form networks with non-clan members, non alma-mater members, and non-home-town people, leading to low levels of social capital and slow diffusion of tacit knowledge, which hampers the creativity of people. Some experts argue that strong networks among clan members, alma-mater members, and home-town people, which are known as '*hak yeun, chi yeun, hyul yeun*' in Korean, are substitutes for the lack of social trust. They worry about deterioration of this '*hak yeun, chi yeun, hyul yeun*' without social trust since informatization is generating continuous change in the traditional culture.

Social trust may be improved not by policies themselves, but by trust building of organizations such as levels of governments, public organizations and private firms. Building social trust may take a long time, perhaps a generation, even in a stable society. Thus, honesty and transparency should be emphasized more in managing organizations and planning and implementing policies. Discussions among stakeholders and rational decision-making are necessary for shortening the time to build social trust up to the level of advanced countries.

Labor Unrest

Nowadays, unionists' demands are increasingly moving into management and political issues as democratization increases. Labor unrest is shifting the center of quality from large *chaebol* firms to the public sector, including government offices, and public companies engaged in social infrastructure such as transportation, ports, railways, and so on. Strong labor unions in public companies are likely to weaken competitive advantage arising from the economy of speed that Korean firms have enjoyed in the past. Furthermore, 'economy of speed' which is regarded as an important source of innovation, may be weakened so that speed of innovation is expected to slow down too.

Labor unions also demand more leisure and less work time without a reduction of income level. In this situation, the government recently introduced the five-workday per week law and is adjusting the number of holidays to compensate for the loss of workdays. This movement may have a positive or negative impact on industrial innovation. It may have a positive impact if the creative capability of people increases after having more leisure and relaxation time, but a negative impact may result if it does not. Much study on how to increase the creative thinking of people seems to be required at the micro level in order to design proper policies.

Education: Lack of Flexibility

Korea's education system at all levels is prepared for massive training and imitative learning, but is vulnerable to creative learning. There also has been a lack of a creative learning mechanism at all levels of education. Teachers from primary schools to high schools generally agree that they do not have the teaching techniques and systems to nurture the creativity of students. University education at the graduate level is even worse, except for some research-oriented universities. Education in Korea has focused more on learning by reading textbooks and listening to lectures rather than learning by creative thinking as most graduate schools impose heavy course work on master's and research students. Besides the lack of nurturing creativity, university education has also been criticized for its failure to meet industrial needs. It has often been pointed out by industrialists that university graduates do not meet a standard that industries require due to a theory-oriented education and research system in universities. It is rare to see research students who take technological problems of firms as their thesis topic. This kind of environment erodes the confidence and trust between universities and industries. University–industry R&D cooperation is, therefore, weak, being one of the problems of Korea's innovation system (Lee and Song, 1998a and 1998b).

Making matters worse, there is a trend for the entries of the young generation of students into science and technology to be rapidly declining, causing concern for the future of the national technological capability. The academic quality of students in engineering departments has deteriorated year by year. Students admitted both to engineering schools and medical schools tend to prefer medical schools to engineering schools. It is often said that this phenomenon came too early at the US$10 000 per-capita income level, and ultimately weakened the national innovation capability of Korea. The Korean government regards this issue as serious and is eagerly searching for solutions.

Lack of Mobility of Professional Manpower

Korea has a culture highly segregated by social clan and family with low levels of geographical and occupational mobility, which is a critical weakness affecting strengthening the national innovation capability (Lee, 1998). A social hierarchy among vocations was formed by the Confucian philosophy during the late Chosun dynasty period. It has been handed down to the current Korean society in which teaching positions are regarded as more prestigious than those of research positions in universities. Professionals tend to move from research positions to teaching positions, and conducting research projects is not regarded as an important job in universities. This tendency obviously exerts a negative impact on creative research work and weakens the role of the university in strengthening national innovation capability.

STEPI research (Ko et al., 2001) found that R&D personnel to a large extent moved from public research institutes to universities, but rarely moved in the reverse direction or from universities to industries. The survey showed that government-sponsored research institutes (GRIs) recruited

648 researchers with Ph.D. degrees for four years from 1998 to 2001. However, 674 experienced researchers, more than the number that entered, left GRIs during the same period, of which 19 per cent moved to universities and 17 per cent to industries. Movements of professors from universities to GRIs or industries are rare since the university has provided the incentive of lifelong tenure to faculty members. There has been a one-sided movement of professionals from industry and public R&D institutes to universities which may hinder the innovation of both industry and public R&D institutes in the long run.

Unbalanced Regional Innovation

A distinctive characteristic of Korea's innovation system is that its R&D resources are geographically concentrated, resulting in an unbalanced regional innovation capability. R&D resources have been mostly allocated to the Seoul and Daejon areas (34.7 per cent of national R&D expenditure and 40 per cent of R&D human resources). This imbalance in R&D resources has been strategically formed as the result of the government's policies. The central government intentionally concentrated government R&D institutes in Daejon in order to increase the efficiency of R&D investment before regional innovation was emphasized for economic growth of regional economies.

The imbalance in the allocation of R&D inputs is clearly one weakness of Korea's innovation system. That has led to gaps in innovative capability by regions, giving rise to differences in the growth potentials of regional economies. As the present government geared government policy direction toward the balanced development of regions, its correction emerged as the main agenda of the science and technology policy. Various policy tools to form innovative clusters by varying regional levels have been designed and are being implemented for the balanced development of regions.

Underdevelopment of Innovation Management

The World Bank (2002) examined to what extent Korean firms developed their innovation management after the economic crisis of 1997, and concluded that they are by and large at a stage that adopts a strategy to innovate their products and processes. The examination also revealed that Korean firms have a well-developed sense of the need for technological change, are highly capable of implementing new projects, and take a strategic approach to the process of continuous innovation. This is, however, true only for some selected firms that have gained competitive advantage.

Korean firms generally lag behind with regard to taking a creative and proactive approach to exploiting technology for competitive advantage. In particular, small and medium sized firms do little R&D work that will enable them to carry out product innovation or innovation of process technologies. They lack the management techniques to deal with innovation-related matters, a flow of complex knowledge within organization as well as with outside organizations. Unfortunately, the government does not have any policy tools to improve the innovation management of SMEs. Universities do not invest much in research on innovation management so that the associated expertise is not provided to students or business people.

Rigid Management of Public R&D Institutes

Government-sponsored R&D institutes have undergone radical reforms twice since their origin: the first one was initiated in 1980 by the so-called military government, and the other was made in 1999 by the people's government. In the first reform, the government consolidated the 15 GRIs under various ministries into nine large research centers under the auspices of the Ministry of Science and Technology (MOST). Policy-makers judged that, in comparison with total national R&D funding, there was more than the optimal number of GRIs. It was advocated that through restructuring, overlapping R&D investment should be avoided.

In the second reform, the people's government adopted the European style of public R&D system by establishing research councils under the office of the Prime Minister. Nineteen science and technology GRIs, including KIST, belonged to one of three research councils by research field. The Korea Research Council of Fundamental Science & Technology included four institutes; the Korea Research Council of Industrial Science & Technology included seven institutes; and the Korea Research Council of Public Science & Technology included eight institutes.

Although the second reform aimed to improve the autonomy of GRIs, they are still controlled by the budget office and related ministries having their R&D budget. GRIs have conducted research projects that come from governments, and some research work contracted with industries. Therefore, the R&D results of GRIs do not effectively spill over to the private sector. GRIs even compete with private R&D centers for national R&D projects.

The two radical reforms may not have fundamentally cured the problems of GRIs. Policy analysts point out that the governance system of GRIs is too complex, and has resulted in rigid management, lack of flexibility, and constrained research activities. GRIs may have a long way to go in order to find their proper position in Korea's innovation system.

CONCLUSIONS AND LESSONS

We have briefly looked over the performance and sources of industrial innovation from the perspective of the national innovation system. This chapter has implicitly argued that industrial innovation in Korea's innovation system has been active in terms of indicators at both the macro and the mid level. Sources of industrial innovation at the macro level are hard-working people and aggressive learning, an export-oriented strategy, sequential capability, heavy investment in R&D activities, overcoming crises as opportunistic learning and the role of government.

Sources of innovation at the meso level vary by industry. In the semiconductor industry, exploitation of 'economy of speed' by speedy decision-making, shortening learning time and development period can be regarded as the most significant sources along with large R&D investment and an effective R&D strategy. Meanwhile, government-sponsored research institutes and the government played relatively important roles in the innovation of the mobile telecom service industry. In the automobile industry, aggressive learning of foreign technology and in-house development of production technology are emphasized as important sources of innovation.

Although Korea's innovation system has showed good performance in the area of industrial innovation, it is weak in many aspects. First of all, low social trust seems to be problematic in advancing the knowledge base of Korea's innovation system. Labor unrest, lack of flexibility in the education system, lack of mobility of professional manpower, unbalanced regional innovation, underdevelopment of innovation management and rigid management of public R&D institutes are obstacles to the Korean innovation system's further strengthening of its learning capability.

In order to achieve continuous development of the economy, the problems currently confronting Korea's innovation system should be solved and there should be an improvement toward a more creative innovation system. The private sector as well as the public sector need to work diligently to overcome drawbacks and strengthen the capability of the national innovation system. Private firms, including *chaebol* companies, should continue to show commitments to technological innovation. It is vital for top management leadership to have long-term vision and technological insight in order to create a virtuous circle of innovation. Technological learning and absorptive capacity remain important as future industrial innovation elements, but agile strategies and an effective system are likely to be necessary.

The government is also likely to be important not only in strengthening the national innovation system, but also in building the innovation capability of industries as in the past. A demand side stimulus, especially a foreign impetus, will be important for the continuation of aggressive learning in private firms. Dynamic technological learning can ensue when firms turn to the export market. The Korean government should therefore bear in mind that protection of domestic markets fails to create competitive domestic markets and fails to stimulate technological learning.

This chapter has been written as a type of essay for understanding Korea's innovation system. It has focused on performance and sources of industrial innovation. Discussions and arguments in this chapter may or may not be lessons for other countries, and are obviously subject to further empirical investigation. They should be discussed further and examined by experts who are interested in the Korean innovation system and modified in order to apply them to other developing countries wishing to build their own national innovation capability.

REFERENCES

Cho, Hyun-Dae and Lee, Dal Whan (1997), *An Analysis of Firms' Participation Strategy on Government R&D Program and Its Performance in Korea*, Seoul: STEPI (in Korean).

Choi, Y. (2003), 'Sources of corporate growth – experiences of Korean enterprises', paper presented to the Portland International Conference on Management of Engineering and Technology, *Technology Management for Reshaping the World*, Portland: Portland State University.

Chung, KunMo and Kong-Rae Lee (1999), 'Mid-entry technology strategy – the Korean experience with CDMA', *R&D Management*, **29** (4), 1–11.

Dosi, G., K. Pavitt and L. Soete (1990), *The Economics of Technical Change and International Trade*, London: Harvester Wheatsheaf.

Fransman, M. and K. King (eds) (1984), *Technological Capability in the Third World*, London: Macmillan Press.

Freeman, C. (1982), *Economics of Industrial Innovation*, London: Frances Pinter.

Fukuyama, F. (1995), *Trust: Social Virtues and the Creation of Prosperity*, NewYork: The Free Press.

Hobday, M. (1995), *Innovation in East Asia – The Challenge to Japan*, Aldershot, UK and Brookfield, US: Edward Elgar.

Kim, Linsu (1993), 'National systems of industrial innovation: dynamics of capability building in Korea', in R. Nelson (ed.), *National Innovation Systems: A Comparative Analysis*, Oxford: Oxford University Press.

Kim, Linsu (1997), *Imitation to Innovation – The Dynamics of Korea's Technological Learning*, Boston: Harvard Business School Press.

Ko, Sangwon (1998), 'Manpower training and retraining system of Korea', in KongRae Lee and Wizin Song (eds), *National Innovation System of Korea* (in Korean), Seoul: STEPI, pp. 215–35.

Ko, Sangwon, Min Chulkoo, Jaemin Park and Eunkyong Lee (2001), *Mobilizing S&T Personnel for Innovation: Main Policy Challenges*, Seoul: STEPI.

Korea Industrial Technology Association (KOITA) (2003), *White Papers on Industrial Technology*, Seoul (in Korean).

Lee, KongRae (1998), 'Technology diffusion policy of Korea', in KongRae Lee and

Wizin Song (eds), *National Innovation System of Korea* (in Korean), Seoul: STEPI, pp. 119–43.

Lee, KongRae (2000), 'Technological learning and entries of user firms for capital goods in Korea', in Linsu Kim and Richard R. Nelson (eds), *Technology, Learning, & Innovation – Experiences of Newly Industrializing Economies*, Cambridge: Cambridge University Press.

Lee, KongRae and Song, Wizin (1998a) (eds), *National Innovation System of Korea* (in Korean), Seoul: STEPI.

Lee, KongRae and Song, Wizin (1998b), 'Structure and characteristics of the Korean National Innovation System' (in Korean), *Kisul Hyukshin YeonKu*, **6** (2), 1–31.

Lundvall, B.-Å. (1992), *National System of Innovation towards Theory of Innovation and Interactive Learning*, London: Pinter Publishers.

Ministry of Science and Technology, Korea Institute of S&T Evaluation and Planning (2003), *Report on the Survey of R&D in Science and Technology*, Seoul: KISTEP.

National Computerization Agency (2003), *National Informatization White Paper*, Seoul.

Nelson, R. and N. Rosenberg (1993), 'Technical innovation and national systems', in R. Nelson (ed.), *National Innovation Systems: A Comparative Analysis*, Oxford: Oxford University Press.

Pavitt, K. (1988), 'International patterns of technological accumulation', in N. Hood and J.-E. Vahline (eds), *Strategies in Global Competition*, London: Croon Helm.

Song, J.K., J.W. Lee, D.W. Lee and M.K. Kim (2002), *Changes in Technology: Strategy of Firms and Policy Implications*, Seoul: STEPI (in Korean).

Utterback, J.M. (1979), 'The dynamics of product and process innovation in industry', in T.C. Hill and J.M. Utterback (eds), *Technological Innovation for a Dynamic Economy*, New York and Oxford: Pergamon Press.

von Hippel, E. (1988), *The Source of Innovation*, New York: Oxford University Press.

Westphal, L.E. (1978), *Korean Industrial Competence: Where It Came from*, Washington, DC: World Bank.

Westphal, L.E. et al. (1984a), *Exports of Capital Goods and Related Services from the Republic of Korea*, Washington, DC: World Bank.

Westphal, L.E., K. Linsu and C.J. Dahlman (1984b), *Reflections on Korea's Acquisition of Technological Capability*, Washington, DC: World Bank.

Westphal, L.E., Y.W. Rhee, L. Kim and A.H. Amsden (1984c), 'Republic of Korea', *World Development*, **12** (5/6), 505–33.

World Bank (2002*), Technology, Skills and Internet Services in Korea: Moving towards a Knowledge-based Economy*, mimeo, Washington, DC.

9. Advance of science-based industries and the changing innovation system of Japan

Hiroyuki Odagiri

INTRODUCTION

With the decline of demand in existing industries, intensifying technological competition on a global scale, and the rapid progress of scientific knowledge, Japan now aims at advancing *science-based industries*. In 2001, based on the recommendation of the Council of Science and Technology Policy, the Japanese Government drew up the 'Science and Technology Basic Plan', in which four areas were given strategic priorities. They are life sciences, information and telecommunication, environmental sciences, and nanotechnology and materials. It is hoped that the promotion of these sciences will foster the development of industrial technologies, such as biotechnology, IT technology and nanotechnology-based materials, thereby stimulating the development of related industries.

Accordingly, Japan's national innovation system is changing. In part, it is a spontaneous change that is occurring in response to changing market needs. Also, it is a consequence of conscious policy efforts because the advance of such industries made the existing institutional, legal and policy framework obsolete. In this chapter, I intend to describe such changes in Japan, occasionally taking biotechnology as a case, and show how technological changes, socio-economic changes, and institutional changes interact with each other, creating a new and yet path-dependent national innovation system.

In the first section, I begin by describing Japan's national innovation system up to the 1980s, followed by the comparison of its experience with those of Korea and Taiwan in the next section. Then, in the third section, I discuss how the conditions underlying Japan's system have recently changed. In the fourth section, I will describe peculiar features of science-based industries. I will focus on four such features: (i) the need for close university–industry collaborations, (ii) the impact of the intellectual property rights system, (iii) the role played by new start-up firms, and (iv) the changing and diversifying boundary of the firm. How Japan is making changes in these four aspects will be discussed in the following section. The final section will conclude the chapter.

JAPAN'S NATIONAL INNOVATION SYSTEM UP TO THE 1980S

Goto and I have earlier discussed the technological and industrial development of Japan from the Meiji Restoration of 1867 to the post-war high growth era by using the framework shown in Figure 9.1 (Odagiri and Goto, 1996).

Technologies imported from overseas played a critical role, particularly in the early period of Japan's development. They were brought to Japan through the movement of goods, people, information and capital. Final products were imported and disassembled for the purpose of 'reverse engineering'. Capital equipment, such as plants, machines and tools, was also imported, bringing advanced technologies with it. It is noted, however, that Japanese engineers often redesigned and modified imported equipment to make it suitable for the Japanese geographic, climatic, economic and social conditions. People moved in both directions. The Japanese government and industries not only invited many experts from abroad despite the heavy cost at the time but also sent Japanese scientists, engineers, managers and government officials abroad to learn the advanced technologies and management methods of the West.

Information was also brought into Japan through licensing of patents and know-how. On many occasions, alliances with foreign partners helped Japanese firms to acquire technologies from abroad. Capital inflow as a means of technology transfer was also common before World War II. A number of electrical and communications equipment producers, such as NEC and Fuji Electric (from which Fujitsu was later hived off), were established as joint ventures with American or European advanced firms. Several domestically established firms, such as Toshiba and Mitsubishi Electric, later invited foreign partners to become their major shareholders in return for technologies and management know-how. The automobile industry was dominated by the Japanese subsidiaries of General Motors and Ford until 1935, and these firms played a major role in Japan's starting the car component industry.

However, this role of capital inflow as a means of technology transfer became relatively unimportant in the post-war period. In fact, inward foreign direct investment (FDI) was restricted after the war until the early

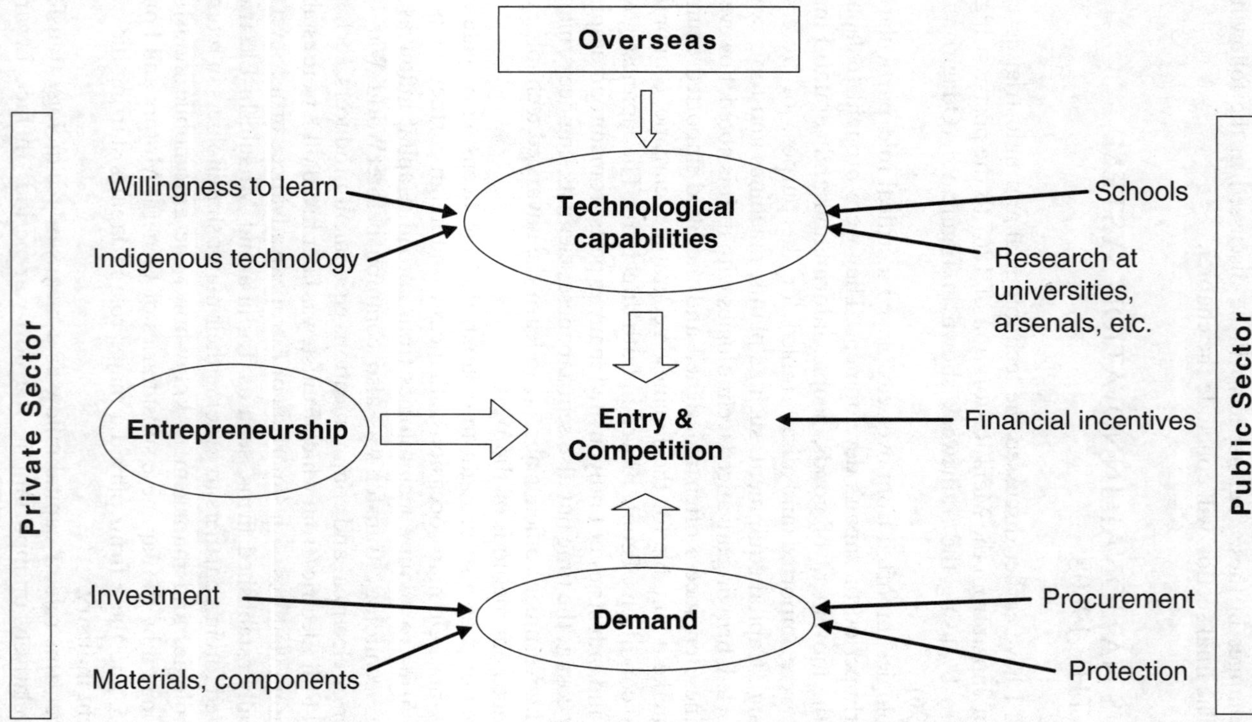

Source: Odagiri and Goto (1996).

Figure 9.1 *Technology and industrial development in Japan: the basic view*

1970s, with only a few exceptions. Even after the capital liberalization of the 1970s the rate of inward FDI remained low. Accordingly, FDI was not the major channel of technology transfer in post-war Japan (Goto and Odagiri, 2003).

Nevertheless, it had an important indirect effect through a potential threat of multinationals making investment in Japan. Japanese firms were aware that capital liberalization was inevitable. Also they were keenly aware from their own pre-war experience that American and European multinationals were far ahead of them not only technologically but also in terms of size, financial power, and marketing and management capabilities. Although such a fear may appear unwarranted today, it was certainly relevant in 1965 when General Motors was 26 times larger than Toyota in sales, and similarly for IBM versus NEC in computers and General Electric versus Toshiba in electrical equipment. With this threat of formidable competition expected after capital liberalization, Japanese firms made every effort to catch up technologically and raise productivity.

The threat of competition came not only from potential and existing foreign rivals but also from domestic rivals, which, in our view, was a central force in Japan's technological development as shown in Figure 9.1. Abundant entrepreneurship was there, which resulted in a number of entries, despite the risk and financial burden of investment in R&D and technology acquisitions and to sell the products. In the electrical equipment industry for instance, Toshiba entered by developing incandescent bulbs in the 1880s, Sony by developing transistor radios in the 1950s, and Sharp by developing calculators in the 1960s. Entry also occurred as a result of diversification efforts of existing firms; for example, Suzuki, originally a manufacturer of weaving machines, diversified into the production of motorcycles and then automobiles, and Toray, originally a manufacturer of rayon, expanded their product line to include nylon and other high-polymer synthetic fibre and then various fine chemicals. Challenges to market leaders were also made by smaller fringe firms with their innovations; in the steel industry for instance, Kawasaki Steel, hived off from Kawasaki Heavy Industries in 1950, increased their market share rapidly after building an innovative integrated steel mill, making itself a serious competitor for the dominant Nippon Steel.

The restriction of FDI, together with import restriction, meant that foreign firms could neither export products to Japan nor invest to manufacture within Japan. Thus, they could exploit their technological superiority only by licensing the technologies, making it easier for Japanese firms to import technologies. Technology importation was also regulated during the 1950s and 1960s because the firms had to apply to the Ministry of International Trade and Industry (MITI) for the allocation of foreign exchanges needed for royalty payment. Often a large number of firms applied to import a technology, even when the technology was still at a commercially untested stage. In such cases, MITI tended to permit just one firm to import the technology, thereby improving the Japanese firm's position *vis-à-vis* that of the foreign licensor in a licensing bargaining. As a result, as some authors argued (for example, Peck and Tamura, 1976), Japanese firms may have paid a royalty rate lower than the international rate. This by no means implies that MITI had the ability to select the most appropriate licensee. Kiyota and Okazaki (2005) suggests that MITI tended to select large but low-productivity firms.

Technology importation increased rapidly. Domestic R&D also increased rapidly. As a percentage to GNP, domestic R&D expenditures increased from 0.84 in 1955 to 1.73 in 1961, 2.14 in 1980, and 3.26 in 1998, surpassing the US in 1987. We emphasize therefore that Japanese firms invested heavily not only for licensing but also for own R&D to assimilate and apply imported technology, although, gradually, the weight of R&D shifted from the improvement of imported technologies to the invention of original technologies.

Besides the control of technology importation, several government policies played important roles. Firstly and probably most importantly, the investment in the early period to establish both a compulsory elementary education system and a higher education system laid the foundations for Japan's economic development. Second, the provision of infrastructure including both hard infrastructure, such as transportation and communication networks, and soft infrastructure, such as Commercial Code, Patent Law, and other legal systems, was essential. Third, the government secured demand to domestic firms through procurement; for instance, military procurement before the war and the procurement of communication equipment by Nippon Telephone and Telegrams, and through the protection of the domestic market by means of import restriction during the 1950s and 1960s. And, fourthly, there were also cases of the government financially supporting target industries by providing tax concessions, low-interest loans, and subsidies, such as the subsidies given by MITI through research associations in the 1960s and 1970s. In our view, however, these policies were not as successful or effective as the provision of infrastructure or the support of demand: see Odagiri and Goto (1996) for details.

In sum, as shown in Figure 9.1, three factors contributed to the technological and industrial development of Japan, with the consequence of active entry and competition. They are (1) the accumulation of technological capabilities, which, particularly in the early period of industrialization, were acquired from abroad by the private sector through its willingness to learn and on the basis of its inherited indigenous technology, and by the public sector, which then diffused the information to the private sector through schools and public research institutions, (2) strong entrepreneurship of the private sector, and (3) the presence of sufficient domestic demand, which was supported not only from the large population but also from protection and procurement.

COMPARISON WITH KOREA AND TAIWAN

One may compare this experience of Japan to that of other countries that have achieved similarly rapid economic growth in the post-WWII period, most notably Korea and Taiwan.[1] There are several similarities; for instance, the promotion of supposedly infant 'target' industries by the government (whether or not such targeting was pursued consistently or effectively), the encouragement of technology importation, and yet limited reliance on FDI as a means of technology transfer. There are, however, marked differences. The first concerns the initial condition. When Japan started the post-war reconstruction effort in 1945, it already had some eighty years of experience in industrialization since the Meiji Restoration of 1867. Indeed, even during the Tokugawa Era that preceded the Restoration, commercial and industrial activities prospered despite the limited inflow of Western technologies caused by the seclusion policy of the Tokugawa Shogunate government. Hence, notwithstanding the devastating bombing and other damages during World War II, Japan had a set of accumulated physical assets and human capital. More than 60 per cent of plants and equipment survived the war and a large labour force with knowledge, skills and experience returned to industries. By contrast, because both Korea and Taiwan were under Japanese rule before the war and Japan had little interest in raising the managerial and technological capability of occupied people, the lack of such capability was a serious handicap when these countries started their development effort after their independence.

The second is the different size of the economy. With a population of about 81 million, Japan in 1949 was more than four times larger than Korea and ten times larger than Taiwan. In terms of economic size, the difference was even larger. Thus, with a domestic economy large enough for most industries to achieve economies of scale, Japanese firms could accumulate experience in domestic markets before entering into export markets. Moreover, this large domestic market provided opportunities for profitable introduction of new products and for profitable entry by new firms. In consequence, several firms competed intensely against each other as discussed earlier. The automobile industry gives a good example.

Although, in the beginning, the quality of Japanese cars was much lower than that of American and European cars, the Japanese carmakers improved both their products and production processes to survive the domestic competition, before starting export to the US. By contrast, Korean and Taiwanese firms had to target export markets almost from the beginning owing to the limited size of their domestic markets, with their government pursuing an export-oriented policy. The Japanese government also promoted export until around 1970 but import substitution was as important as export. In this regard, China, with its huge population, may resemble Japan more than Korea and Taiwan. China's regulation of FDI may also resemble Japan during its high-growth era. Obviously, however, the global environment today is quite different from that of the 1950s and so is the importance of intellectual property rights. And, of course, Japan had a very different political system from China: in fact, there were many occasions on which Japanese firms refused (or neglected) to follow the government's 'administrative guidance', which would not be tolerated under the Chinese political scheme. Differences are hence more pronounced than similarities.

The third difference concerns the industrial structure. In Korea, several big company groups (*chaebols*) were established and the government supported and utilized them to foster the development of heavy and chemical industries. In Taiwan, by contrast, large firms hardly existed and, to support small and medium enterprises (SMEs), the government played important roles by, for instance, funding R&D at government research institutions and transferring the technology to SMEs. Between these two extremes, Japan was more balanced. Pre-war *zaibatsus* were disbanded after the war, with all the formerly *zaibatsu*-controlled firms becoming independent.[2] A number of non-*zaibatsu* large firms, such as Toyota, survived and started to grow, and so did a vast number of SMEs. Some of these SMEs were independent while some were dependent on large firms as suppliers, subcontractors, or subsidiaries and tended to rely on the technologies provided by large firms. Yet several suppliers, such as Denso, started to invest in own R&D and grew to become multinationals themselves.

As a consequence of these differences, the role of the government was more limited in Japan. The proportion of public R&D expenditure in total R&D expenditure was 64 per cent in Korea in 1980 (Kim, 2003) and 60 per cent in Taiwan in 1986 (Aw, 2003), whereas it was only 27 per cent in Japan in 1981.[3] We may therefore conclude that Japan's development was essentially industry-led with the government providing the necessary infrastructure and occasional (but not necessarily successful) intervention. This is why we put 'entry and competition' at the centre of Figure 9.1. In comparison, the governments played more active roles in the economic development of Korea and Taiwan.

It is out of the scope of this chapter to discuss which of these three countries would give a best lesson to current developing countries. Presumably, since these developing countries tend to have neither inherited industrial, managerial and technological bases nor well-balanced size distribution of firms, the Japanese experience is less applicable than that of Korea or Taiwan.

THE JAPANESE BUSINESS SYSTEM AND THE CHANGE

In addition to the macro and industrial factors discussed above, several features of the business system also influenced Japan's national innovation system during its high-growth era. First, owing to the presence of friendly shareholders (for example, banks and group firms) and the practice of appointing executives through internal promotion, the management could pursue long-run projects more easily. Second, the proportion of top management with a science or engineering background was higher in Japanese firms in comparison to large American firms (in the pre-Silicon Valley era), which helped Japanese top management to have a better understanding of the potential and limitation of R&D. Third, because of the Japanese internal labour system in which long-term company– employee relationships were common, coupled with internal training and rotation programmes, the linkage among R&D, production and sales departments used to be tight, fostering the manufacturing and marketing applications of innovations. Moreover, such linkage tended to expand to suppliers and other affiliated firms, prompting the sharing of information among them. And fourthly, the introduction of new technologies to production lines was easier both because of the above-mentioned interaction between R&D and manufacturing departments and because of the flexibility in rearranging workshops and the broader skills of workers nurtured through internal training and rotation: see Odagiri (1992) for more details.

Since 1990, however, some of these advantages weakened. Banks (excluding trust banks) reduced their share ownership of public companies from 15.7 per cent in 1988 to 7.7 per cent in 2002 while the percentage of trust funds, pension funds, and foreigners together increased from 8.4 to 27.5.[4] Even hostile mergers and acquisitions, which had been rare except for those aiming at greenmailing, took place. Bankruptcy is now more common and so is *de facto* dismissal of workers. The loss of production skills is now a concern owing to the ageing of the working population together with the shift of plants to overseas, particularly China and Southeast Asia.

Changes also occurred as regards the mechanism shown in Figure 9.1. Even though, on the one hand, the inflow of technologies from abroad has increased because of globalized business activities and the use of advanced information technologies, the end of catch-up means that Japan can no longer rely on one-way technology acquisition from abroad. In addition, intellectual property rights (IPR) have become more strictly enforced by foreign firms, as exemplified by the IPR (and trade secret)-related lawsuits filed, for instance, by IBM and Corning against Japanese firms during the 1980s.

It is difficult to say if entrepreneurship of Japanese firms has been weakening. Still the fact is that the rate of new company establishment has declined. The rate of entry (the number of new enterprises as a percentage of the initial number of enterprises) dropped from 5.9 per cent of 1975–1978 to 3.1 per cent in 1999–2001 and is now lower than the rate of exit, which was 3.8 per cent in 1975–1978 and 4.5 per cent in 1999–200.[5] Of course it is easy to imagine that the loss of market demand owing to the business stagnation of the 1990s caused this drop in business start-ups. Yet, it is noteworthy that the entry rate started to drop not after 1990 but in the early 1980s when business conditions were still favourable.

In addition to these changes in the business environment of Japanese industries, the emergence of new scientific and technological environment calls for a significant change in Japan's national innovation system. Most importantly, the increasing importance of science-based industries has been causing a profound impact on the R&D strategies of Japanese firms and Japan's science and technology policy. To this topic we now turn.

SALIENT FEATURES OF SCIENCE-BASED INDUSTRIES

Science-based industries (SBIs) are the industries in which the development is pursued by means of innovations based on sciences. Scientific knowledge, we say, 'is used' in innovations in two senses. First, the scientific research outcome would be applied and developed for industrialization. Second, sciences would be used to solve the bottlenecks that may arise in the course of R&D and production. Also, any discovery during R&D or production would be fed back to scientific research. Therefore, the flow of information is not only from science to development and commercialization as the so-called 'linear model' of innovation implies, but also from development or production to science (Kline and Rosenberg, 1986). This bi-directional interaction between scientific activities and industrialization is an important characteristic of SBI.

Four issues are relevant regarding SBIs. They are (1) science linkage, (2) intellectual property rights (IPRs), (3) the boundary of the firm, and (4) widespread use of the technology across industries. We will discuss these in turn.

Science Linkage

Science linkage is commonly measured by the number of citations to scientific papers per US patent (Narin and Olivastro, 1992). When applying for patents in the US (but not in Japan), applicants are required to list any prior papers and patents that are related to the technologies to be patented. It can be assumed that, if the application cites many scientific papers, then the invention has benefited greatly from scientific discoveries. Thus, the per-patent number of citations to scientific papers is used as a measure of science linkage.

Table 9.1 shows the trend in science linkage in six major fields in the US and Japan. There are three major findings. First, there is an increasing trend in any field and in either country. Second, it is higher in the US than in Japan. This difference may be overstated because American inventors are more likely to cite papers by American authors in American journals, and they will be careful not to miss citing related papers for fear of receiving complaints from the authors. Japanese inventors may have benefited from Japanese papers but they are probably less careful in citing them than American inventors, because Japanese authors are less likely to notice the lack of citation, and likewise will be the USPTO (US Patent and Trademark Office) examiners.

Third, the six fields that are closely related to SBIs have higher scores of science linkage than that in all fields. This tendency is strongest with biology/microbiology, followed by organic chemistry, suggesting that biotechnology is the most science-based of all industries.

This high and increasing level of science linkage implies that, to foster innovations in SBIs, collaboration between universities – the main players of scientific research – and industries – the main players of development and commercialization – is essential. Moreover, as stated earlier, the relationship should not be unidirectional from scientific discovery to development, because the feedback to the scientific sector of information gained in the process of development is also essential. Here rests an opportunity as well as a need for university–industry collaboration.

Table 9.1 Trends in science linkage in Japan and the USA

	All fields	By six major fields						
		Biology/ microbiology (1)	Organic	Inorganic	Medicine/ veterinary medicine (2)	Agriculture, forestry, fishery (3)	Computation/ counting	Basic electronic circuitry
Japan								
1985	0.17	1.48	0.85	0.32	0.45	0.16	0.19	0.35
1990	0.27	2.42	1.04	0.93	0.53	0.40	0.31	0.62
1995	0.56	5.15	1.95	0.85	1.27	0.89	0.83	0.91
2000	0.51	5.84	2.99	0.55	2.31	1.41	0.69	0.69
2002	0.49	6.80	2.99	1.19	1.83	1.16	0.63	0.61
USA								
1985	0.39	5.13	1.38	0.72	1.08	0.27	0.50	0.55
1990	0.72	8.07	2.54	1.86	1.74	0.59	1.08	0.99
1995	1.61	15.53	6.54	2.72	3.58	1.33	1.91	1.47
2000	2.82	23.20	14.20	3.08	7.15	5.63	2.01	1.44
2002	3.23	24.32	15.83	3.53	8.24	6.03	1.97	1.62

Notes:
(1) Biology, Beer, Alcohol Spirit, Wine, Vinegar, Microbiology, Enzyme, Mutation or Genetic.
(2) Medicine and Veterinary Medicine, Hygienic.
(3) Agriculture, Forestry, Stockbreeding, Hunter, Capture, Fishery.

Source: National Institute of Science and Technology Policy, 'Kagaku Gijutsu Shihyo 2004' ('Science and Technology Indicators 2004'), NISTEP Report No. 73. The original data is from CHI Research Inc.

Intellectual Property Rights

Intellectual property rights (IPRs), such as patents and copyrights, have strategic importance in SBIs. This is primarily because huge R&D expenditures are required in SBIs, and IPRs are considered to be the most effective means of appropriating the returns to R&D investment. Besides, IPRs are particularly effective in some SBIs. In the US, the effectiveness of patents as an appropriability mechanism is highest in medical equipment and drugs among industries, followed by special purpose machinery, auto-parts and computers (Cohen et al., 2000). Also in Japan, the effectiveness is highest in drugs, followed by computers (Goto and Nagata, 1996).

This fact suggests that the design of an IPR system and its implementation profoundly affects the development of SBIs. As just stated, IPR is considered to provide an incentive for innovation by protecting rights to invented technologies. Particularly in small start-up firms, technologies can be virtually the single source of income and, without IPR protection, they may not be viable. From this viewpoint, the so-called pro-patent policy of strengthening IPR may appear to be the right policy proposal for promoting SBIs.

Yet, stronger IPRs may actually hurt technological progress because IPRs restrict the usage of invented technologies (Merges and Nelson, 1990). In particular, an increase in 'research tool patents' can hinder technological progress as the need to clear permission with the owners of these patents can make R&D more costly and time-consuming. Examples of research tool patents in biotechnology include those on the PCR (polimerase chain reaction) method, DNA chips, and transgenic mouse. If each patent-holder acts aggressively, then many R&D projects would become economically infeasible, hindering technological progress as a result. This is what Heller and Eisenberg (1998) called the 'tragedy of anticommons'. In electronics, it is often the case that hundreds of patents have to be cleared before a product is to be commercialized. Pooling of patents and cross-licensing among patent-holders are common. Again, however, the cost of searching for patents that have to be cleared and of making agreements with the patent-holders can be huge. For instance, large electronics firms employ hundreds of people to deal with IPR issues.

These two effects of IPR, that is, the positive incentive effect and the negative usage-restriction effect, have to be balanced in any design of an IPR system.

The Changing R&D Boundary of the Firm

The issue of the boundary of the firm has been discussed widely in relation to the make-or-buy decisions on parts and materials. For instance, the close and long-term relationship with suppliers has been considered to be the strength of the Japanese automobile producers, which is contrasted with a higher proportion of in-house part production of American firms until the 1990s combined with arm's-length transaction with independent suppliers (Odagiri, 1992). However, even in the US, more use of outside suppliers (including those spun off from the assemblers) and more collaboration with them have become common.

Similarly, in R&D, the use of outside suppliers and partners has become prevalent. It is now impracticable to perform all R&D works in-house, and how to incorporate and utilize outside capabilities has become the key for successful innovation. Such utilization is made in several fashions. The firm may outsource routine R&D services, such as software development, supply of order-made samples, manufacture of prototypes, and animal tests (in the case of pharmaceuticals). They may form R&D alliances by commissioning research to other established firms, new start-up firms, universities, or public laboratories, or starting joint research projects with them. They may also acquire technologies by licensing-in. In such a diverse manner, firms today are extensively utilizing outside capabilities (Odagiri, 2003).

As a consequence, it is indispensable for the development of SBIs that the economy is equipped with a wide variety of potential partners and opportunities that the firms can use for outsourcing, R&D alliances, technology acquisitions, and so on.

Widespread use of the Technology across Industries

Another feature common to SBIs is that a technology is used not in a single industry but in a variety of industries. Biotechnology is a good example. Table 9.2 shows the shipment of biotechnology-related products in Japan in 2000. The food and beverage industry has the largest shipment, accounting for nearly two thirds of the total shipment, followed by pharmaceuticals. Still, these are not the only biotechnology-related industries and, as shown in the table, biotechnology is used in a wide range of industries from chemicals to machinery, electronics, informatics, and environmental remediation. In addition, a wide variety of technologies is used. In food and beverages, almost all the technologies are the so-called 'traditional' biotechnology, such as fermentation and cultivation. By contrast, about half of the technologies used in the production and R&D of pharmaceuticals are the 'new' biotechnology, such as cell fusion, recombinant DNA, and bioreactors, and if one only considers new biotechnology, pharmaceuticals (including medical equipment) are the largest user of biotechnology. In the US also, the health care industry is the dominant user of biotechnology.[6]

That is, the range of the biotechnology industry is not only wide, but is also dependent upon the definition of 'biotechnology'. The same can be said about information technology and nanotechnology as these technologies are used in a wide array of industries.

REFORM

With the declining demand for traditional products and in search of high value-added businesses, Japan is now in the process of building up science-based industries. However, in view of the above-mentioned four characteristics, the national innovation system discussed earlier may no longer be appropriate for SBIs. As a consequence, the Japanese innovation system has been in the process of transforming itself and the government policy to foster this transformation is under way. Let us now discuss these changes in detail.

Towards a Closer University–Industry Collaboration

In SBIs, since the largest performer of scientific research is universities (and national laboratories), a close university–industry collaboration (UI collaboration) is called for. Of course, as the measurement of science linkage implies, the most common channel of information from academic research to industrial innovation is published papers; for instance, Branstetter's recent research (2004) confirms a great contribution of papers published by university faculties on industrial patenting.

Universities can also contribute to industrial innovation in other diverse ways because, as discussed earlier, the unidirectional flow of science to innovation is insufficient. Industrial R&D teams may face technological difficulties and, to solve them, they may seek the advice of academic scientists or propose to start joint research with such scientists. Invented technology may be transferred from universities to industries by means of licensing of university patents. However, the development of a commercially viable product out of a patented invention need not be straightforward. The licensed patent may not cover all the necessary technology and know-how, which may be smoothly transferred only when the university inventor is actively involved. Also, as the term 'absorptive capacity' implies, a sufficient capability is needed on the licensee's side and, even with such capability, unexpected bottlenecks may arise in the course of development.

Table 9.2 Shipment of biotechnology-related products, 2000

Product field	Domestic shipment		Composition of shipment by the type of technologies (%)				
	Billion yen	%	Traditional fermentation, cultivation, modification, etc.	Traditional environmental remediation with organisms	Cell fusion, recombinant DNA, bioreactors, etc.	Biomaterials, and electronic and other equip. and software making use of biological knowledge	Non-response
Food and beverages	4474	66.0	100.0	0.0	0.0	0.0	0.0
Misc. food	155	2.3	90.0	1.2	6.3	0.0	2.5
Agricultural	24	0.4	24.8	0.0	15.2	0.0	60.1
Livestock and fishing	32	0.5	24.0	8.4	9.9	5.3	52.4
Pharmaceuticals and medical equip.	1098	16.2	38.4	2.8	48.3	3.1	7.6
Laboratory samples and reagents	23	0.3	32.9	0.0	60.3	6.4	0.4
Textiles	2	0.0	60.6	3.4	11.1	10.7	14.2
Chemicals	439	6.5	47.0	0.0	52.4	0.0	0.6
Bio-electronics	33	0.5	0.0	0.0	0.2	73.0	26.7
Environment-related equip. & materials	212	3.1	0.3	92.8	1.0	1.8	4.1
Laboratory and plant equip.	159	2.4	8.8	0.0	5.6	84.7	0.8

Misc. manufacturing	61	0.9	88.7	1.6	1.2	8.5	0.0
Informatics	12	0.2	0.0	0.0	0.0	91.8	8.2
Services	54	0.8	9.4	20.6	26.1	43.1	0.7
Total	6779	100.0	75.9	3.6	14.5	3.2	2.9

Source: Ministry of Education, Culture, Sports, Science and Technology, et al. (2002), 'Heisei 13 Nendo Baio Sangyo Souzou Kiso Chosa Houkokusho' ('Report on the Basic Survey of Biotechnology Industries, 2001').

Advice by university inventors or other academics may help the industry to acquire a necessary capacity or to solve the bottlenecks.

There are also cases in which UI collaborations are called for at a pre-invention stage. Thus, industries often commission research to universities or propose joint research with them. Joint research is an attractive option because supposedly complementary capabilities of university scientists (who are good at, say, theorizing) and industry engineers (who are good at, say, experimenting and building prototypes) can be combined.

To promote SBIs, therefore, there must be an environment in which UI collaborations of various forms are feasible and encouraged. Accordingly, Japan is now shifting gear towards this direction.

It is not that UI collaboration was absent in Japan and, on the contrary, universities played an important role in Japan's industrial and technological development (Odagiri, 1999). As was somewhat common with the US, another late-developing country at the time, Japan in the mid-nineteenth century was desperate to catch up with the then state-of-the-art technologies of advanced European nations.[7] Thus, its higher education system emphasized the acquisition of practical technological knowledge and skills. Technologically knowledgeable people were scarce and mostly in universities; hence, industries actively sought information and advice from university faculties.

Unfortunately, particularly after World War II, a uniform and rigid regulation began to be applied to the conduct of university faculties. Such regulation was strictly enforced because most of the major universities in Japan were national and their professors were civil servants.[8] Hence, professors could receive funds from industry only in a limited manner and after much tedious paperwork, and their time spent working for industries was similarly restricted. They were not encouraged to apply for patents and could not become a director of a private company.

In the past few years, however, there has been a drastic shift towards deregulation and encouragement of UI collaborations. Professors can now join boards of directors of private companies. Several policies have been adopted to promote joint research with industries. First, red tape was relaxed regarding the acceptance of research funds from industries and of researchers dispatched from companies to university laboratories. Second, many universities built special facilities for UI joint research. Third, universities can now offer their space to start-ups at a low rent, if these start-ups were established for the purpose of commercializing technologies originating from the university. Fourth, many universities have founded technology licensing offices (TLOs), which help faculties in applying for patents and licensing them and help companies in finding suitable university patents to be licensed and suitable faculties with which to start joint research. Fifth, patent fees have been reduced for applications by university researchers or TLOs. Sixth, special tax concessions are now given to company R&D expenditures used for UI collaborations.

Furthermore, with the National University Corporation Law, every national university in Japan was incorporated into a semi-independent corporation in April 2004. Although the majority of its budget will continue to be supported by the government, this reform is expected to promote UI collaborations further for several reasons. First, incorporated universities can now hold patents, whereas in the past patents belonged to the nation. Second, as the faculties are no longer civil servants, a more flexible employment arrangement is now possible, making it easier for the faculties to work for companies part-time and receive industry funds. Also, the recruitment of specialists to support patenting, licensing, spinning-off and other activities should become easier. Third, naturally, each university will have more incentive to increase its revenue, not only by offering more up-to-date courses but also by attracting industry funds for UI collaborations and promoting patenting and licensing of university inventions.

With these reforms, UI collaborations have been increasing rapidly.[9] The number of UI joint research by national universities increased from 1139 in 1990 to 4029 in 2000 and 6767 in 2002. The number of start-ups based on university-invented technologies increased from 11 in 1995 to 135 in 2002 and, in 2005, the number of such companies in operation is more than 1000. Though this figure is smaller than in the US, the increase is impressive.[10] Thirty-nine TLOs have been set up, and several cases of licensing have been already reported, even if they are still few and the TLOs are all suffering from loss. Also, as of September 2003, 280 cases were reported in which the professors of national universities were acting as directors or auditors of companies.

This rise in UI collaborations and, more generally, the change in the expected role of universities has been significantly transforming the national innovation system of Japan.

Intellectual Property Reform

In 2002, Japan enacted the Basic Law on Intellectual Property. With this law the government established the Intellectual Property Policy Headquarters within the Cabinet, for the purpose of 'providing stipulations on the development of a promotion program on the creation, protection and exploitation of intellectual property'. There is a strong pro-patent feeling among the policy-makers as well as the industries, particularly in response to the pro-patent shift (at least at some point in time) of the US, as indicated by its wider acceptance of EST (expressed sequence tags) patents and business model patents. However, because stronger patent rights may hinder diffusion as discussed earlier, the Japanese Patent Office (JPO) has not explicitly shifted its stance towards an extension of patentable inventions. Still, such a shift seems to have gradually occurred in reality.

The new IP policy emphasizes, first, a wider utilization of patented technologies and, second, a stronger enforcement of patent rights.

According to JPO's survey, among the patents held by companies, only 27 per cent were actually used by the patent owners.[11] Although no comparable US statistics are available, this proportion is considered to be much lower than in the US. This difference partly comes from the larger number of applications in Japan. Owing to the first-to-file rule adopted in Japan (as opposed to the first-to-invent rule in the US), firms tend to make an application for any invention they make, even if many of them will later be found to be commercially useless.

The effectiveness of patent protection depends on the enforceability of patents. Patent litigation is a notoriously costly and time-consuming process in Japan, mainly because there has been no court specialized in patent-related litigations and, hence, there was no judge (and only a few lawyers) with technological knowledge. To remedy this situation, a special court was established in 2005 within the Tokyo High Court to deal with patent litigation.

Also, to supply more personnel with knowledge on patent law and patent management to industries as well as the courts and the government, the government encouraged universities to establish necessary courses, and JPO has been making efforts to hold public lectures on related issues. Since small and medium enterprises in particular suffer from the lack of necessary personnel, the government has also been establishing a scheme to introduce to SMEs those people having retired from large companies after years of experience in patent management.

Another policy emphasis has been on the production and utilization of patents by universities and national institutes. As discussed in the previous section, the government has been promoting more active involvement of universities in industrial innovation. As regards patents, an important policy change was made in 1999 with a law dubbed 'the Japanese Bayh-Dole Act' after the 1980 Bayh-Dole Act of the USA. With this law, researchers who made inventions out of the R&D projects commissioned and funded by the government can now claim the ownership of the inventions. This new policy aimed to give more incentive for researchers to patent and also to promote commercial application of the patents by the researchers themselves or by licensing. As discussed already, patent fees were reduced for academic inventions and TLOs have been set up in many universities.

Consequently, although the rise in university patenting and licensing may not be spectacular yet, a gradual change has been occurring and, together with the incorporation of national universities as discussed in the previous section, a big impact on Japan's national innovation system is anticipated in coming years.

Promotion of Start-ups

The promotion of start-ups has been another major policy issue, because the aforementioned decline in the rate of entry of new enterprises is worrisome and because the advance in the US of biotechnology and IT industries is considered be largely due to the activity of high-tech start-ups, such as those in Silicon Valley.

Thus, several policy measures have been taken besides those regarding university-based start-ups discussed above. In 1999 the Law for Facilitating the Creation of New Business (dubbed the Japanese SBIR programme after the US Small Business Innovation Research Program) was enacted, with which the government started to provide subsidies and debt guarantees to support the investment by SMEs (existing SMEs, new start-ups, or individuals) to start new businesses and to develop and commercialize new technologies. In 2002, the government started to give tax advantages to individuals investing in start-up companies (called the Angel Tax System), and reduced the minimum amount of capital required to found a stock company from 10 million yen to a mere 1 yen, provided the company is established to start a new business, on the condition that the capital should be increased to a minimum of 10 million yen within five years of the establishment. Start-up firms were also allowed to use stock options as a compensation scheme to its directors and employees, because these firms often face a cash-flow shortage as they have to spend money on R&D and make other investments in the early stage when their revenue remains still low.

With these and other policies to promote start-ups, the number of high-tech start-ups has actually been increasing. For instance, the per-annum number of newly established biotech start-ups increased from less than 20 in the latter half of the 1990s to more than 40 after 2000. As a result, the number of existing biotech start-up companies increased from 60 in 1998 to 387 in 2003.[12]

Financing for these start-ups has been made easier. Three stock markets, called JASDAQ, MOTHERS and HERCULES, were opened or reorganized to make it easier for start-ups to trade their shares. The number of new initial public offerings (IPOs) has accordingly increased: in 2003, about 100 firms made IPOs in these markets. Much venture capital has also been established.

Yet, there remain a number of problems. In 2001–2002, Odagiri and Nakamura (2002) conducted interviews and carried out a survey by questionnaire of 65 Japanese biotech start-ups. When we asked the firms if they felt each of thirteen probable obstacles in the list to be a significant barrier in founding their firms, 54 per cent of them answered yes to 'the difficulty in recruiting technological staff'. This was followed by 'difficulty in financing' (49 per cent), 'difficulty in recruiting non-technological staff (e.g., finance, accounting, and legal)' (23 per cent), and 'difficulty in securing wet laboratories' (23 per cent). Evidently, recruitment of technological and non-technological staff is a big hurdle for Japanese start-ups, together with financing.

A part of this difficulty comes from the lower number of specialists, such as lawyers and certified accountants, in Japan in comparison to the US. More important in our view is the lower mobility of workers in general in Japan. As discussed earlier, long-term attachment between workers and their companies has been the norm in Japan. Although *de facto* dismissal did occur even in Japan and the so-called 'lifetime employment' may have been adopted only in the public sector and large firms, most managers took it as a norm towards which they should make efforts (Odagiri, 1992). Likewise, most workers tended to assume that they can and will stay with the same company until retirement. True, with the collapse in the past decade of a number of firms that were once considered infallible, many people started to regard the concept of lifetime employment with suspicion. And, in fact, there now appear to be more cases of workers changing jobs and firms hiring mid-career workers, part-time and temporary workers. Still, according to the government survey, 40 per cent of the firms with 1000 employees or more replied that they intend to maintain the lifetime employment system.[13] The same survey, however, also reveals that those firms replying that they no longer have a lifetime employment system accounted for less than 10 per cent in 2003 but had increased compared to four years earlier.

That is, Japan's labour system has been gradually moving towards a more mobile one and yet, at least at this time, mobility is lower compared to other countries, particularly the US. This tendency is most evident in large firms and these large firms tend to have talented people both because they can recruit better workers and because their workers tend to receive more in-company training and wider experience. This situation makes it difficult for start-ups to recruit good scientists and engineers as well as management staff including those in accounting, finance, legal affairs, intellectual property management, and administration, as shown in the abovementioned result of our survey to Japanese biotech start-ups.

Nevertheless, a gradual change is occurring towards more mobile labour markets and more recruitment by start-ups of talented people from established companies. For instance, the first university-spinoff biotech company to have made an IPO, called Anges MG, first had a CEO who had had experience of leading a start-up in Silicon Valley in the US but they were succeeded by a person who had left one of the biggest chemical companies in Japan. Another university-spinoff biotech company is led by a former employee of one of the biggest securities firms. Thus, the move of people from large companies to start-ups has been occurring and, we expect, is going to be more common in the coming years.

R&D Boundaries of the Firm

It is now not only inefficient but also impractical for firms to perform all R&D-related works in-house. Through R&D alliances, licensing, outsourcing and such, they have to incorporate and utilize outside capabilities to achieve innovations efficiently and swiftly. As a consequence, how to set a boundary between in-house R&D and external R&D has become a key factor for successful innovation. From a national viewpoint, the presence of opportunities for R&D outsourcing and alliance is a key factor for a successful national innovation system.

I trust that this increasing importance of inter-organizational collaboration applies to any industry today. Still, it probably applies best in SBIs, such as biotechnology and pharmaceuticals. The number of research alliances (including those with firms and universities, in Japan or elsewhere) by the ten largest pharmaceutical firms in Japan increased three-fold in ten years, from 65 in 1989 to 189 in 1999, and, during January 1999 to August 2001, 103 cases of alliances by these firms were reported by the press (Odagiri, 2003). Forty-three of them were technology acquisitions (that is, licensing-in), and 50 were joint or commissioned R&D, with the rest being access to databases and so forth. As a partner of these alliances, new biotech firms (NBFs), particularly those in the US, were as popular as established firms. These tendencies, that is, an increase in the cases of R&D alliances with many of them being those with NBFs, are found among all major pharmaceutical firms across the world (see, for instance, Henderson et al., 1999).

In a survey conducted by Japan's National Institute of Science and Technology Policy (NISTEP), among the 146 firms who replied that they have conducted biotechnology-related businesses in 2000, 97 performed R&D alliances and/or technology acquisitions (Odagiri et al., 2002). Asked about the reasons why they perform R&D alliances, they gave the highest score to the 'utilization of the partner's technological knowledge and capabilities (particularly non-patented ones)' and the next highest to 'speed', 'utilization of capital equipment', and 'cost reduction' that can be gained through alliances. This result illustrates the importance of utilizing outside assets (tangible or intangible) and capabilities, and of combining them with internal ones.

Of course, firms cannot relegate all R&D works to external sources because they have to maintain capabilities that are indispensable not only for their own development and commercialization but also to evaluate potential alliance partners, monitor them, and understand and absorb the results supplied by them. In the NISTEP survey, many firms reported that they have had cases in which they could have found reasonable alliance partners but nevertheless had decided to perform the R&D themselves. Besides the fear of ambiguity in the ownership of the outcome, these firms cited 'utilization of internal human and other resources and capabilities' and the 'need to nurture them internally' as the main reasons for this decision. That is, firms are keenly aware of the need to accumulate their internal capabilities, not just for in-house R&D but also to perform more efficient R&D alliances.

Utilization of outside resources and capabilities also occurs in the form of outsourcing of more routine R&D-related services. In such outsourcing, the contract specifies the details of the work to be outsourced and all the output from the work is to be handed over to the outsourcer. Examples are, in the case of biotechnology and pharmaceuticals, animal tests, supply of specific samples (such as knockout mice), production of test products, software development, genome analyses, and clinical tests. The amount spent for outsourcing reached 25 per cent of R&D expenditures among pharmaceutical firms, according to the NISTEP survey.

I have earlier discussed the wide application of biotechnology across industries. Bio-related informatics and services, as well as the provision of laboratory equipment, bio-electronics, and samples and reagents, constitute an important part of biotech-related industries. Many firms in these fields are active outsourcees, whether they are large or small and established or new. The presence of such firms is a prerequisite for an SBI-oriented innovation system.

That is, an increasing importance of the issue of the R&D boundaries of the firm in SBIs is closely connected with the important role played by universities and by start-ups and with the widespread use of the technology across industries. Together, they constitute a background without which SBIs can never grow.

CONCLUSION

After explaining the innovation system behind the industrial and technological development of Japan from the Meiji Restoration of 1867 until the

1980s, I have discussed that the end of catch-up, together with the depressed market demand and the weakened financial power of the banking sector, necessitated a significant change in Japan's national innovation system.

Scientific advance has been playing a central role in the emergence of new science-based industries (SBIs), as exemplified by those based on biotechnology, nanotechnology and information technology (IT). The advance of life science promoted its industrial application in the form of biotechnology, which transformed pharmaceuticals and other industries and gave rise to new industries, such as bio-informatics and other bio-related services. Nanotechnology has been changing the material-related industries, and IT has been changing the electronics and communications industry. Also, technologies are interrelated as exemplified by the application of nanotechnology in biotech devices and the application of IT in bio-informatics.

The Japanese government, therefore, has designated four key areas (life sciences, information and telecommunication, environmental sciences, and nanotechnology and materials) and increased the science and technology budget allocated to these key areas. In addition, the government has been making efforts to promote industries based on these sciences by, for instance, promoting university–industry collaborations and the start-up of new high-tech firms.

Changing a national innovation system is by no means an easy task because the factors constituting a national innovation system are complementary. The financial system of Japan, characterized by a close bank–firm relationship and the presence of stable shareholders, was complementary to the labour system characterized by a long-term worker–employer relationship. And this system was conducive, for instance, to the accumulation of firm-specific human skills and the close intra-firm (and intra-group) information sharing, which made cumulative technological innovation easier. Such advantage should not be disposed of easily. Many such advantages have been actually exploited even in science-based fields; for instance, a large brewery applied its fermentation process technologies to the production of biotech drugs. Still, to promote new industries and new firms, the economy needs to foster the reallocation of talented people through external markets (as opposed to internal labour markets) and the supply of more venture funds (for which banks lack comparative advantages). Probably, these needs do not conflict with the traditional Japanese system because, for instance, many established firms today have redundant middle-aged workers and hence their reallocation must be mutually beneficial, and venture funds and the risk-averse banking sector should be able to coexist. The search for a right balance between the traditional system and the, say, more Silicon Valley-type system is in process, from which a new national innovation system is hopefully to emerge.

NOTES

1. For Korea, see Kim (1993, 2003). For Taiwan, see Hou and Gee (1993) and Aw (2003).
2. It is true that former *zaibatsu* members later formed *kigyo shudan* (business groups) with cross shareholding, etc. However, *kigyo shudan* is merely a loose federation of independent firms and essentially differs from *zaibatsu* (Odagiri, 1992).
3. The proportion has since decreased in both Korea and Taiwan. More recently, it is about the same between Japan and Korea, but still higher in Taiwan.
4. The average of 2661 firms listed in five Stock Exchanges in Japan. Source: Tokyo Stock Exchange, *Kabushiki Bunpu Jokyo Chousa*.
5. Source: Small and Medium Enterprise Agency, *White Paper on Small and Medium Enterprises in Japan*, 2003.
6. Source: US Department of Commerce, *A Survey of the Use of Biotechnology in US Industry*, 2003.
7. For the US, see Rosenberg and Nelson (1994).
8. Actually, in terms of the number of universities or of students, private universities dominated, accounting for 75 per cent of universities and 74 per cent of students. However, prestigious universities (e.g., Tokyo, Kyoto, Osaka, Hitotsubashi, and Tokyo Institute of Technology) were all national with only a few exceptions (e.g., Keio and Waseda).
9. The following statistics are available at the website of the Ministry of Education, Culture, Sports, Science and Technology (http://www.mext.go.jp/), although few of them are given in English.
10. In the US, 450 start-ups were formed in 2002 and the accumulated number during 1980–2002 was 4320, of which 2741 were still in operation. Source: The Association of University Technology Management, *AUTM Licensing Survey: FY2002*.
11. The 'use' here includes both own use and licensing. See the website of Japan Patent Office (http://www.jpo.go.jp/indexj.htm) for this and most of the following statistics and facts. Most of them are in Japanese only.
12. Source: Japan Biotechnology Association, *2003-Nen Baio-Bencha Toukei Houkokusho*.
13. Source: Ministry of Health, Labour and Welfare, *White Paper on Labour, 2003*.

REFERENCES

Aw, Bee-Yan (2003), 'Technology acquisition and development in Taiwan', in Sanjaya Lall and Shujiro Urata (eds), *Competitiveness, FDI and Technological Activity in East Asia*, Cheltenham, UK and Northampton, MA, USA: Edward Elgar, pp. 168–90.

Branstetter, Lee (2004), 'Is academic science driving a surge in industrial innovation? Evidence from patent citations', presented at the NISTEP Research Seminar on R&D Strategy and Science and Technology Policy, February 2004, Tokyo.

Cohen, Wesley M., Richard R. Nelson and John P. Walsh (2000), 'Protecting their intellectual assets: appropriability conditions and why US manufacturing firms patent (or not)', *NBER Working Paper*, No. 7552, National Bureau of Economic Research.

Goto, Akira and Akira Nagata (1996), 'Sabei deta ni yoru inobeshon purosesu no kenkyu' (A study of innovation process by survey data), unpublished report, Tokyo: National Institute of Science and Technology Policy.

Goto, Akira and Hiroyuki Odagiri (2003), 'Building technological capabilities with or without inward direct investment: the case of Japan', in Sanjaya Lall and Shujiro Urata (eds), *Competitiveness, FDI and Technological Activity in*

East Asia, Cheltenham, UK and Northampton, MA, USA: Edward Elgar, pp. 83–102.
Heller, Michael A. and Rebecca S. Eisenberg (1998), 'Can patents deter innovation? The anticommons in biomedical research', *Science*, **280**, 698–701.
Henderson, Rebecca, Luigi Orsenigo and Gary P. Pisano (1999), 'The pharmaceutical industry and the revolution in molecular biology: interactions among scientific, institutional, and organizational change', in David C. Mowery and Richard R. Nelson (eds), *Sources of Industrial Leadership*, Cambridge: Cambridge University Press, pp. 267–311.
Hou, Chi-Ming and San Gee (1993), 'National systems supporting technical advance in industry: the case of Taiwan', in Richard R. Nelson (ed.), *National Innovation Systems*, Oxford: Oxford University Press, pp. 384–413.
Kim, Linsu (1993), 'National systems of industrial innovation: dynamics of capability building in Korea', in Richard R. Nelson (ed.), *National Innovation Systems*, Oxford: Oxford University Press, pp. 357–83.
Kim, Linsu (2003), 'The dynamics of technology development: lessons from the Korean experience', in Sanjaya Lall and Shujiro Urata (eds), *Competitiveness, FDI and Technological Activity in East Asia*, Cheltenham, UK and Northampton, MA, USA: Edward Elgar, pp. 143–67.
Kiyota, Kozo and Tetsuji Okazaki (2005), 'Foreign technology acquisition policy and firm performance in Japan, 1957–1970: micro aspects of industrial policy', *International Journal of Industrial Organization*, **23**, 563–86.
Kline, Stephen J. and Nathan Rosenberg (1986), 'An overview of innovation', in Ralph Landau and Nathan Rosenberg (eds), *The Positive Sum Strategy*, Washington, DC: National Academy Press, pp. 275–305.
Merges, Robert P. and Richard R. Nelson (1990), 'On the complex economics of patent scope', *Columbia Law Review*, **90**, 839–916.
Narin, Francis and Dominic Olivastro (1992), 'Status report: linkage between technology and science', *Research Policy*, **21**, 237–49.
Odagiri, Hiroyuki (1992), *Growth through Competition, Competition through Growth*, Oxford: Oxford University Press.
Odagiri, Hiroyuki (1999), 'University–industry collaborations in Japan: facts and interpretations', in Lewis M. Branscomb, Fumio Kodama and Richard Florida (eds), *Industrializing Knowledge: University–Industry Linkages in Japan and the United States*, Cambridge, MA: The MIT Press, pp. 252–65.
Odagiri, Hiroyuki (2003), 'Transaction costs and capabilities as determinants of the R&D boundaries of the firm: a case study of the ten largest pharmaceutical firms in Japan', *Managerial and Decision Economics*, **24**, 187–211.
Odagiri, Hiroyuki and Akira Goto (1996), *Technology and Industrial Development in Japan*, Oxford: Oxford University Press.
Odagiri, Hiroyuki and Yoshiaki Nakamura (2002), 'Nihon no baio bencha kigyo: Sono igi to jittai' (Biotechnology-related startup firms in Japan: Lessons from a survey study), *NISTEP Discussion Paper* No. 22, Tokyo: National Institute of Science and Technology Policy.
Odagiri, Hiroyuki, Tadahisa Koga and Kenta Nakamura (2002), 'Baio tekunoroji kenkyu kaihatsu to kigyo no kyokai: Kenkyu teikei, gijutsu dounyu, autosousing, kaigai kenkyu ni kansuru chousa houkoku' (Biotechnology R&D and the boundaries of the firm: results from a survey study on R&D alliance, technology acquisition, outsourcing, and overseas R&D), *NISTEP Research Material* No. 90, Tokyo: National Institute of Science and Technology Policy.

Peck, Merton J. and Shuji Tamura (1976), 'Technology', in Hugh Patrick and Henry Rosovsky (eds), *Asia's New Giant*, Washington, DC: The Brookings Institution, pp. 525–85.
Rosenberg, Nathan and Richard R. Nelson (1994), 'American universities and technical advance in industry', *Research Policy*, **23**, 323–48.

10. National innovation systems and India's IT capability: are there any lessons for ASEAN newcomers?

Nagesh Kumar and K.J. Joseph

1. INTRODUCTION

The Information and Communication Technologies (ICTs) and the associated innovations are considered to be instrumental in bringing about wide-ranging socio-economic transformation of the developing world during the 21st century. No wonder today there is hardly any country in the developing world which has not initiated policy measures and institutional interventions to harness the powers of new technology for development. The focus of these initiatives has involved among other things developing a domestic ICT production base on the one hand and facilitating the diffusion of the new technology into different sectors of the economy and society including governance. The new ASEAN countries such as Cambodia, Laos, Myanmar and Vietnam, though lagging behind their counterparts in ASEAN, have not been left behind as manifested in the series of policy measures initiated by them during the last decade. One of the distinguishing characteristics of the new general purpose technology, unlike its predecessors, is that a developing country such as India is acclaimed for her competence. Hence an obvious issue before the new ASEAN countries aiming at harnessing the new technology for development arises; are there any lessons for these countries from India, which is known for its IT success? The present chapter seeks an answer to this question.

India's ICT capabilities and its presence in the world market for ICT software and services are well documented (D'Costa, 2003; Heeks, 1996; Arora et al., 2001; Arora and Athreya, 2002; Joseph, 2002; Joseph and Harilal, 2001; Kumar, 2001a). The ICT software and service sector has sustained an annual compound growth rate of over 45 per cent during the last decade which has been unprecedented in any of the sectors of the Indian economy. As a result the ICT sector contributed about 22 per cent of total export earning and provided employment to over one million in 2004. India exports ICT software and services to about 133 countries and over 300 Fortune companies outsource ICT services from Indian companies.[1] What is more, the majority of the ICT firms in the world with CMM-level 5 certification are from India. While some studies have shown that Indian firms by and large operate at the low end of the value chain (Arora et al., 2001) and have lagged behind in innovative effort (Parthasarathi and Joseph, 2002) another study (Joseph and Abraham, 2005) estimating the firm-level technological competence index has shown that Indian firms are building up technological competence. Similar conclusions were drawn by Kumar (2001a) which explored the issue by analysing the value-added per employee, profitability and net foreign exchange outflow. Since the focus of this study is to draw lessons from India for ASEAN newcomers one needs to have a better understanding of the underlying forces that brought about India's IT success. There are a number of studies (Arora et al., 2001; Singh, 2003), which tend to argue that India's ICT success has been an outcome of the free play of the market and of benign state neglect. Such conclusions not only conceal more than they reveal but are of little relevance for other countries because there are many countries in the developing world that have assigned a prime role to the market yet their success in the IT sector has been limited. Hence the present study makes an attempt towards exploring the underlying forces of India's ICT success using the framework of a national system of innovation.

This chapter is organized in the following sections: the second section analyses the role of different actors in India's national innovation system which contributed to the growth of India's ICT sector. The third section examines the present state of ICT production and use as well as the present state of human capital availability in the new ASEAN countries (Cambodia, Laos, Myanmar and Vietnam) followed by the last section, which presents certain concluding observations along with lessons for the ASEAN newcomers towards developing an ICT base.

2. THE NATIONAL INNOVATION SYSTEM AND INDIA'S IT DEVELOPMENT

The concept of a System of Innovation introduced by Lundvall (1985) has been further articulated analytically and empirically into the national system of innovation (NIS) in the hands of Freeman (1987). With further contributions by Freeman (1988, 1995), Nelson (1988, 1993) and Lundvall (1988, 1992) the NIS has emerged as a framework for the growing body of literature that addresses the process of innovation both at the national, regional and even sectoral level. While the earlier literature has been focusing mostly on the developed countries, with limited empirical content, of late there have been a number of studies focusing on developing countries and providing empirical content. In its broad sense the NIS refers to the national network of institutions, both public and private, and the policy initiatives for the development and diffusion of various technologies (Freeman, 1987; Nelson, 1993; Lundvall, 1992). The NIS in India has evolved over time and has been instrumental in the creation of an extensive infrastructure base for the development of innovative and skill-intensive activities such as ICT. This, *inter alia*, includes one of the largest and most expanding mass of technically trained manpower, a network of centres of international reputation in specific sciences such as the Indian Institute of Science, the Indian Institute of Technologies (IITs) and national laboratories and a number of Software Technology Parks to facilitate the export of ICT software and services. What follows is a brief account of the various state initiatives towards the development of the ICT and software sector in India.

Institutional Infrastructure and Policy Initiatives[2]

As early as in the mid-1960s the government recognized the critical importance of the electronics industry and computing for national development in view of their 'pervasive' applications, and has consciously strived to build a local institutional infrastructure for development of local capabilities (India, Electronics Commission, 1975). As will be seen below, these early initiatives have provided a base for rapid development of the IT software industry in the 1980s and 1990s.

In contrast to the general perceptions the importance of promoting software development, particularly for export, had been recognized by the erstwhile Department of Electronics and suitable policies and programmes were put in place as far back as 1972 (Parthasarathi and Joseph, 2002). In a period when very high tariff and non-tariff barriers were the rule, the import of computer systems on a custom duty-free basis and without reference to indigenous angle clearance was permitted for software export. Moreover in a period when there were a series of restrictions on FDI, 100 per cent foreign-owned companies were permitted to set up software export operations provided they located in the Santacruz Electronics Export Processing Zone (India, Department of Electronics, 1972).

A series of government committees and policy measures (see Table 10.1) have contributed to the evolution of the NIS in the IT sector. The early initiatives include the Bhabha Committee of 1963, the Electronics Committee chaired by Dr V.A. Sarabhai in 1966 and the National Conference on

Table 10.1 Milestones in electronics policy

Milestones	Remarks
1. Bhabha Committee (1966) Report	Recommended development of an integrated electronics sector to achieve self reliance with minimal recourse to foreign capital and dominant role to public and small-scale sector.
2. Formation of Department of Electronics (1970)	The Department was endowed with the responsibility for developing an electronics industry in the country.
3. Formation of Electronics Commission (1971)	This was mainly the policy formulating body relating to the electronics industry in the country.
4. Sondhi Committee (1979)	Recommended dismantling of controls in general and MRTP and FERA in particular.
5. Menon Committee (1979)	Recommended liberalization of import of foreign capital and technology and duty free import of capital equipment.
6. Components Policy (1981)	Delicensing of component manufacture except for MRTP and FERA companies. Provision of 74% foreign equity to companies in high tech. areas. No clearance required under section 21 and 22 of MRTP Act except for LSI and VLSI circuits. General reduction in duty on components and liberal import of capital goods for component manufacture.
7. Telecommunication Policy (1984)	Telecommunication equipment manufacture was opened to private sector.
8. Computer Policy (1984)	All Indian companies, including FERA, were allowed to enter all segments of the computer industry with no restriction on capacity. Most of the components needed were put under OGL to facilitate import.
9. Integrated Policy (1985)	Dereserved certain components of small-scale sector. Introduced broadbanding and liberal approach towards foreign companies even with more than 40% equity in high technology areas.
10. Computer Software Policy (1986)	Reduction in the import duty on all imports meant for software exports and no duty for

Table 10.1 (continued)

Milestones	Remarks
	100% export. Provision of special financing schemes and permission for foreign companies (with more than 40% equity) in 100% export projects.
11. National Taskforce on ICT (1998)	Made 104 recommendations on software and 87 on hardware development in the country.
12. Telecommunication Policy (1994, 1999)	Opening up the Telecommunication Services for the private sector.
13. Formation of MIT (1999)	Brought together different actors involved in IT to form a separate Ministry of Information Technology.

Source: Adapted from Joseph (1997).

Electronics of March 1970. As a follow-up of their recommendations a separate Department of Electronics (DoE) was set up to coordinate and implement policies for development of electronics industries including computer software in 1970. In 1971 the government constituted the Electronics Commission as a policy formulation body with a heavy emphasis on R&D and technology development. In 1973 the Technology Development Council was set up to assist the Electronics Commission on the recommendation of the National Seminar on R&D Policy in Electronics (India, Electronics Commission, 1975). The DoE has spearheaded a number of programmes of human resource development for software engineers, a technological and communication infrastructure for software development and other facilitating measures over the past three decades in tune with the recommendations of the Electronics Commission's perspective plan.

The Computer Policy of 1984 gave a thrust to software development by underlining the need for institutional and policy support on a number of fronts. The policy, for example, called for the setting up of a separate Software Development Promotion Agency (SDPA) under the Department of Electronics (DoE). Imports of inputs needed for software development were made more liberal. However, the policy also emphasized that

> Effective software export promotion on a sustained basis can be effective in the long run only if it is planned as a part of an overall software promotion scheme covering both export and internal requirements including import substitution. Also, planning for software development is integrally connected with the

plan for hardware development and system engineering (India, Department of Electronics, 1985).

The accelerated growth of the computer industry following the above policy posed numerous problems for the software activities, calling for a rationalization of the policy for import and manufacture of software, and using this base for promoting software exports. At the same time world trade in computers was expected to be of the order of US $100 billion by 1990 wherein more than half was estimated as software. The Seventh Plan (1985–90) had a software export target of US $300 million, accounting for about 0.6 per cent of the world trade in software. Against this background an explicit software policy was announced in 1986, and software was identified as one of the key sectors in India's agenda for export promotion. The policy underlined the importance of an integrated development of software for the domestic and export markets (India, Department of Electronics, 1986).

The policy emphasized the need for simplifying the existing procedures pertaining to all the aspects of software development and production for both domestic and export markets and provided various commercial incentives to software firms, such as tax holidays, tax exemption on the income from software exports, export subsidies, and duty free import of hardware and software for 100 per cent export purposes.

With the initiation of economic reforms in the early 1990s an assessment was made by the finance ministry that apart from the general orientation of all industries towards export markets, India's comparative advantage was in software and not in hardware. Therefore a major thrust was consciously given to software exports. Accordingly new policy measures have been initiated which *inter alia* included: removal of entry barriers for foreign companies; removal of restrictions on foreign technology transfers; participation of the private sector in policy making; provisions to finance software development through equity and venture capital; measures to make available faster and cheaper data communication facilities; and reduction and rationalization of taxes, duties and tariffs and so on[3] (Narayana Murthy, 2000).

Along with the policy measures initiated by the National Government, various state governments also enacted IT policies with a view to promote ICT growth in the respective states. As of now 18 state governments have enacted such policies. These policies generally focus on the key issues of infrastructure, electronic governance, IT education and providing a facilitating environment for increasing IT proliferation in the respective states.[4]

Recognizing the potential of IT-related industries and software for India's development, the Prime Minister appointed a National Taskforce on Information Technology and Software Development (NTITSD) in May 1998 under the chairmanship of the Deputy Chairman, Planning Commission. NTITSD submitted its report outlining a National IT Plan comprising 108 recommendations for software and 87 recommendations for hardware (India, NTITSD, 1998). These recommendations have since been notified by the Government in the Gazette of India dated 25 July 1998. NTITSD has set before the country an ambitious target of $50 billion software export by 2008. The DoE was upgraded into a fully fledged Ministry of Information Technology (MIT) in October 1999 to coordinate the promotional role of the government in the industry.[5]

Supply of Trained Manpower for Software Development

The National Conference on Electronics in 1970 had estimated the need for about 300 M.Techs and 50 Ph.D.s in computer science and technology and had recommended launching of specialized Masters level programmes at the IITs and other major institutions. In addition it made proficiency in computer programming mandatory for undergraduates of IITs and science postgraduates of all major universities in the country (India, Electronics Commission, 1975). As a follow-up of these recommendations M.Tech (2-year postgraduate) and B.Tech (4-year graduate) courses in computer science were started in 1974 and 1977 respectively with DoE support at the IITs. In 1982 two new courses, viz. a 3-year Master of Computer Applications and a Diploma of Computer Applications were started, besides expansion of M.Tech/B.Tech courses as a follow-up of the Rajaraman Committee of 1978. These facilities were further expanded, and new polytechnic diplomas were started in 1984 further to the Computer Manpower Development Programme launched in 1983. In 1984 the Sampath Committee reviewed the training needs and in 1985 a Standing Committee on Computer Education was set up to plan further actions. The new courses, introduced under the Computer Manpower Development Programme supported by DoE at about 400 institutions, had produced some 15 000 software personnel by 1996 (Heeks, 1996). The DoE's support has not been restricted to financial grants, but has also involved curricula development. In addition to the courses started at the educational institutions a number of enterprises and other institutions promoted by the DoE have also been providing training in software development. These include NCST and C-DAC, running advanced software engineering courses and CMC Ltd, ETTDC, NIC, running routine software application training.

Besides these, the government has permitted private investment in IT training since the early 1980s.[6] These privately run centres offer diploma courses varying in duration ranging from short-term specialized courses to longer-term basic courses. However, the quality of the training imparted by these institutions has been uneven. The DoE has stepped in to provide accreditation of the courses as a step towards their standardization. A scheme called DOEACC was started in 1990 jointly with the All India Council of Technical Education (AICTE) to provide accreditation to a specified level of courses viz., O-foundation course, A-Advanced Diploma, B-MCA Level, C-M.Tech Level. By January 2004 a total of 850 institutes had been accredited by the DOEACC Society. The Society conducts examinations for all the four levels twice a year and grants certificates/diplomas (India, Department of Information Technology, 2004).

The demand for software personnel, and especially engineering graduates, has grown rapidly since the mid-1990s due to the expansion of the software development activity in India as well as the growing brain drain. In view of this, easing the supply of IT professionals has been one of the challenges faced by the country. In a survey conducted during the late 1990s, 57 per cent of the firms interviewed indicated manpower and skills shortage as the major problem (Arora et al., 2001). The NTITSD has made a number of recommendations dealing with augmenting the quantity and quantity of trained manpower for the software industry. In tune with these recommendations the capacity of the higher education system in engineering in the country has been expanded in addition to the setting up of new institutions.[7]

Infrastructure for Technology Development in Software

The DoE, as the government arm, has played an instrumental role in creating the necessary infrastructure for the development of the software industry. In the 1970s the DoE set up Regional Computer Centres run like public utilities attached to educational institutions that were following the recommendations of the National Conference on Electronics of 1970. These centres were set up in Bombay, Delhi, Bangalore, Calcutta, Kanpur and Hyderabad. A Computer Maintenance Corporation (CMC) was set up in 1976 for maintaining these and other computer centres. Over time the CMC has grown into a full fledged software company with a strong R&D capability. The National Informatics Centre (NIC) was set up to facilitate automation and networking of government offices at the centre, state and district levels.

Since the late 1980s the DoE has concentrated on providing a data communication and networking infrastructure to the educational and research community and to the software industry. This infrastructure has played a critical role in the development of the industry in the 1990s. The Education and Research Network (ERNET) project was initiated in 1986 with participation of NCST Mumbai, IISc Bangalore, five IITs, and support of the

DoE and the UNDP, with the objective of enhancing the national capability in the area of computer communication by progressively setting up a nationwide computer network for the education and research community. ERNET has evolved into a separate institution now providing networking services to over 80 000 users in 750 academic and research institutions with its dedicated satellite data transfer backbone.

A notable institutional intervention has been the establishment of Software Technology Parks[8] (STP) to provide the necessary infrastructure for software export. The first ones to come into being were those at Bangalore, Pune and Bhubaneshwar in August, October and December 1990 respectively. In 1991 four more STPs were set up by the DoE at Noida, Gandhinagar, Trivandrum and Hyderabad.[9] As of now there are 39 Software Technology Parks set up in different parts of the country and they play a significant role in the export of software from the country. The infrastructure facilities available in these STPs include, among other things, modern computers and a communication network. The STPs also envisage a transparent policy environment such as single window clearance, and a package of concessions such as tax holidays, no value addition norms, exemption of export profit from income tax, and permission to sell in the domestic market. The total number of units registered with the STPs increased from 164 in 1991 to 5582 in 1999 and accounted for about 68 per cent of India's IT exports (see Table 10.2). By 2002–03 the number of units increased to 7000 and they accounted for 80 per cent of the software exports (India, Department of Information Technology, 2004).

Table 10.2 Trend in IT export from units registered with STPs

Year	Number of units registered with STPs	Total exports from India ($ million)	Share of STP units in total export
1991–92	164	164	NA
1992–93	227	225	8
1993–94	269	330	12
1994–95	364	485	16
1995–96	521	734	29
1996–97	667	1085	46
1997–98	844	1750	54
1998–99	1196	2650	58
1999–2000	5582	3900	68

Source: Joseph (2002).

In June 2000 a new STP was set up in Silicon Valley, composed of a Business Support Centre and an India Infotech Centre with a view to facilitate software export by small and medium firms to the US.[10] The centre also fosters business relationships by providing access to financial institutions in the US, venture capital funds and specialized trade bodies to promote partnerships and strategic alliances between the US and Indian ICT software and service companies.

R&D Capability Building

The DoE has put heavy emphasis on R&D activity relating to, among other fields, the development of computer software by supporting R&D in the area at different institutions such as TIFR, IITs, IISc, select universities (such as Jadavpur University), ISI, and CSIR Laboratories since the early 1970s. The Technology Development Council has been supporting R&D projects since its inception in 1973 (India, MIT, 2000a). These programmes of technology development have led to a build-up of capabilities and have provided experienced manpower for the rapid development of the industry. For instance the capabilities built in the process of early work on data communication at TIFR started in the late 1970s and anchored at the DoE-supported National Centre for Software Technology (NCST) set up in Bombay in 1984, proved instrumental for the development of countrywide networks and for the Internet in the country in the 1990s. The National Aeronautical Laboratory (NAL) also developed a supercomputer, Flosolver. The government S&T agencies have set up a parallel Supercomputer Education and Research Centre (SERC) and Department of Computer Science and Automation at IISc, which provides high end expertise and manpower to the industry in software. Besides NCST, DoE also set up another institution for technology development in the 1980s, the Centre for Development of Advanced Computing (C-DAC). C-DAC has developed India's first supercomputer, Param and has developed software for Indian languages' script. The Electronics Research and Development Centre (ER&DC) is another new R&D institution set up by the DoE. The ER&DC has research facilities at Thiruvanathapuram, Calcutta and Noida near Delhi. The government has also stimulated and supported R&D activity of industry through tax incentives and direct funding by the DoE.

Procurement Policies and other Promotional Measures

The government has also helped in the development of the software industry by generating large and complex assignments that have given confidence to the local firms. These include automation of railways reservation and bank automation, among others (Heeks, 1996 and Parthasarathi and Joseph, 2002). The government also assisted firms in standardization and quality control, for instance in their effort to obtain ISO 9000 certification. The small and medium enterprises have been assisted in their overseas expansion by subsidies for their participation in industry fairs such as CebIT, venture capital and tax breaks for export activity.

Financial Support: the Role of the Exim Bank of India

The Exim Bank, which is fully owned by the Government of India, has had a number of schemes to help promote the software industry from 1986 onwards. To list a few: the Exim Bank has entered into a memorandum of understanding in 1998 with NASSCOM for promoting software service export. Under the 1986 software export policy, the Exim Bank operated a special window for export-oriented software companies. It also supported market research quality satisfaction, a buyers' visit to India, participation in specialized fairs and so on. The bank also provided term loans to Indian companies to finance their equity contribution in overseas ventures. To support the industry move up the value chain towards products, the bank had a special programme to finance software product development.[11]

Measures to Address Software Piracy

Until recently the weak copyright regime in the country facilitated the proliferation of software piracy, which in turn acted as a disincentive for firms to develop software products. The magnitude of the problem has been illustrated by an estimate from Lotus Development Corporation that in the early 1990s, of 150 000 copies of Lotus 1-2-3 in use, 140 000 were pirated (Schware, 1992). To address this problem the government initiated a series of measures. The copyright of computer software has been protected under the provisions of the Indian Copyright Act of 1957. Major changes were made to the Copyright Law in 1994. Accordingly it is illegal to make or distribute copies of copyrighted software and therefore punishable. Section 63 B of the Act stipulates a minimum jail term of seven days, extendable to up to three years. The Act further provides for a fine ranging from Rs 0.05 million to Rs 0.2 million. In addition the government, in cooperation with the NASSCOM, conducts regular anti-piracy raids to discourage software piracy. As a result the piracy rate in the country has come down from 89 per cent in 1993 to 60 per cent in 1997 (NASSCOM, 1999).

Role of Private Sector and Industry Associations

It may be myopic to attribute the observed dynamism of the industry entirely to the initiatives made by the state (Joseph, 2002). While the state initiatives laid the foundation for faster growth, the industry associations,[12] particularly the National Association of Software and Service Companies (NASSCOM), played an important role. In addition to lobbying at the Central and State government levels, the NASSCOM also played a key role in projecting India's image in the world IT market. For example, in 1993 NASSCOM appointed a full-time lobbying firm in Washington. It facilitated the participation of Indian firms in a large number of international IT exhibitions, projecting India's capabilities in the sphere of IT. The role that NASSCOM played in getting the visa rules relaxed by the developed countries, especially USA, is well known. Also in 1994, NASSCOM initiated anti-piracy initiatives in India when IPR was becoming a major issue in Indo–US relations. It took up the campaign against software piracy and conducted a number of well publicized raids.[13]

NIS and Patterns of Spatial Agglomeration of the IT Sector

IT and software development in different parts of the world is characterized by a strong tendency towards clustering because of agglomeration economies. In India the software industry developed initially in Mumbai (formerly Bombay). Subsequently, especially after the entry of Texas Instruments in the mid-1980s, Bangalore emerged as a centre of software industry development. Besides Bangalore and Mumbai, Delhi along with its suburbs, namely Noida and Gurgaon, has emerged as the third most popular concentration of software units (see Table 10.3). Hyderabad and Chennai have also emerged as alternative locations in the South after the saturation of Bangalore in terms of available infrastructure and scarcity of space. The state government's promotional role has also contributed to the emergence of Hyderabad as the fourth most important centre of concentration of software companies. The top five cities together account for 80.5 per cent of the 600 top companies. But other cities such as Calcutta, Pune, Thiruvanantpuram, Ahmedabad and Bhubaneswar are coming up as increasingly popular locations. One important facilitating factor has been the availability of high-speed data communication links and facilities provided in the Software Technology Parks (STPs).

The pattern of concentration of software development industry in and around select cities does corroborate the key importance of NIS for the activity. Table 10.4 shows that the cities of high concentration of software development, Bombay, Bangalore, Delhi and Hyderabad, have shared a

Table 10.3 Patterns of clustering of Top 600 software companies

City	Number of company headquarters located	Percentage share
Mumbai	131	21.83
Bangalore	122	20.33
Delhi and around	111	18.50
Hyderabad	64	10.67
Chennai	55	9.16
Calcutta	25	4.16
Pune	23	3.83
Thiruvanathapuram	14	2.33
Others	55	9.16

Source: Adapted from NASSCOM (2000b).

disproportionate portion of national innovative infrastructure, skills base, and other resources for technology development. Because of significant agglomeration economies present in skills and knowledge-intensive activities such as software development, this disproportionate share of national innovative infrastructure has crowded in the software development activity to these cities.

The presence of higher educational enterprises in engineering and technology thus ensuring the supply of engineering manpower, centres of excellence, a nucleus of local software industry companies and clustering of high technology enterprises together with telecommunication infrastructure in the form of Software Technology Parks in these cities has facilitated agglomeration of software industry in these cities.

Are there any lessons for the new ASEAN countries aspiring to develop an ICT base? In seeking an answer to this issue it is important to have a fair idea of the state of the ICT sector in the new ASEAN countries. The next section seeks to draw the broad contours of the ICT sector in the new ASEAN countries.

3. NEW TECHNOLOGY IN THE NEW ASEAN COUNTRIES[14]

Information Technology Use

Let us begin with an examination of the present state of ICT infrastructure and ICT use in new ASEAN countries in comparison with the old ASEAN

Table 10.4 Illustrative S&T infrastructure in four IT clusters in India

Type of NIS infrastructure	Bombay	Bangalore	Delhi and around	Hyderabad
Institutions of Higher Technical Education and Excellence	IIT-B; Bombay University; SNDT Women's University; Bajaj Institute of Management and several other engineering and management institutes	IISc; University Visvesraya College of Engineering; SKSJ Technology Institute; and 28 private engineering colleges; Indian Institute of Management-B	IIT-D; Delhi College of Engineering; Delhi University Department of Computer Sciences, Roorkee University of Engineering (within 200 kms); J.N. University; Jamia Milia Islamia Engineering College; FMS; IIFT; plus several private institutions.	J.N. Technological University; Hyderabad University; Osmania University; Kakatiya University;
Public Funded Research Laboratories and Institutions	TIFR; NCST; BARC; UDCT; SAMEER	ISRO; NAL, CMTI; Electronics and Radar Development Establishment; Aeronautical Development Establishment; Gas Turbine Research Establishment; Centre for Aeronautical Systems Studies and Analysis; ER&DCI	NIC; NPL; Institute for Systems Studies and Analysis; SPL; C-DOT	National Remote Sensing Agency; RRL; NGRI; IICT; Defence Electronic Research Laboratory; DRDL

Local Software Champions	TCS; PCS; Tata Infotech; Mastek; L&T ITL; APTECH; COSL; Datamatics; Silverline	Infosys Technologies Ltd.; WIPRO Information Technologies	HCL Technologies; NIIT Ltd. CMC Ltd.	Satyam Computer Services Ltd.
High Speed Data Communication Facilities	Earth Station of STPI	Earth Station of STPI	Earth Station of STPI	Earth Station of STPI
High Technology Enterprises, mostly public sector	L&T; Godrej; Tata group and a large number of engineering and electronics enterprises	ITI; BEL; HAL	Central Electronics Ltd.; NRDC; EIL; RITES; ETTDC; ET&T; TCIL	ECIL; BHEL

Source: Kumar (2001b).

member countries and also with other low-income countries (see Table 10.5). With respect to IT infrastructure it may be noted that the number of fixed telephone lines per 1000 people even in the largest city of the new ASEAN member countries is lower than the national average for the old ASEAN countries. When it comes to mobile telephones, computers and the Internet, the divide between the old and new ASEAN countries is much wider. In general in terms of ICT infrastructure and use, while the old ASEAN member countries are found to be either on a par with or at a higher level than the middle-income countries, the new ASEAN member countries lag not only behind their counterparts but also behind the low-income countries in general. Here Vietnam appears to be an exception. More importantly it has been found that IT use in these countries (with the possible exception of Vietnam) in terms of telecommunication network (fixed or mobile) and Internet use is confined to the urban areas leading to what is called the 'intra-national digital divide'.

To address these issues there have been initiatives at the individual country level and at the regional level. At the country level all the new ASEAN (CLMV) countries have made a series of institutional arrangements and policy measures, which in general aim at building up ICT infrastructure, an ICT production base, human resource development and promoting the use of ICT in different sectors of the economy and society.

Policy Initiatives

While the economies in the new ASEAN countries in general were faced with making the difficult choice of 'investing in Pentium or in Penicillin' they have undertaken a series of bold policy reforms and institutional interventions towards developing an ICT production base and promoting the use of new technology for addressing their development needs. Nonetheless, given the gigantic task at hand and the rocky road which they have to traverse, the destination still remains far away.

In the case of Cambodia the present policy towards IT lays emphasis on promoting IT use in different sectors of the economy and also for promoting e-governance. To achieve this, policy calls for, among others, the development of infrastructure, promoting computer literacy, standardization of the Khmer language in computers and a greater role for the private sector. However, a comprehensive IT policy is yet to be framed and NiDA is responsible for undertaking initiatives to promote information technology in the country.

In Lao PDR as early as in 1996 the Science Technology and Environment Agency (STEA) was given approval by the Prime Minister's office to implement the overall policy for monitoring and controlling information

Table 10.5 Indicators of ICT infrastructure and use in ASEAN countries (2001)

ASEAN member countries	Telephone mainline				Mobile phone per 1000 people	PCs per 1000 people	Internet users (000)	Information and communication expenditure	
	Per 1000 people	In largest city per 1000 people	Waiting time year (2000)	Cost of local call per 3 minutes ($)				% of GDP	per capita $
Old ASEAN									
Brunei									
Indonesia	35	261		0.02	31	11	4000	2.2	17
Malaysia	196		0.7	0.02	314	126.1	6500	6.6	262
Philippines	42	265		0	150	21.7	2000	4.2	41
Singapore	471	471	0	0.02	724	508.3	1500	9.9	2110
Thailand	99	452	1.6	0.07	123	27.8	3536	3.7	76
New ASEAN									
Cambodia	2	19		0.03	17	1.5	10	–	–
Lao PDR	10	65	1.1	0.02	5	3	10		
Myanmar	6	32	5.3	0.01	0	1.1	10	–	–
Vietnam	38	–	–	0.02	15	11.7	1010	6.7	26
Low income	30	130	1.4	0.05	10	6.1	15932	–	–
Low–middle income	93	270	2	0.04	72	21.6	112591	–	
High income	593	–	0	0.08	609	416.3	388888	–	–

Source: Adapted from Joseph and Parayil (2004) based on The World Bank, *World Development Indicators*, 2003.

technology. Given this mandate, STEA proposed a 4-year plan (1996–2000) which dealt with, among other things, developing IT infrastructure (including human capital, IT industry base, communication network), promoting IT application (in government, business and industries as well as the economy at large) and devising policies for promoting ICT development which also included policies relating to promotion of FDI in IT. Achievements by the year 2000 appear to have fallen short of the targets on account of various reasons which *inter alia* included lack of resources and lack of coordination among different agencies involved. Today the government is in the process of making an integrated IT policy for the country, and five working groups have been appointed for this purpose.

In the case of Myanmar an examination of the computer science law which governs the use of computers and the Internet in the country and IT production to a great extent revealed that though the objectives were highly laudable, the series of restrictions has had the effect of mitigating the positive effect of various initiatives that the government has undertaken to promote the use of IT for development. Hence the study highlighted positive outcomes of phasing out various restrictions on the use of information technology in general and the Internet in particular. In the case of Vietnam, IT production and use is governed mainly by an eminent Vision Document concerning new technology prepared by the Party, and subsequently by an Action Plan developed by the government. Both these documents are unambiguous in their approach, exhaustive in their coverage and ambitious but realistic in their targets (Government of Vietnam, 2001; 2002).

On the whole the policy initiatives and institutional interventions tend to suggest that there is a high degree of awareness among the policy-makers of new ASEAN countries regarding the need for harnessing new technology for development. The initiatives notwithstanding, the present state of IT production is in its infancy in the new ASEAN countries as is evident from the forthcoming discussion.

IT Production

Conceptually IT production could be divided broadly into ICT goods and IT services. Each of these broad product groups comprises a wide range of goods and services with varying levels of entry barriers, and incorporates varying levels of technology. Hence the central issue is to find ways and means of evolving a national system of innovation that will enable the new ASEAN countries to enter into IT production.

In Cambodia as of now there is only one firm engaged in the production of IT goods in the country. The company began its operations in 1992 as a joint venture and became 100 per cent foreign owned by 2000. In addition to producing television sets and VCRs it also had the dealership for leading computer companies. In the initial years the company used to employ more than 70 people. Over the years, various reasons such as high import duty and VAT (import duty plus VAT put together are about 26.5 per cent) leading to large scale smuggling and poor infrastructure, have led to a situation wherein the firm was forced to scale down its operations in the country. Today the company employs only about 20 people in its IT factory and focuses more on computer and software service-related activities. Given the fact that the present level of IT production in the country is negligible, the entire domestic demand is being met entirely through imports. This has had the effect of adversely affecting the overall trade balance of the country on the one hand and forgoing the potential opportunities for employment and income generation in the country through ICT production on the other. The situation is not much different in Lao.

In Myanmar, too, the IT goods production in the country is at a low level. While there are two local producers of computers, almost 70–80 per cent is accounted for by the so-called 'grey market'. While looking at the major areas of operations of the members of the Myanmar Computer Industry Association it was discerned that hardly any are engaged in computer production. Most of them are engaged in hardware/software supply and service, with limited software production. In consumer electronics, MNCs such as Toshiba and Daewoo have operations in the country. In the public sector Myanmar Machine Tool and Electrical Industries (MTEI) has one electrical and electronics factory located at South Dagon with the following product lines – fluorescent lamps and incandescent bulbs, electric rice cookers, electric irons, electric hot plates and dry cells.

While the production base in IT goods appears to be limited, Myanmar has already initiated some bold steps towards creating software/service production in the country. This is manifested in the setting up of Software Technology Parks at the instance of Myanmar ICT Development Corporation (a consortium of 50 private companies) with the active support and cooperation from the Government of Myanmar, with the necessary communication infrastructure being provided by the publicly owned Bagan Cybertech. The project, set up in the Hline University Campus with a total investment of about Ks 2.5 billion, was initiated in March 2001. The first phase was completed within a very short span of about 10 months and the park was inaugurated in January 2002. The first phase of the project, covering a developed area of about 11 acres, has 32 rooms (100ft×50ft). As of June 2003 the occupancy rate is 100 per cent. The park has been able to attract two foreign companies; it is also the home for an incubation centre for promising local software programmers and the Japan–Myanmar e-learning Centre. The activities in the park include; software development, human resource development, national level projects, data processing services, consultancy services and it provides employment for about 700 people. On the whole the technology park experiment is a testimony to the positive outcome of public–private participation. The government is making an effort to replicate this success by setting up another park at Mandalay.[15]

Vietnam is perhaps the only country in the region having various explicit policy resolutions for promoting IT production. In addition, Software Technology Parks have been set up in HCMC, Da Nanag and Hanoi to attract investment into the software development and export sector and to achieve the target of US $500 million worth of software production, including an export of US $200 million by 2005. It is understood that Vietnam's ICT industry (covering software, hardware, network services and systems integration) was about $337 million in 2000. It has been estimated to reach US $417 million in 2002 and $690 million by 2003. Surveys conducted by the International Data Group indicate that the IT market in Vietnam has been recording an annual growth rate of 25 per cent, and that this rate of growth is expected to continue through 2010.[16] Another survey by *PC World* has shown that by the end of 2002 there were about 260 software companies, employing about 5000 specialists, and the recorded growth rates in sales were of the order of 23 per cent. On average these firms employ 20 people and the turnover per person for the year 2000–01 was $6400 and $11 000 for software companies and software outsourcing companies respectively.

When it comes to IT goods production in Vietnam, it is understood that about 0.4 million computers have been produced in the year 2002 and the industry has been recording a growth rate of 30–40 per cent in recent years. About 60 per cent of this has been accounted for by the non-branded sector. With a view to gain a better understanding of the growth of IT goods sectors we have analysed the data on the production of IT goods (TV, radio communication equipment – ISIC 32; computing machines – ISIC 30) and compared the share as well as growth of the industrial sector as a whole and other industries (2-digit level) and reached the following conclusions.

The industry segment, Office accounting and computing machinery recorded a very high growth rate of over 72 per cent during 1995–2002, albeit from a very low base, which is in tune with the observations made by the industry associations. When it comes to the industry segment, Radio, TV and communications equipment, the recorded growth rate has been a little over 15 per cent, only marginally higher than the growth rate recorded by the manufacturing sector as a whole (14.13 per cent). More importantly there have been a number of industries which recorded a higher growth rate than this industry. If we look at the IT sector as a whole, notwithstanding the importance attached to the IT sector, its share in the manufacturing output increased only by about 0.5 per cent (from 2.5 per cent in 1995 to 3.0 per cent in 2002) and the recorded growth rate has been only of the order of 17.6 per cent. Thus it appears that there is the need for greater focus on developing the IT goods sector, which could not only be a source of income and employment in the domestic economy but also a source of foreign exchange.

Human Capital Constraint for IT Use and Production

If the experience of India is any indication, the availability of human capital, which plays a dual role as both producer and user, is a necessary condition for developing an ICT production and user base. Against this background, what follows is a brief discussion of the institutional arrangement for creating IT human capital in the new ASEAN.

The higher education system in Cambodia is comprised of five public universities, three semi-independent specialized institutes of faculties and six recognized private higher education institutes. It is estimated that all these institutes together turn out a total of 25 000 students. Of these only two, the Royal University of Phnom Penh and Norton University, offer degree programmes in Information Technology-related fields. The total number of IT graduates from these institutes is estimated at about 200 to 300 per year. Recently CISCO, at the instance of NiDA, also started undertaking IT training. Also a number of NGOs are found to be involved in capacity building and IT training along with a large number of private training centres offering short-term courses in IT. Thus there are multiple actors involved in the generation of human capital for IT.

In Laos, computer and IT-related education[17] is provided not only by the National University of Lao PDR, but also by the private sector. The computer science programme in the NUL began in the year 1998 under the faculty of Science, Department of Mathematics.[18] The number of graduates is only 29 (UNDP, 2002). The faculty of Engineering and Architecture (FEA) is considered as best equipped with IT facilities in the NUOL system.[19] The main component of these facilities is the Lao–Japan Technical Training Center (LJTTC). The courses offered at the LJTTC are a combination of general application courses; computer-aided engineering courses and a course on network software. LJTTC even offers a course on Internet café set-up and maintenance.

Given the limited IT education facility in the public sector the vacuum is filled at least partly by the private sector. The following private colleges are currently providing IT education, albeit at a very preliminary level. The

Vientiane College, a private Institution with the academic and financial support of the Monash University in Australia, was established in 1992. Other institutions involving foreign investment are the Micro Info Centre (Joint Venture) and Lao American College, another joint venture. The Lao American College has established working relations with the National University of Lao PDR, City University of Washington State, USA, the Ohio University, USA and the Bangkok University, Thailand. In addition to these educational institutions with foreign investment there are three local initiatives, Rattana Business Administration College, Com Centre and PVK Computer Center. Parallel to these educational institutions, there are a number of computer dealers who provide short-term training in computer operations.

In Myanmar the government has taken proactive steps in promoting education in science and technology in general and IT in particular. Major institutions of higher learning have been kept under the administrative control of the S&T ministry to enable them to be given better focus and attention. These are:

a. Yangon University of Technology
b. Mandalay University of Technology
c. Pyay Technology University
d. Yangon University of computer studies and technology
e. Mandalay University of computer studies and technology

It could be observed that out of the five institutions, two of them are specialized universities focusing exclusively on IT (d and e above). The Yangon University of Computer Science and Technology, the leading university in IT education, provides 12 courses in IT education (Kyaw, 2002).

In addition to these universities there are 24 government colleges and 80 university colleges, and all of them have IT departments and offer diplomas or degrees in IT education. All these universities have access to computers with LAN. By 2003 the number of graduates with IT qualifications was as high as 3000 and the government has set a target of reaching 25000 by 2010.

Vietnam has a target of training over 50 000 IT specialists at different levels, of whom 25 000 are high-level programmers fluent in English. For IT training there are 20 IT faculties in Vietnam's various universities and colleges, 45 technical colleges with IT programmes, and about 67 vocational schools with IT subjects. It is estimated that at present there are approximately 20 000 IT professionals (with bachelor's degrees) in Vietnam, with 10 000 working directly in the IT industry on research and development or in education services (USAID, 2001).

There are a number of bilateral agreements to promote IT training in Vietnam. A joint working group on IT and electronics between India and Vietnam has been in existence since 1999. The Prime Minister of India, during his visit to Vietnam in January 2001, announced a grant of Rs 100 million for a software and IT training centre in Vietnam. In November 2001 the Government of Vietnam allocated this grant to the Hanoi People's Committee for utilization in a US$ 4.4 million Hanoi IT Transaction Centre project.[20] In November 2001 a Vietnam–Japan e-learning centre was opened in HCMC as part of the cooperation between the two countries in IT. Under a memorandum of understanding signed between the two countries in August 2001, Japan's Ministry of Economy, Trade and Industry had agreed to provide US $150 000 as non-refundable aid to Vietnam for IT courses. The Republic of Korea plans to send about 50 government officials to Vietnam to provide training in the latest IT techniques.

The government has also been successful in attracting investment in the field of IT training. A number of private schools are entering the local Vietnamese market with a specific aim of developing IT professionals. The Royal Melbourne Institute of Technology (RMIT) has opened a school in HCM City with plans to build a large campus near there.[21] APTech, Tata Infotech and NIIT, the three leading private training institutes from India with operations in many countries, are also providing IT-related programmes in Vietnam. In addition to these formal educational institutions there is considerable IT-related training provided by IT-related business associations. The Vietnam Association of Information Processing (VAIP) has been instrumental since the early 1990s in providing IT-related awareness over public TV and in setting up 90 IT training centres throughout Vietnam. These centres issue approximately 1000 certificates a month. In addition, the VAIP sponsors an annual IT Olympiad with participation from each university in an effort to promote student IT learning.

While all goes well with quantity, what matters in a highly-skilled intensive and competitive field like IT and software development, is quality. A study by USAID (2001) quoted the results of another study conducted by the Political and Economic Risk Consultancy Ltd, in which a Human Resource Index for Asian countries was developed. Vietnam ranked low in virtually all categories (for example, none were above 3.50 on a scale of 0–10), with the high-tech proficiency ranked 2.50 – the lowest of all countries included in the survey! English proficiency was also ranked the lowest of any country (including China), with a ranking of 2.62 out of 10.

4. CONCLUDING OBSERVATIONS AND LESSONS FOR NEW ASEAN COUNTRIES

Our analysis of the contributory factors towards the development of the ICT sector in India has shown that the National System of Innovation, which evolved over time as an outcome of the policies initiated by the government, played a key role. These included development of a system of higher education in engineering and technical disciplines, creation of an institutional infrastructure for S&T policy-making and implementation, and building centres of excellence and numerous other institutions for technology development, among other initiatives. We have seen that the new ASEAN economies are also in the process of making various initiatives towards building up the three pillars of ICT – IT infrastructure, human capital and an IT production base. Against the backdrop of this discussion we shall now reflect on the plausible lessons that may be learned by the new ASEAN countries from the Indian experience. Here we must hasten to add that India's National Innovation System has been built up over nearly five decades, and the road ahead for the new ASEAN countries is long as well as rocky. Nonetheless, much could be learned from India, which in turn could facilitate their leapfrogging.

Discussion on the policy initiatives and institutional interventions by the ASEAN newcomers tends to suggest that there is a high degree of awareness among the policy-makers on the need for harnessing new technology for development. While comprehensive ICT policies have already been formulated in Vietnam, other countries such as Cambodia, Laos and Myanmar are yet to come up with a comprehensive ICT policy. The institutional arrangements are also found to vary from country to country. While there is a ministry exclusively for ICT in Vietnam, in other countries, ICT issues are handled either by the Ministry of Science, Technology and Environment or an Independent Agency such as NiDA in Cambodia. There are also instances of more than one agency dealing with ICT-related issues, leaving room for coordination failures. Given the pre-eminent role that the new technology plays today, cutting across different ministries and administrative departments, and the imperative need for evolving a National System of Innovation, it may be worthwhile to have an exclusive Ministry for Information Technology in those countries, which are yet to set up a separate ministry.

Also the policies in almost all the countries seem not to be assigning an appropriate role for the provincial governments in developing and harnessing ICT for development. We have seen that as of now most of the provincial governments in India have their own IT policies to promote the production and use of IT in the respective states. While the role of private and public sectors and the coordinated effort has been underlined in the policies of all the countries, there are other stakeholders, such as the Civil Society Organisations that could play a very constructive role especially in addressing the issue of the 'intra-national digital divide' and harnessing ICT for the rural sector in general and the agricultural sector in particular, the mainstay of all the new ASEAN economies.

Given the fact that affordability is a major issue in promoting the use of IT in the new ASEAN countries, which in turn arises on account of the high price of hardware and software in relation to income levels, there appears to be the need for greater focus on promoting the use of open source software. In this regard there is great potential in cooperating with each other as well as with other developing countries such as India, known for her IT capabilities.

A major issue being confronted by the new ASEAN countries relates to the human capital constraint. It may be argued that the present approach of 'training the trainers' adopted by most of the countries has its obvious limits and underscores the need for targeted measures to attract more investment into the field of IT education and training. While ambitious targets coupled with concerted actions have been made towards developing IT manpower by most of the countries, the focus so far appears to have been on 'quantity', leading to mushrooming of private training institutions. This has the potential danger of creating a pool of 'unemployable' human power. Hence drawing from the Indian experience there appears to be the need for an accreditation system on the one hand and evolving deliberate policies on the other, to nurture strong linkages between academia and industry. In this way teachers are exposed to the real world environment through consultancy and other means, students take up internship with the private sector and the private sector participates in the teaching and development of the curriculum in the academic institutions, ultimately resulting in an overall improvement in the quality of manpower. Here again much could be gained by joining hands with countries like India known for its institutional arrangements conducive to bringing out high quality IT manpower.

Since the present IT production base in the new ASEAN countries is limited, there is an urgent need to devise appropriate policies to facilitate the establishment of an IT production base, perhaps initially focusing on low technology products. While the IT policy in Vietnam and Myanmar lays emphasis on IT production and could claim some success, there is an urgent need for other countries in the region to build an IT production base which needs to be incorporated as an integral part of the IT policy. Here, there is a need for appropriate trade and investment policies and the creation of a facilitating environment such that the new ASEAN countries appear in the radar screen of 'flying geese' and those firms operating in high-cost countries planning relocation. In the sphere of ICT services, again there appear to be real opportunities to enter into some of the relatively less skill-intensive services such as IT-enabled services including medical transcription, call centres, data entry and so on wherein the required skilled labour could be developed in the short run. But such IT-enabled services also call for better communication infrastructure at affordable prices.

In making efforts towards developing an IT production base, it is important to keep in mind the lessons offered by the experience of India. To begin with, the strategy might be to make available a large pool of IT manpower at different levels such that the primary condition for the establishment of an IT goods/service production base is satisfied. Here the strategy needs to be one of pooling together the resources of different actors such as civil society organizations, the private sector and so on. Also the strategy should be not one of spreading the resources thinly across the country, instead the investment needs to be undertaken in such a way as to take advantage of the agglomeration economies. This might be possible through the setting up of Technology Parks wherein office space, a communication infrastructure and other facilities, which are beyond the reach of an individual entrepreneur, are provided along with a 'single window clearance' system. Secondly, such technology parks needs to be close to and have constant interaction with the centres of learning such that mutual learning and domestic technological capability is built up in the long run. Thirdly, there is also the need for conscious efforts towards skill empowerment such that the economy does not get locked into low technology activity, and an upward movement along the skill spectrum is ensured. It needs to be noted that the investment policy of new ASEAN countries, given their commitment towards developing an ICT base, needs to focus on developing a National System of Innovation, which in turn facilitates the creation and strengthening of the basic pillars of ICT, such as IT infrastructure, human capital and an IT production base.

NOTES

1. See in this context, the three-part article on the Indian IT industry by India's IT Minister Arun Shourie (2004a, 2004b, 2004c).
2. For a more detailed discussion see Kumar (2001b).
3. Mention needs to be made of the substantial reduction in the duties and tariffs across the board for components and sub-assemblies, zero duty of software import and zero income tax on profits from software exports.
4. A detailed comparative analysis of the policies initiated by different state governments against the backdrop of the national policies would be highly rewarding, but it falls

beyond the scope of the present chapter and is reserved for future work. For the details of policies enacted by different state governments the interested reader may visit the home page, www.NASSCOM.org.

5. See India, MIT (2000a) and http://www.mit.gov.in for programmes of the Ministry.
6. Dataquest, 31 May 2000, 15 June 2000.
7. See Kumar, 2000, for details.
8. A Software Technology Park (STP) is in all respects similar to a free trade zone exclusively for software. The specific objectives of the STPs are:

   - To establish and manage the infrastructural resources such as data communication facilities, core computer facilities, office space, common amenities, etc.
   - To provide services (import certification, software valuation, project approvals, etc.) to the users who undertake software development for export purposes.
   - To promote development and export of software and software services through technology assessments, market analysis, marketing support, etc.
   - To train professionals and to encourage design and development in the field of software technology and software engineering (Government of India, 1995).

9. In 1991 there was also a policy change as regards the management of the STPs. The earlier autonomous societies for managing each park were dissolved and a new society, called the Software Technology Park of India registered in June 1991, was given the charge of managing all the STPs in the country through individual executives in each of the parks. Under the new scheme the participating companies have the advantage of being fully involved in all decision-making, including fixing of rent, selection of hardware etc. The companies are represented in the executive board which manages the park under the overall supervision of the governing council.
10. 'STPI now opens office at Silicon Valley, USA', *The Economic Times*, New Delhi, Special Supplement on Software Technology Parks of India, 11 June, 2000. Also see, 'Software exports and role of Exim Bank', (mimeo), Exim Bank, Mumbai (undated).
11. Based on discussion with Mr T.C. Venkat Subramanian, Chairman and Managing Director, Export–Import Bank of India.
12. To begin with, there was the Computer Society of India, which is essentially an association of academics and professionals, and did not address many of the issues faced by the industry. Hence a new association called the Manufacturers Association of Information Technology (MAIT) was formed in 1982. This consisted of both hardware and software firms. Later an association, currently known as NASSCOM, was formed to address specific issues being faced by the software and service companies. The Electronics and Software Export Promotion Council, an autonomous body under the MIT, through its various initiatives, also made a significant contribution towards India's IT export growth.
13. For a detailed account of the NASSCOM activities in promoting IT and the role played by the late Mr Dewang Metha, see 'Power lobbying', *Business India*, 19 February to 4 March, 2001.
14. This section draws heavily on Joseph (2004).
15. See for details, 'Billions of Kyats, millions of Dollars spent in developing ICT infrastructure and facilities' opening address given by Secretary-I at the Second Annual Myanmar ICT week.
16. Quoted from www.emich.edu/ict_usa/Vietnam.htm.
17. For a detailed account of higher education in Lao PDR, see John C. Weidman (undated), 'Reform of higher education in the Lao People's Democratic Republic', paper presented at the 1995 Annual Meeting of the Association for the Study of Higher Education, Orlando, Florida, 1 November. Revised version published under the title, 'Lao PDR' (Weidman, 1997).
18. For a detailed account of the IT education in Laos, see UNDP (2002).
19. The UNDP study reported that the FEA intends to offer a course on Computer Engineering Program in the near future.

20. Sudhir Kumar, 'Foreign presence in the field of IT in Vietnam', Indian Embassy in Vietnam (mimeo), undated.
21. The Melbourne-based Royal Melbourne Institute of Technology (RMIT) has announced a US $30 million expansion plan. RMIT, which currently has about 700 students at both graduate and postgraduate level (in IT and other fields) in its facilities in HCMC and other places has a target to reach 10 000 students at its new educational centre in HCMC.

REFERENCES

Arora, A. and S. Athreye (2002), 'The software industry and India's economic development', *Information Economics and Policy*, **14** (2), 253–73.

Arora, A., V.S. Arunachalam, J. Asundi and F. Ronald (2001), 'The Indian software services industry', *Research Policy*, **30** (3), 1267–87.

D'Costa, A.P. (2003), 'Uneven and combined development: understanding India's software exports', *World Development*, **31** (1), 211–26.

Dosi, G., C. Freeman, R. Nelson, G. Silverberg and L. Soete (eds) (1988), *Technical Change and Economic Theory*, London: Pinter.

Edquist, C. (1997), 'Systems of innovation approaches – their emergence and characteristics', in C. Edquist (ed.), *Systems of Innovation: Technologies, Institutions and Organizations*, London: Pinter/Cassell.

Freeman, C. (1987), *Technology Policy and Economic Performance: Lessons from Japan*, London: Pinter.

Freeman, C. (1995), 'The National Innovation Systems in historical perspective', *Cambridge Journal of Economics*, **19** (1).

Government of India (1985), 'New computer policy', *The Gazette of India*, 31 August.

Government of India (1995), *STP Info Data Base*, New Delhi: Department of Electronics.

Government of Vietnam (2001), 'Policies and measures on the acceleration of information technology application and development in Vietnam', Hanoi: Ministry of Science Technology and Environment.

Government of Vietnam (2002), 'Policies and Legal Documents on Information and Telecommunication Technology of Vietnam', Hanoi: The National Steering Committee on Information Technology.

Heeks, R. (1996), *India's Software Industry: State Policy, Liberalization and Industrial Development*, New Delhi, Thousand Oaks, London: Sage Publications.

India, Department of Electronics (1972), 'Annual report', New Delhi: Department of Electronics.

India, Department of Electronics (1985), *New computer Policy*, New Delhi: Departement of Electronics.

India, Department of Electronics (1986), 'Policy on computer software exports, software development and training', New Delhi: Department of Electronics.

India, Department of Information Technology (2004), *Guide to Electronics Industry in India*, New Delhi: Department of Information Technology.

India, Electronics Commission (1975), 'Perspective report on electronics in India', New Delhi: Electronics Commission.

India, MIT (2000a), *Annual Report 1999–2000*, Ministry of Information Technology, Government of India.

India, MIT (2000b), *Action Taken Report of the National Task Force on Information Technology and Software Development*, IT Action Plan: Part I, New Delhi, Ministry of Information Technology.

India, NTITSD (1998), *IT Action Plan* (in three volumes), New Delhi, National Taskforce on Information Technology and Software Development, http://it-taskforce.nic.in/.

India, Planning Commission (1999), *Ninth Five Year Plan 1997–2002: Development Goals, Strategy and Policies*, Volumes 1 and II, New Delhi: Government of India Planning Commission.

Intarakumnerd, P. et al. (2002), 'National innovation systems in less successful developing countries: the case of Thailand', *Research Policy*, **31**, 1445–57.

Joseph, K.J. (1997), *Industry under Economic Liberalization: the Case of Indian Electronics*, New Delhi: Sage Publications.

Joseph K.J. (2002), 'Growth of ICT and ICT for development: realities of the myths of the Indian experience', Discussion Paper No. 2002/78, WIDER-UNU August, http://www.wider.unu.edu/publications/publications.htm.

Joseph, K.J. (2004), 'Development of enabling policies for trade and investment in the IT sector of the greater Mekong subregion', UNESCAP, http://www.unescap.org/tid/projects/gms.asp.

Joseph, K.J. and Abraham Vinoj (2002), 'Moving up or lagging behind in technology? An analysis of India's ICT sector', paper presented at the International Seminar on ICTs and Indian Development – Processes, Prognoses and Policies, organized by the Institute for Human Development, New Delhi and Institute of Social Studies, The Hague, 9–11 November, Bangalore.

Joseph, K.J. and V. Abraham (2005), 'Moving up or lagging behind? An index of technological competence in India's ICT sector', in A. Saith and M. Vijayabaskar (eds), *ICTs and Indian Economic Development*, New Delhi: Sage Publication.

Joseph, K.J. and K.N. Harilal (2001), 'Structure and growth of India's IT exports: implications of an export-oriented growth strategy', *Economic and Political Weekly*, **36** (34), 3263–70.

Joseph K.J. and G. Parayil (2004), 'India–ASEAN Cooperation in Information and Communications Technologies (ICTs): issues and prospects', RIS Discussion Paper No. 70. 2004, New Delhi: Research and Information System for the Non-Aligned and Other Developing Countries.

Kumar, N. (2000), 'Developing countries in international division of labour in software and service industry: lessons from Indian experience', background paper for *World Employment Report 2001*, Geneva: ILO.

Kumar, N. (2001a), 'Indian software industry development: international and national perspective', *Economic and Political Weekly*, **36** (44).

Kumar, N. (2001b), 'National innovation system and Indian software industry development', background paper for World Industrial Development Report, New Delhi: Research and Information System for the Non-Aligned and Other Developing Countries (RIS).

Kyaw, Aye (2002), 'IT needs and readiness assessment of the Union of Myanmar, Ministry of Communications, Posts and Telegraphs', (mimeo) Yangon.

Lateef, Asma (1997), 'Linking up with the global economy: a case study of the Bangalore software industry', NIOP, DP/96/97, Geneva: International Institute for Labour Studies.

Lundvall, B.-Å. (1985), *Product Innovation and User–Producer Interaction*, Aalborg: Aalborg University Press.

Lundvall, B.-Å. (1988), 'Innovation as an interactive process: from user-producer interaction to the national innovation systems', in G. Dosi, C. Freeman, R. R. Nelson, C. Silverberg and L. Soete (eds), *Technology and Economic Theory*, London: Pinter.

Lundvall, B.-Å. (ed.) (1992), *National Systems of Innovation: Towards a Theory of Innovation and Interactive Learning*, London: Pinter.

Metcalfe, S. (1995), 'The economic foundations of technology policy: equilibrium and evolutionary perspectives', in P. Stoneman (ed.), *Handbook of the Economics of Innovation and Technological Change*, Oxford and Cambridge: Blackwell Publishers.

Narayana Murthy, N.R. (2000), 'Making India a significant IT player in this millennium', in Romila Thapar (ed.), *India: Another Millennium*, New Delhi: Viking, Penguin Books.

NASSCOM (1999), *The Software Industry in India: A Strategic Review*, New Delhi: National Association of Software and Service Companies.

NASSCOM (2000a), *Directory of Indian Software and Service Companies*, New Delhi: National Association of Software and Service Companies.

NASSCOM (2000b), *The IT Software and Services Industry in India: Strategic Review 2000*, New Delhi: National Association of Software and Service Companies.

NASSCOM (2004), *The IT Software and Services Industry in India: Strategic Review 2004,* New Delhi: National Association of Software and Service Companies.

Nelson, R.R. (ed.) (1993), *National Innovation Systems: A Comparative Study*, Oxford: Oxford University Press.

Parthasarathi, A. and K.J. Joseph (2002), 'Limits to innovation with strong export orientation: the experience of India's information communication technology sector', *Science, Technology and Society*, **7** (1).

Schware, R. (1992), 'Software entry strategies for developing countries', *World Development*, **20** (2), 143–64.

Sen, Pronab (1995), 'Indian software exports: an assessment', *Economic and Political Weekly*, **30**, Nos. (7) & (8), 2053–8.

Shourie, A. (2004a), 'In infotech, we have a headstart so let's not put our feet up', *Indian Express*, 3 January, New Delhi.

Shourie, A. (2004b), 'Indian infotech needs to partner east Europe, target China', *Indian Express*, 3 January, New Delhi.

Shourie, A. (2004c), 'IT's written, now just dot the Is, cross the Ts', *Indian Express*, 5 January, New Delhi.

Singh, Nirvikar (2003), 'India's information technology sector: what contribution to broader economic development?', Technical paper No. 207, OECD Development Centre, Paris.

UNDP (2002), 'E-readiness assessment in the Lao PDR', UNDP/UNV Vientiane, Lao PDR.

USAID (2001), '*Vietnam: ICT assessment final report*', USAID Asia Near East (ANE) Bureau Sponsored Study, September.

Weidman, J.C. (1997), 'Lao PDR', in Gerard A. Postiglione and Grace C.L. Mak (eds), *Asian Higher Education: An International Handbook and Reference Guide*, Westport, CT: Greenwood Press.

11. Innovating for global competition: Singapore's pathway to high-tech development

Henry Wai-chung Yeung

INTRODUCTION

The territorialization of firm-specific competitive advantages has been featured strongly in recent research in urban and regional development studies. Localized *clusters* become very significant spatial formations for understanding processes of economic development and technological innovation (Storper, 1997; Porter, 1998a; 1998b; Scott, 1998; Fujita et al., 1999; Scott and Storper, 2003). Clusters are defined as 'geographic concentrations of interconnected firms, specialised suppliers, service providers, firms in related industries, and associated institutions in particular fields that compete but also cooperate' (Porter, 1998a: 197). The emergence of clusters is not a recent phenomenon and has been recognized and explored in a wide range of literature (Hirschman, 1958; Amin and Thrift, 1994; Ellison and Glaeser, 1999; Schmitz and Nadvi, 1999; Becattini et al., 2003; Schmitz, 2004; see a recent critique in Martin and Sunley, 2003). The concept is also well explored in the geographic literature on spatial agglomerations (Malmberg and Maskell, 1997; 2002; Maskell and Malmberg, 1999).

This geographic literature argues that cluster development is significantly embedded in networks of relational assets and geographical proximity particularly on local and regional scales. Such social processes as norms and conventions, collective learning, and localized capabilities help to define these specific interdependencies and subsequent territorial tendencies. When tapped by firms in specific localities, these interdependencies provide a significant source of location-based advantages through both competition and cooperation. Other researchers have developed further the concept of innovation and learning systems on regional and national scales (Lundvall, 1992; Asheim, 1996; Braczyk et al., 1997; Simmie, 1997; 2003; Keeble and Wilkinson, 1999; Boekema et al., 2000; Cooke, 2001; Narula,

2003; see recent critiques in Hudson, 1999; Bunnell and Coe, 2001; MacKinnon et al., 2002). Coupled with the 'new institutionalism' literature (Amin, 1999; MacLeod, 2001), this line of enquiry has led to a resurgence of research interest in the role of *institutions* in promoting firm growth and regional transformation. The broader urban and regional structures in which localized clusters are embedded are recognized as far more important determinants of economic development (Gertler et al., 2000; Bunnell and Coe, 2001; cf. Sternberg and Arndt, 2001).

Drawing upon the empirical materials in my earlier work on Singapore's economic development (see Wang and Yeung, 2000; Chew and Yeung, 2001; Phillips and Yeung, 2003; Yeung, 2002), this chapter aims to show the pathways taken by Singapore to achieve high-tech development. In particular, I pay special attention to the changing post-war economic development strategies in Singapore, thereby showcasing how state institutions matter in shaping the national system of technological innovation and in chartering a unique pathway to economic development. Through the empirical analysis of Singapore's science parks and chemical clusters, I also aim to demonstrate how the city-state of Singapore has harnessed the benefits from cluster development which offers significant economic synergies and economies of scale and scope to enhance high-tech development potential in Singapore through a peculiar combination of institutional support, foreign investment and local supplier development. Unlike their counterparts elsewhere in industrialized economies, industrial clusters in Singapore represent a deliberate and state-driven attempt to attract the location of high-tech activities by transnational corporations (TNCs) and local enterprises. Aiming to create specific places to ground globalizing R&D activities, the Singapore government has contributed to cluster formation through various initiatives to generate agglomeration economies for R&D activities (for example superior physical infrastructures, generous financial incentives, and the nearby location of universities and research institutes).

This chapter is organized into five sections. The next section briefly charts Singapore's changing pathways to economic development since the 1980s. This historical review is then followed by three empirical sections respectively on (1) the development of R&D capabilities in Singapore through science parks; (2) the harnessing of Singapore's SME advantage through reverse technology flows to foreign TNCs and (3) the nurturing of cluster development in the chemical industry. In the concluding section, I offer some lessons and policy implications from Singapore's experience in promoting high-tech development.

INNOVATION AND HIGH-TECH IN SINGAPORE: CHANGING PATHWAYS TO ECONOMIC DEVELOPMENT SINCE THE 1980S

Singapore has grown from a British colonial entrepôt in the late 19th century and early 20th century to a modern city-state specializing in high value-added manufacturing activities and international financial and business services (Régnier, 1991; Huff, 1994; Perry et al., 1997; Low, 1998; Pereira, 2000; Yeung, 2002). With its independence and changing global economic systems, Singapore was able to attract a huge influx of foreign investment that took advantage of Singapore's explicit policies towards export-oriented industrialization. This reliance on foreign capital worked very well in the first two decades of Singapore's industrialization and plugged Singapore into the so-called 'new international division of labour' (Fröbel et al., 1980). This strategy of courting foreign capital was perceived as 'essential in view of the weak domestic technological base and the long lead-time needed to transform domestic entrepôt traders and small-scale entrepreneurs into a dynamic industrial entrepreneurial class able to compete in the global market' (Chia, 1997: 32).

As shown in Table 11.1, manufacturing was still a relatively minor sector of the Singapore economy by 1960, accounting for only 13 per cent of total GDP at 1968 prices and 16 per cent of total employment. By 1970, this ratio rose respectively to 24 per cent and 22 per cent as a consequence of rapid industrialization. The ratios increased further to 28 per cent and 30 per cent in 1980. Since 1970, manufacturing has been the single most important sector in Singapore. Given this early developmental strategy, the Singapore economy was, and still is, heavily dependent on foreign investment, particularly in the manufacturing sector. As shown in Table 11.2, the share of foreign investment in Singapore's GDP rose steadily from 5.3 per cent in 1965 to 17 per cent in 1970 and 52 per cent in 1980. By the early 1970s, for example, Singapore had become a preferred offshore assembly location for foreign semiconductor manufacturers (Henderson, 1989; McKendrick et al., 2000).

The next two decades during the 1970s and the 1980s witnessed a massive expansion of foreign direct investments (FDI) from the US, Japan and other European countries (Mirza, 1986; Huff, 1994; McKendrick et al., 2000). Net foreign investment commitments in Singapore's manufacturing sector grew tremendously from S$88.6 million in 1963 to S$6.3 billion in 1999, representing a more than seventy-fold increase over a period of three and a half decades (Department of Statistics, various years a). The cumulative foreign direct investment in Singapore reached S$196 billion in 2000 (http://www.singstat.gov.sg, accessed on 15 October 2003). Throughout the

Table 11.1 Key macroeconomic indicators for Singapore, 1960–1999

Singapore (S$)	Annual growth rate (%)				Annual figures				
	1960–70	1970–80	1980–90	1990–99	1960	1970	1980	1990	1999
Population (,000)	2.4	1.5	2.3	2.9	1646	2075	2414	3016	3894
GDP (at 1990 prices)	12.7	12.4	9.3	9.1	5.8bn	13.9bn	32.9bn	66.5bn	126.8bn
Manufacturing[1]	26.7	16.7	8.6	7.8	0.29bn	3.29bn	9.26bn	18.0bn	32.1bn
Trade	10.7	8.5	8.1	10.2	0.72bn	2.74bn	5.29bn	10.0bn	20.2bn
Finance & business	16.8	12.4	13.9	14.3	0.27bn	2.56bn	6.04bn	15.3bn	29.6bn
Other	9.5	12.0	8.1	9.4	1.02bn	5.31bn	12.3bn	23.2bn	44.9bn
GNP per capita	10.2	22.7	11.8	7.5	$1330	$2825	$9882	$22693	$39721
Gross fixed capital formation (1990 prices)	–	16.0	8.1	10.9	–	4.6bn	12.7bn	21.6bn	45.2bn
Exchange rate (US$)	–	−2.9	−1.5	−0.5	–	3.094	2.094	1.745	1.666
Inflation rate (%)	–	–	–	–	0.3	0.4	8.5	3.4	0.4
Total exports[2]	6.5	70.1	17.9	14.1	2.77bn	4.76bn	34.1bn	95.2bn	229.4bn
Re-exports	–	–	12.2	17.1	–	2.92bn	13.9bn	32.5bn	88.0bn
Domestic exports	–	118.9	19.2	14.1	–	1.83bn	20.2bn	62.8bn	151.3bn
Total imports	10.6	52.9	12.3	9.6	3.48bn	7.53bn	46.5bn	109.8bn	215.4bn
Total labour force (,000)	3.5	6.5	3.4	2.4	471.9	650.9	1115.3	1537.0	1911.6
Manufacturing	8.5	12.5	2.9	−0.9	74.1	143.1	339.2	447.4	409.0
Trade	3.1	5.4	3.5	2.0	114.2	152.6	243.0	337.5	404.5
Finance & business	1.7	20.9	8.8	8.8	21.7	25.8	85.0	167.2	314.9

Notes:
[1] Data for 1960 are at 1968 prices. Annual growth rates for the 1960–1970 period are calculated based on 1968 prices for both 1960 and 1970.
[2] Data for 1960 refer to 1964. Data for 1964 and 1970 are at current prices. Data for 1980, 1990 and 1998 are at 1990 prices.

Sources: http://www.singstat.gov.sg; accessed on 17 May 2000; Department of Statistics (various years a).

Table 11.2 *Cumulative equity investments in Singapore by country of origin, 1965–1999 (in S$million)*

Country	1965	1970	1974	1980	1985	1990	1991	1992	1993	1994	1995	1996	1997	1998	1999
USA	23.0	343.0	1082.0	2551.5	6170.0	9951.8	11 108.4	11 731.0	13 356.7	14 325.9	17 048.6	19 812.9	24 922.1	26 095.9	29 511.1
Australia	–	–	–	403.9	593.9	3033.8	3122.8	3158.0	3315.8	3552.6	3380.5	3531.2	3632.7	–	–
EUROPE	85.0	406.0	997.0	4814.9	7688.7	16 272.0	18 414.7	18 299.5	20 455.2	25 307.2	29 781.1	37 368.2	44 038.8	–	–
United Kingdom	–	199.0	424.0	3432.8	4365.2	6951.4	8238.1	7310.7	8086.1	9149.3	10 453.5	13 063.5	16 083.2	19 446.0	13 857.0
Germany	–	3.0	107.0	421.6	565.2	902.1	1008.1	1054.0	1093.0	1339.1	1936.0	2031.0	1868.0	–	–
Netherlands	–	183.0	420.0	253.1	877.2	4349.6	4623.2	4498.2	4470.6	4631.7	4765.7	7335.3	8259.3	10 473.6	23 819.9
Switzerland	–	–	–	505.1	1415.4	2362.3	2537.6	2766.4	3480.4	5834.8	7250.0	9180.9	10 960.3	13 877.4	15 854.9
ASIA	49.0	–	–	4679.0	9068.2	20 324.3	21 435.3	23 448.8	26 202.8	30 033.7	33 364.7	37 883.0	41 938.0	–	–
Japan	–	68.0	354.0	1420.6	3261.3	11 271.6	12 449.1	13 813.4	14 721.5	16 919.2	18 817.0	21 645.8	23 518.6	26 378.5	26 322.7
Hong Kong	–	–	–	1707.0	2352.8	4220.8	4187.4	4522.9	5021.6	5018.2	5348.0	5998.4	6038.2	6173.0	5831.0
Taiwan	–	–	–	61.6	82.0	254.1	284.7	390.2	571.4	790.9	1006.8	1214.7	2024.9	–	–
ASEAN	–	–	–	1361.0	3165.7	4338.7	4164.7	4561.5	5408.6	6622.1	7139.0	8041.4	9073.5	–	–
Malaysia	–	–	–	1171.4	2784.8	3286.6	3183.6	3525.6	3791.9	4331.1	4712.5	5610.6	6575.3	7612.5	7059.4
OTHER COUNTRIES	–	199.0	667.0	553.1	1981.9	8353.3	8829.3	8713.7	9954.7	12 624.1	15640.8	17 052.8	24429.0	–	–
Total (Foreign)	157.0	995.0	3054.0	13 002.4	25 502.7	57 935.2	62 910.5	65 351.0	73 285.2	85 843.5	99215.7	115 648.0	138 960.6	156 859.5	178 019.9
Ratio to GDP at current prices (%)	5.3	17.1	24.3	51.8	65.5	87.2	83.6	80.7	78.9	80.6	83.7	89.7	98.4	111.1	123.6

Table 11.2 *(continued)*

Country	1965	1970	1974	1980	1985	1990	1991	1992	1993	1994	1995	1996	1997	1998	1999
Percentage in non-manufacturing sectors (%)	0.0	0.0	0.0	46.7	53.3	60.9	63.7	67.3	67.4	65.2	62.8	63.6	63.4	63.6	65.5
Total (Foreign and local)	157.0	995.0	3054.0	34 010.8	74 644.5	155 748.7	173 347.6	188 915.3	220 527.9	262 076.9	315 421.2	365 925.6	418 655.1	455 379.5	48 6367.5
Percentage of foreign (%)	100.0	100.0	100.0	38.2	34.2	37.2	36.3	34.6	33.2	32.8	31.5	31.6	33.2	34.4	36.6

Note: Data on 1965, 1970 and 1974 refer to foreign investment in Singapore's manufacturing industries in terms of gross fixed assets.

Sources: Economic Development Board (various years b) and Department of Statistics (various years b).

1980s and into the early 1990s, Singapore attracted over 10 per cent of all FDI received by destinations outside the OECD countries (Perry et al., 1997: 15). As an assessment of the importance of foreign capital in Singapore's economy, Huff (1995: Table 6) estimates that foreign investment contributed to some 22 per cent and 26 per cent of the gross domestic fixed capital formation (GFCF) during the 1970–1979 and 1980–1992 periods respectively. This ratio of foreign investment to GFCF is certainly one of the highest among the Asian Newly Industrialized Economies (NIEs). In 1997, Singapore played host to some 16 190 foreign TNCs, over 300 of which were in the manufacturing sector (Department of Statistics, 2000: xi).

In terms of industrial distribution, foreign ownership is most conspicuous in Singapore's manufacturing sector. As mentioned earlier, foreign capital contributed significantly to net investment commitments in Singapore's manufacturing sector from the 1970s onwards. In 1966, foreign investment represented some 45 per cent of total gross fixed assets in manufacturing (Rodan, 1989: 99). By 1975, foreign-controlled firms in Singapore's manufacturing sector were responsible for 52 per cent of total employment, 55 per cent of remuneration, 71 per cent of gross output, 63 per cent of value-added, 84 per cent of direct exports, and 65 per cent of capital expenditure (Table 11.3). In the next two decades (the 1980s and the 1990s), Singapore's industrialization was characterized by a shift towards high value-added manufacturing activities, particularly in the electronics industry (Wong, 1995; Chia, 1997; Perry and Hui, 1998; Mathews, 1999; McKendrick et al., 2000; Chew and Yeung, 2001) and the chemical industry (Wang and Yeung, 2000). This trend is indicated in the growing domination of foreign firms in Singapore's manufacturing sector in terms of gross output, value-added, and direct exports from 1975 to 1998. In 1998, although they accounted for only 21 per cent of total establishments, foreign firms contributed to 77 per cent of gross output, 73 per cent of value-added and 88 per cent of direct exports.

DEVELOPING R&D CAPABILITIES THROUGH SCIENCE PARKS

My first empirical example is the Singapore Science Park, which is a specific state-driven exercise to bring R&D to Singapore. As a newly industrialized economy and dependent fast-follower of technology, Singapore has invested significantly in R&D since the 1980s (Wong, 1995; Goh, 1998). Its national expenditure on R&D has been increasing steadily from 0.54 per cent of GDP in 1984 to 1.47 per cent in 1997 and 1.89 per cent in 2000

Table 11.3 Local and foreign ownership of selected sectors in Singapore, 1960–1998 (50% equity as cut-off and in S$million unless otherwise specified)

Industries	1960	1970	1974–5		1980–81		1990		1997–8	
	Total	Total	Total	Foreign (%)	Total	Foreign (%)	Total	Foreign (%)	Total	Foreign (%)
Manufacturing			**1975**		**1980**				**1998**	
Establishments[1]	548	1747	2385	22.0	3355	24.9	3703	23.4	4004	20.8
Workers (,000)	27.4	120.5	191.5	52.0	285.2	58.5	351.7	58.9	352.3	50.5
Compensation	66.8	397.6	1180.5	55.0	2526.9	58.5	6852.2	61.3	11 768	56.2
Gross output	465.6	3891.0	12 610	71.3	31 658	73.7	71 333	75.9	121 433	77.3
Value-added	142.1	1093.7	3411.1	62.7	8521.9	67.4	21 607	72.7	29 627	73.3
Direct exports	164.3	1523.0	7200.7	84.1	19 173	84.7	47 000	85.8	75 530	88.3
Capital expenditure[2]	9.8	421.3	622.6	64.6	1861.9	74.6	4184.4	70.7	37 246	67.8
Trade[3]			**1975**		**1981**				**1997**	
Establishments	–	–	21 208	–	35 251	4.7	40 147	6.0	51 337	13.1
Workers (,000)	–	–	124.5	–	172.8	15.6	211.0	21.0	265.3	29.1
Compensation	–	–	698.2	–	1676.0	34.3	4191.5	39.6	9055.1	49.3
Turnover	–	–	22 334	–	64 264	42.5	179 856	63.1	371 978	74.1
Value-added	–	–	2369.4	–	4993.0	39.2	10 338	50.4	19 249	52.3
Number of companies	–	–	–	–	8196	20.7	15 259	23.9	37 225	19.3
Shareholders' equity	–	–	–	–	5491.9	34.5	15 871	38.6	36 730	41.8
Fixed assets	–	–	–	–	3491.5	36.4	11 552	36.3	26 705	36.1
Equity investments[4]	–	–	1406.7	25.4	4553.7	40.1	13 195	51.0	37 550	44.4

Finance			**1974**		**1980**				**1997**	
Establishments	–	–	1327	–	2330	18.3	2724	18.0	5499	26.4
Workers (,000)	–	–	19.3	–	28.1	40.2	11.7	41.0	18.7	25.7
Compensation	–	–	223.0	–	541.9	49.8	386.4	44.4	1256.5	35.9
Receipts	–	–	801.5	–	19 346	68.4	10 970	74.2	22 909	19.2
Value-added	–	–	272.0	–	2177.2	57.0	1006.0	62.6	2859.9	41.3
Number of companies	–	–	–	–	2010	22.3	3763	30.5	11 914	20.7
Shareholders' equity	–	–	–	–	9867.2	17.5	62 578	24.7	178 458	27.8
Fixed assets	–	–	–	–	1237.2	34.9	6424.6	20.7	15 348	22.5
Equity investments[4]	–	–	1306.1	14.4	10 515	22.2	72 552	36.9	182 498	29.2

Notes:
[1] In 1964 and 1970, 7.9% and 11.7% of respectively 965 and 1626 total establishments were majority or wholly foreign-controlled.
[2] Data for 1998 refer to net fixed assets.
[3] Data for 1980 refer to 1981.
[4] Data for 1974–5 refer to paid-up capital.

Sources: Department of Statistics (various years b; c; d; 1992; 2000). Data on trade and finance for 1980–81, 1990, and 1997 are from unpublished data supplied by the Department of Statistics, 9 June 2000.

(http://www.a-star.gov.sg; accessed in July 2002). Table 11.4 shows that Singapore's R&D activities during the 1985–1995 period were not too far from those in industrialized economies. Heralding the 'Second Industrial Revolution' in Singapore (Rodan, 1989), the year 1979 saw the first sign of establishing a national R&D programme (*Business Times*, 8 June 1979). Singapore's state philosophy was simple and clear: to prepare a conducive R&D environment that will help Singapore 'maintain

Table 11.4 R&D activities in selected countries ranked by R&D expenditures

Country	Expenditures on R&D as % of GNP, 1985–95	Number of RSEs per million people, 1985–95	High-technology exports	
			$millions	% of manufactured exports
Sweden	3.4	3714	21 969	34
Japan	2.9	6309	152 431	38
Korea	2.8	2636	44 433	39
Finland	2.5	2812	8797	26
US	2.5	3732	197 657	44
France	2.4	2584	68 655	31
UK	2.2	2417	95 755	41
Denmark	1.9	2647	8174	27
Norway	1.8	3678	2703	24
Australia	1.7	3166	6415	1.7
Belgium	1.7	1814	–	–
Canada	1.6	2656	33 608	25
Italy	1.1	1325	32 747	15
Singapore	**1.1**	**2728**	**74 585**	**71**
India	0.8	149	2654	11
Indonesia	0.8	–	4474	20
South Africa	0.7	938	–	–
Brazil	0.6	168	5175	18
China	0.5	350	33 344	21
Argentina	0.4	671	1355	15
Malaysia	0.4	87	39 490	67
Mexico	0.4	213	29 692	33
Hong Kong China	0.3	98	7392	29
Philippines	0.2	157	6249	56
Thailand	0.1	119	17 758	43

Note: Not all countries are reflected in the table here as there are more than 250 countries in the original source.

Source: http://www.statistics.com; accessed in July 2002.

a competitive environment' in the light of increasing competition in the global economy (Ministry of Trade and Industry, 1991: 60). From its conception, the Singapore government duplicated the successfully tried-and tested Singapore style of industrial estate development in Jurong: (1) the nature of R&D according to guidelines is development, or 'mission-oriented' research (*The Straits Times*, 14 September 1992); (2) the directives will be government-led, and (3) there will be a heavy reliance on foreign R&D firms. The establishment of such institutions as the National Science and Technology Board (NSTB)[1] was meant to support R&D activities in Singapore.

The National Technology Plan (NTP) in 1991 maps out a technology corridor along the southwestern area of Singapore in line with the Strategic Economic Plan (Ministry of Trade and Industry, 1991). In Figure 11.1, the concept plan for a technology corridor has contributed to the spatial integration of science habitats, business parks and tertiary institutions. There is thus nothing coincidental about the location of the Singapore Science Park in this corridor. Figures 11.1 shows the links that the Park can share with tertiary institutions and other major research institutions, as it is located within a one-kilometre radius of the National University of Singapore, the National University Hospital, and such research institutions as the Institute of Systems Science and the Institute of Molecular and Cell Biology. There is no doubt that the Park can potentially benefit from such geographical advantages as physical proximity and agglomeration economies. Funding for R&D frequently rose, with the latest figures at S$2 billion[2] annually for the 1996–2001 period, reflecting the Singapore's government commitment to R&D (*The Straits Times*, 22 July 1996). In 1995, the NSTB targeted R&D expenditure to be at 2 per cent of Singapore's GDP, with the private sector accounting for the majority share (Ministry of Trade and Industry, 1991: 30). In 2000, gross expenditure on R&D (GERD) grew by 13.3 per cent, from $2.66 billion to $3.01 billion. The private sector accounted for 62.0 per cent of the total national GERD (http://www.a-star.gov.sg; accessed in July 2002). Furthermore, an elaborate plan to develop Singapore into an innovation hub was announced in early 1998 (*The Straits Times*, 8 January 1998). Within the same year, the government released a plan to develop a S$5 billion science hub in the South Buona Vista area along the Ayer Rajah Expressway which lies at the lower right-hand area within the technology corridor shaded in Figure 11.1 (*The Straits Times*, 16 September 1998). It will be a place where 'ideas born within a scientific community are cradled, nurtured and developed into commercial products', creating a 'focal point for R&D and developing an innovative milieu' (*The Straits Times*, 16 August 1998).

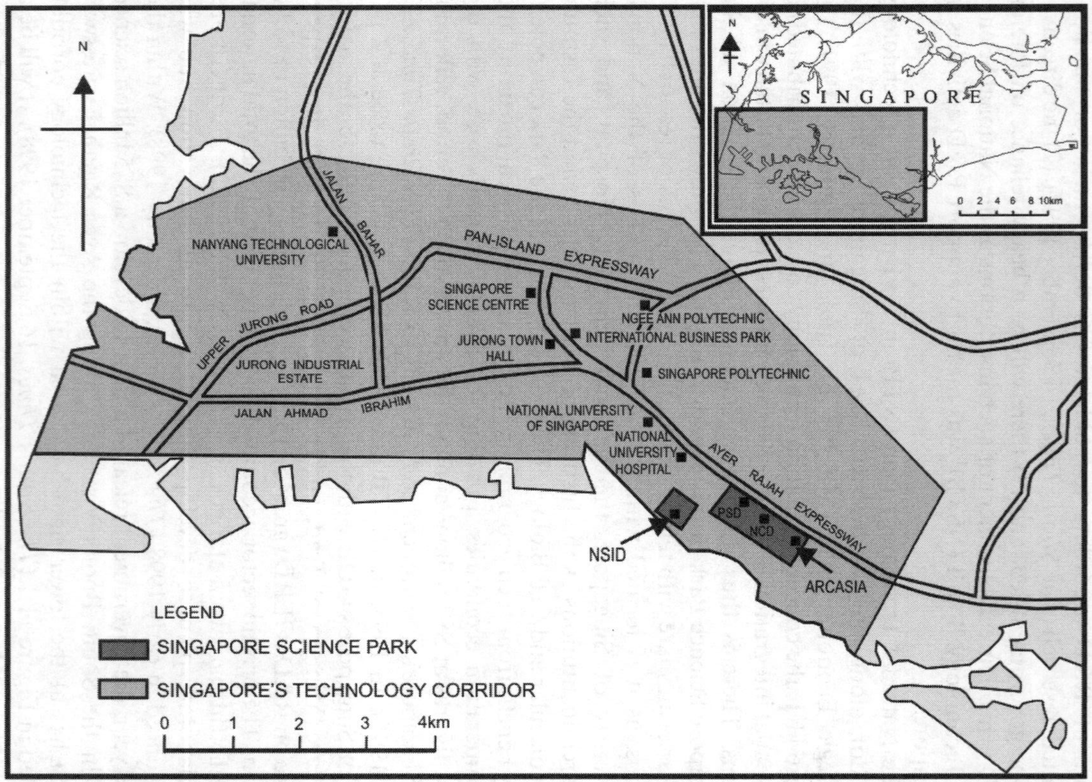

Figure 11.1 The location of Singapore's Science Park and technology corridor

In this national context of promoting research and innovations, the Singapore Science Park was set up in 1980 as a place where R&D can converge and create synergies with institutions and firms alike, and researchers can work any time, meet and share ideas.[3] The Park's establishment can be read as the 'first positive indication' of the government's recognition of linking R&D activities to its economic policy (Goh, 1998: 60). The Economic Development Board (EDB) and the Jurong Town Corporation (JTC), both national economic institutions spearheading economic development since the 1960s, were directly involved in the planning and creation process of the Singapore Science Park. According to the former managing director of Arcasia,[4] the Park was a brainchild of the former chairman of the Jurong Town Corporation. The Jurong Town Corporation continued to manage the Park until the incorporation of Arcasia in 1990. The advantage of transferring the Park's management to its subsidiary is that Arcasia offers 'the flexibility, in terms of ease of changing policies and coming up with innovative products. Government agencies can change things but at a slower pace. We also wish to subject ourselves to more private sector's disciplines'.

However, this idea of a science park in Singapore received some mixed reaction. Initially at its first mention in the early 1980s (again when the NSTB announced its National Technology Plan in 1991 and recently, with the current debate on nurturing technopreneurship; see Coe and Kelly, 2000; 2002), there were many arguments for and against the formation of the Singapore Science Park and the Science Hub (see *Business Times*, 16 April 1984; *The Straits Times*, 27 August 1980; 25 May 1990, 17 September 1998; 23 September 1998). These arguments ranged from poor public perception, to a lack of critical mass in research scientists and engineers, a total transformation of the education system, and a change in attitude towards failure, risk-taking and creativity. The general thread was that Singapore is not ready to move towards a knowledge-based economy. One response highlights the pitfalls on efforts to become the 'Silicon Valley of the East' as it misses out the very spirit of entrepreneurship:

> It would be a mistake to expect a certain output of high-tech entrepreneurs in a predicable, mechanistic manner, just because there has been so much investment, facilities and people put into the effort. The long-term consequence . . . is that this nation will be far less directed by the government and more market-driven (quoted in *The Straits Times*, 29 December 1998).

A report by the Political and Economic Risk Consultancy based in Hong Kong, contrasting Taiwan's environment to Singapore, concluded that although Singapore has better technological infrastructure, Taiwan is more tolerant of failure and has a vibrant entrepreneurial atmosphere. The report went on to criticize Singapore's openness to new technology as 'big on substance but short on flair'. Although a 'master at creating and maintaining very high quality infrastructure', Singapore is finding it 'very difficult to nurture the sort of vibrant, freewheeling atmosphere necessary to transform the country into a centre for technological innovations' (quoted in *The Straits Times*, 13 October 1989).

More than two decades have gone since its inception; the 65-hectare Park (I, II and III) now claims to house more than 200 institutions registered as 'research facilities' (307 companies in total in 2000; http:// www.sciencepark.com.sg, accessed in July 2002). Comprising both local and non-local actors, the tenants constitute the biggest and most influential group of actors involved in the production of Singapore's R&D activities (see detailed analysis in Phillips and Yeung, 2003).[5] The Park's tenants include such global players as Sony, Exxon Chemical, Silicon Graphics, Lucent Technologies, as well as small and medium enterprises (SMEs) and start-ups. Local R&D facilitators, such as the Productivity and Standards Board (PSB) and the Infocomm Development Authority (IDA) make up the second group of diverse, but active, actors in the Park. The function of these organizations is to support R&D activities in the Park and in Singapore. The Economic Development Board provides R&D benefits and incentives to the Park's tenants, while the NSTB encourages all R&D activities in Singapore. Arcasia is responsible for the Park's overall property development, marketing and management (amongst other industrial and business parks in Singapore). The National Technology Plan identifies seven main R&D areas: information technology (IT), manufacturing and engineering technology, pharmaceuticals, telecommunications, chemicals, electronics and, lately, the life sciences (http://www.sciencepark.com.sg; accessed in July 2002).

Table 11.5 summarizes the reasons cited by tenants for choosing to locate in the Park. Two key findings emerge. First, the spatial proximity rationale does not seem to appeal to many tenants located in the Park. The reasons for locating in the Park are based more on such science park perks as 'image' and 'infrastructure', rather than on government incentives for the cluster rationale. 'Attractive infrastructure and support services' has the highest response at 33.3 per cent, followed far behind by such reasons as 'invited to establish in the park' (12.8 per cent), 'convenient location' (12.8 per cent) and 'close to similar activities' (12.8 per cent). None felt that the 'links with suppliers and industries' was an important consideration for establishing themselves in the Park. Second, when asked what aspects of the science park they liked best, most respondents chose 'facilities' and 'physical landscapes'. These responses imply that the Park has only been

Table 11.5 Reasons for tenants' location in the Singapore Science Park

Category	Factors	Responses
Science Park perks	Attractive infrastructure and support services	26 (33.3%)
	Supportive management	3 (3.8%)
	Lower costs	4 (5.1%)
	Total	33 (42%)
Government benefits and incentives	Invited to establish in park	7 (9.0%)
	Supportive government policies	7 (9.0%)
	Total	17 (30%)
Spatial proximity	Close to market/demand	3 (3.8%)
	Convenient location	10 (12.8%)
	Links with suppliers and industries	0 (0%)
	Total	23 (28%)
Total		70 (100%)

Note: Firms were allowed to choose more than one option. All percentages are rounded off.

Source: Author's survey.

successful in appealing to tenants who are attracted to the image of the Park, but not its functioning aspects (for example research links and collaborations). The efforts by state institutions and non-state actors to create an attractive physical place seem to be realized, although the tenants may not be what they had in mind. This unintended outcome may be explained by the overt focus of the planners and developer on the physical aspects of other science parks they visited before building the Singapore Science Park.

When asked to describe the main activity of their firm, 12 respondents (35 per cent) indicated R&D, another 6 (18 per cent) indicated 'marketing & sales and R&D'. This results in a total of 18 firms (53 per cent) that considered R&D to be their main activity (see Figure 11.2). There are tenant firms that are actively involved in R&D activities, and those that are not: 16 (47 per cent) firms indicated no R&D activities and 11 (69 per cent) of them are foreign firms. A seemingly encouraging number of firms described their main activity to be R&D (Figure 11.2). But they are involved in various types of R&D activities. Results from both the surveys and interviews indicate that the nature of R&D activities ranges from mostly applied R&D activities to the organization of these activities. In Table 11.6, 14 firms out of 34 (42.1 per cent) described a total of 34 major developments. These 14 firms have spent an average of 7.3 years in the Park. These developments tend to be new products (n=10 or 71.4 per cent), and fewer

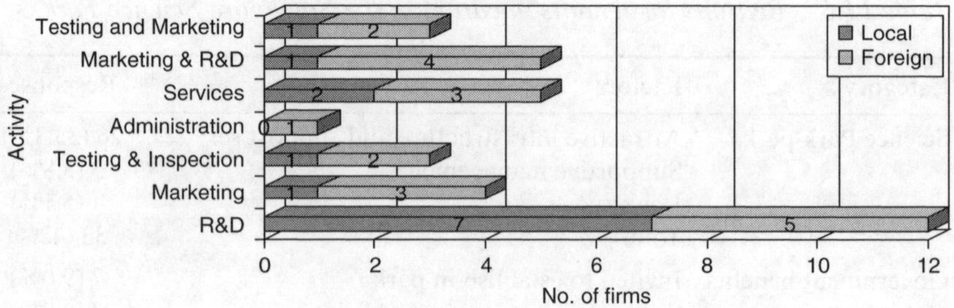

Source: Author's survey.

Figure 11.2 Summary of tenant firms' main activities

Table 11.6 Comparison of local and foreign firms by proportion of R&D expenditure

Type of firms	With major R&D developments	% of expenditure on R&D		
		Above 75%	50–74%	Below 50%
Local firms	8 (N=15) (33.3%)	5	2	1
Foreign firms	6 (N=19) (40%)	3	3	1
Total (%)	14 (N=34) (42.1%)	8 (N=14) (57%)	4 (N=14) (29%)	2 (N=14) (14%)

Note: Total number of foreign firms in the survey = 19; Total number of local firms in the survey = 15.

Source: Author's survey.

are new and improved processes (n=4, or 28.5 per cent). Only 8 (23 per cent) out of 25 are patented. A large number of these firms spent above 75 per cent of their total expenditure on R&D activities. This tentatively implies a positive relationship between R&D expenditure and R&D developments.

Table 11.7 summarizes the collaborative efforts among surveyed tenants. Twenty-nine out of the 34 major developments described earlier were done in collaboration (85.2 per cent). An encouraging number of tenants (n=22, or 65 per cent) have collaborated on R&D projects, but most have done so only on a one-off basis (17 out of 22, or 77.2 per cent). Collaborations occur quite equally among tenants within and outside the Park, highlighting

Table 11.7 Collaborations among tenants in the Park

	R&D project developments	Product development	Testing and inspection
Number of firms (%)	22 (65%)	15 (44.1%)	18 (40.9%)
Firms in Park	Yes (13) No (9)	Yes (4) No (11)	Yes (9) No (9)
Status	Ongoing (5) One time only (17)	Ongoing (9) One time only (6)	Ongoing (12) One time only (6)

Note: Total number of firms in survey = 34.

Source: Author's survey.

the dilemma of R&D firms' strategic desires to be both locally embedded and globally linked (Patel and Pavitt, 1991; Tödtling, 1994; Patel, 1995; Perry and Tan, 1998; Gertler et al., 2000). Ironically, despite these encouraging collaborating efforts, the survey shows that 'linkages with institutions' (31 out of 98, or 32 per cent) is the most agreed upon aspect of the Park that can be improved. This finding implies that there are tenants who wish to collaborate with institutions, but such opportunities have not yet materialized.

HARNESSING THE SME ADVANTAGE: REVERSE TECHNOLOGY FLOWS AND FOREIGN TNCS

While technological collaboration might be fostered by co-location in the Singapore Science Park, such collaborative activities can also be identified elsewhere in Singapore, particularly between foreign TNCs and their local suppliers. Very often, these local suppliers and business partners are relatively small in their size and scale of operations; they are really small and medium enterprises (SMEs) that interestingly can serve as providers of certain locally specific expertise and 'soft' technology to their foreign TNC customers in the context of collaborative supplier–buyer or subcontracting linkages. I argue that this proactive role of SME suppliers is *developmental* rather than *dependent*. Such developmental linkage growth is deemed vital in pushing local supporting industries towards the status of 'technological graduation', defined as the stage when local SMEs are not only suppliers to foreign and local large firms, but also innovative suppliers capable of creating patents and innovative ideas (see Chew and Yeung, 2001). Enduring competitive advantages in a global economy rest increasingly with *localised social capital* – knowledge, relationships, motivation – that distant rivals cannot match (Porter, 1998b: 78).

Local enterprises have been an important player in Singapore's economic development. In fact, they have helped to build up Singapore's industrial base and have been instrumental in moulding Singapore into an attractive international business hub for foreign investments. Today, the local enterprise sector has evolved from its trading and light industry base to undertake a broad spectrum of modern economic activities. They are a key component in the manufacturing, service and commerce sectors, supporting foreign TNCs as well as engaging in the development and utilization of proprietary expertise (Economic Development Board, 1992: 5). In this way, the performance and reliability of local enterprises constitutes one of the key factors in attracting foreign companies. In 1994, they accounted for about 94 per cent of total establishments in the manufacturing, commerce and services sectors, 48 per cent of employment, 34 per cent of value-added and 49 per cent of direct exports (Department of Statistics, 1997: 1). They have increasingly been acknowledged as an integral part of the economy and an important source of employment generation, innovation and economic vitality. This has accelerated the need to promote local entrepreneurship so as to develop Singapore into a major node in global business.

During the 1970s and the 1980s, Singapore's manufacturing growth was largely driven by foreign TNCs (Mirza, 1986; Rodan, 1989; Low, 1998; Rajan, 2003). Local enterprises had only very limited direct contact with foreign TNCs. This was attributed to the offshore-oriented nature of such foreign-dominated industries in petroleum products, machinery and electronics (Mirza, 1986: 258). Inputs and components were usually provided by external suppliers or by TNC subsidiaries in Singapore. Even if supplier–buyer linkages were present, only standardized parts were given to local suppliers. This resulted in relatively low acquisition of product design know-how through subcontracting. In a study of 18 electronics TNCs and 16 of their local SME suppliers in the late 1980s, Wong (1991) explained that TNCs would only devote resources to transfer technological know-how to their suppliers if the expected returns from such efforts outweighed the costs of transfer. However, there is a lack of information on how local SME suppliers could benefit foreign TNC customers through reciprocity. Local SMEs could reciprocate to their foreign TNC customers through a process of *reverse transfer*. It is vital in building up Singapore's technological capability with higher levels of innovativeness and active application of expertise and experience by local SME suppliers. Local SMEs have now moved beyond playing a supporting role since foreign TNCs are now looking for partners to undertake concurrent development activities or to develop complementary capabilities (Economic Development Board, 1996a).

Singapore has about 92 000 SMEs, which are categorized broadly into manufacturing and non-manufacturing sectors. In the manufacturing sector, SMEs have flourished in industrial sectors of electrical/electronic products, transport equipment, precision machinery, and fabricated metal products (Soon, 1995: 67). My emphasis here is on the manufacturing sector due to the desire to relate SME development to the industrial sector as well as the greater proportion of foreign investment channelled into the manufacturing sector. Table 11.2 shows the stock of foreign investment in Singapore, with the manufacturing sector being consistently one of the largest recipients of foreign investment. Prior to the 1997 Asian economic crisis, the number of locally controlled companies grew from 2998 in 1985 to 6465 in 1995, achieving an increase in most of the years except for 1987 and 1991 (Department of Statistics, 1997: Table 6.2). In terms of their performance indicators, the number of establishments, number of workers employed by SMEs and their direct exports revenue have also improved over the years (see Table 11.8). Recent figures have also shown a significant increase in the value-added per worker component. In Table 11.9, the value-added component has increased tremendously since 1994. This is attributed to the efforts, policy measures and initiatives by the Singapore government to promote SMEs. Most of the industries have been performing better since 1994, except for the rubber and plastics industry and electrical machinery and apparatus industry in which output experienced negative and stagnant growth respectively.

My survey of 41 SME suppliers in 1999[6] shows that 14 of them (34.1 per cent) were involved in precision machining such as precision parts for production assembly machinery, optical instruments, disk drives and many others. Another eight (19.5 per cent) were supplying fabricated metal parts that consist of fabrication of stamping tools and making of fixtures; 12 (29.3 per cent) were suppliers of tailor-made plastics components that include plastics mould making for audio, office equipment and computer peripheral products. These three product sectors account for a total of 82.9 per cent of all respondents. Together, they form Singapore's precision supporting industry, supplying custom-made toolings, metal stamped and plastic moulded products to their TNC buyers. Over the past 35 years, this precision sector has grown impressively, especially with the influx of foreign TNCs into Singapore (Economic Development Board, 1996a: 15). Foreign TNCs have brought with them a large market for toolings, parts, components and sub-assemblies. In terms of number of employees, 58.5 per cent of SME suppliers had between 10 and 99 employees. Only six SMEs hired less than 10 employees.

Table 11.8 Performance of local SMEs in Singapore

	1987	1988	1989	1990	1991	1992
Number of establishments	26 598	28 687	31 997	35 443	37 826	41 632
Number of workers	102 219	116 072	118 761	121 666	126 080	127 131
Direct exports (S$ billion)	2.9	3.5	4.1	4.4	4.4	4.8

Source: Economic Development Board (1994), p. 177.

Local SME suppliers in the survey supply a wide range of products and services to foreign TNC customers in Singapore. This diverse range of parts and components supplied by local SME suppliers is either exported or assembled locally. In the SME literature, SMEs are considered to be highly flexible due to their small size. This seems to be quite true for local SMEs in Singapore in terms of their products and services. Generally, SMEs are expected to be financially constrained, resulting in a neglect of employee training and R&D activities. Twenty SMEs in the survey, however, offered formal training courses to their employees. The average percentage of operating costs spent on training was estimated to be 6.6 per cent. The percentages of operating costs allocated to formal training ranged from a low 0.5 per cent to a high 20 per cent. One of the SME suppliers interviewed offered training in ISO9002. Similarly, another SME supplier interviewed sent its workers to attend such part-time courses as 'NTC 3' at the Institute of Technical Education in order to allow workers to upgrade their skills.

In terms of R&D activities, SME suppliers were less active. Only 13 SMEs (32 per cent) in the survey carried out some R&D. The average scale of R&D was relatively small as it constituted only 5 per cent of the company's operating costs. In some cases, R&D took the form of designing the entire product from scratch for TNC customers. R&D by SMEs involved either supporting foreign TNCs or the development of their own niche products (*Singapore Enterprise*, August 1998: 4). Most SMEs were conducting R&D to support their TNC customers that had transferred some of their R&D activities to SME suppliers. Some SMEs were conducting R&D into their niche products. The survey shows that more SMEs in the plastics and material handling industries had R&D. For example, one SME interviewed had already gone into R&D by developing raw polymer materials on their own, instead of getting them from overseas. Singapore has reached a stage where it is competing head-on with some developed nations.

Table 11.9 Manufacturing establishments, output and value-added by industry in Singapore, 1994–1997

Industrial sector	Year	Number of establishments	Output (S$ million)	Value-added (S$ million)
Fabricated metal products	1994	559	5607.5	1509.2
	1995	584	6313.2	1680.6
	1996	605	6297.8	1704.1
	1997	625	6323.0	1733.1
Machinery & equipment	1994	499	4965.1	1392.5
	1995	515	5849.2	1636.7
	1996	565	6452.2	1753.7
	1997	585	6840.6	1942.8
Electrical machinery & apparatus	1994	145	2797.1	746.2
	1995	150	3344.9	834.4
	1996	152	3110.9	808.0
	1997	157	3110.9	812.1
Electronics products & components	1994	237	48 725.3	9543.4
	1995	239	57 872.7	11 987.9
	1996	238	60 912.8	12 672.2
	1997	245	63 410.3	13 181.6
Plastics & rubber products	1994	337	2275.7	732.0
	1995	346	2588.5	781.7
	1996	355	2601.0	796.1
	1997	363	2450.1	771.5
Medical, precision & optical instruments	1994	57	1184.8	472.1
	1995	62	1473.1	536.4
	1996	62	1547.1	578.5
	1997	65	1694.1	621.3

Note: Figures refer to establishments engaging 10 or more workers.

Sources: Economic Development Board (various years a) and Department of Statistics (various years a).

Capabilities to innovate become critical to fulfilling Singapore's Manufacturing 2000 Thrust, a national initiative to develop Singapore's manufacturing capabilities. Another general perception of SMEs was that their customer base is relatively small. This perception seems logical, particularly when SMEs are unable to meet TNC customers' large volume orders with their small scale of operation. My survey shows that the number of TNC customers ranged from merely two to as many as more than 20. There does not seem to be a direct relationship between the sales revenue of SMEs and the number of their TNC customers. The majority of the more profitable SMEs had an annual sales revenue of S$25 million to S$49.9 million and about one to 10 TNC customers.

Why do small manufacturing supplier firms exist? A key reason is that they supply custom goods to other manufacturers under conditions of changing specifications and short notice in order to meet demands of just-in-time inventory control (Young et al., 1994: 37). Several common reasons for establishing supplier relationships with TNC customers were shared by the SMEs in the survey (see Table 11.10). Securing TNC customers was perceived to enhance the company's image and became the most cited reason by sampled SMEs (73 per cent; N=30). The respondents believed that TNC customers allow the SME supplier to build up its credibility since TNC customers usually demand a stringent level of product/service quality. The initial presence of TNC customers can serve as a 'magnet', attracting other TNCs that are sourcing for similar parts and components. Supplying to TNCs was also perceived as a means to gain technological exposure (59 per cent; N=24). SME suppliers perceived supplier–buyer linkages as a convenient means to get exposed to their TNC customers' technology and process knowledge. This is, however, not so easy since access to TNCs' establishments was often confined to preferred suppliers. The third most common reason to supply to TNCs was to offer local

Table 11.10 Rationale for establishing supplier relationships with TNC customers

Reasons for being a supplier*	Number of SMEs	Reasons for TNC buying from SMEs*	Number of SMEs
1. To enhance company's image	30 (73%)	1. Reasonable pricing	39 (95%)
2. To gain technological exposure	24 (59%)	2. Quality of products	31 (76%)
3. To offer local adaptation of their products	22 (54%)	3. Good personal relations	28 (68%)
4. Others	12 (29%)	4. Long-term customers	13 (32%)
5. To gain technical assistance	4 (10%)	5. Others	11 (27%)
6. To add extra features to customer's products	3 (7%)	6. Good product design	10 (24%)
7. To gain assistance from EDB and other intermediaries	2 (5%)	7. Introduction by EDB and other intermediaries	1 (2%)
Total responses	41 (100%)	Total responses	41 (100%)

Note: * A respondent is allowed to select up to three reasons.

Source: Author's survey.

adaptation to TNC products (54 per cent; N=22). This was related to alternatives offered by local SMEs in aspects such as raw materials, dimensions and designs.

TNCs are often very stringent in quality so as to maintain the reputation of their brand names. In order to present the final product in its best quality, the process of assembling and the quality of all parts and components are crucial in contributing to the quality of the final product. Similarly, a final product, bearing the nationality of the company, actually comprises a large number of individual parts and components manufactured by a large pool of suppliers of different nationalities. Local SMEs have been constantly improving the quality of their work since Singapore is beginning to be seen as a supporting hub for high precision manufacturing. They face tremendous pressures in keeping prices reasonably low in order to clinch contracts. However, TNCs do take into account such other non-price criteria as quality of work, promptness of delivery and eagerness to suggest ideas. As shown in Table 11.10, 39 SMEs (95 per cent) were able to provide reasonable prices to their TNC customers. Another 31 SMEs (76 per cent) agreed that their quality of parts and components was good enough to meet TNC requirements. Some local SMEs have risen to play an increasingly important role as partners to TNC customers by undertaking concurrent development activities in product design. The most obvious reason why TNC customers buy from local SMEs is that these suppliers can meet the special needs TNCs cannot meet themselves. In adopting new and sophisticated production technologies, large firms may not want to make the investment as it can jeopardize their operating efficiency and output. Buying from suppliers may be more optimal. Clarke (1994) found that the converse is true when the large firm is unwilling to maintain older technologies that may reduce its production capabilities.

The relations between local suppliers and their foreign TNC customers in Singapore are often reciprocal. The level of technology transfer between either parties and their levels of supplier's participation in customer's product design can determine such relations. Table 11.11 illustrates the ratings of supplier–buyer relations by local SMEs. Twenty-four SMEs rated their involvement in their customers' product design as 'excellent' or 'good'. Instead of manufacturing solely according to customers' specification, the surveyed SMEs were often invited by their customers to offer suggestions concerning the parts and components that they manufacture. Local SMEs also considered trust by TNC customers as 'good'. For instance, one SME interviewed received substantial trust from its TNC customer from the UK. Initially, the customer checked through everything before exporting overseas. Such an arrangement continued for almost a year until the customer fully trusted the SME to the extent of allowing the latter to manufacture,

Table 11.11 Rating of supplier–buyer relation by local SMEs in Singapore

	Rating of relation (number of responses)					
	Very bad	Bad	Neutral	Good	Excellent	N
Involvement in their product design	0	4	13	20	4	41
Trust by TNC customers	0	0	2	26	13	41
Exchange of information	0	0	11	23	7	41
Providing technical assistance to them	0	0	7	23	11	41

Source: Author's survey.

check, pack and export the parts directly on its behalf. The TNC customer saved a tremendous amount of time because it now only had to ensure that it sent the SME good castings or tooling to manufacture the parts, leaving most, if not all, of the job to the SME in Singapore.

NURTURING CLUSTER DEVELOPMENT: THE CHEMICAL INDUSTRY CLUSTER ON JURONG ISLAND

Apart from developing science parks and TNC–SME supplier networks, the Singapore government is also instrumental in nurturing the chemical cluster located on Jurong Island. During the last three decades, Singapore has distinguished itself as a regional centre for trade in petroleum, petrochemicals and chemicals. While Singapore's geographical advantages have provided the basis for the country to secure a role as a regional producer, supportive state policies have been just as important in creating a conducive business environment for the transnational operations of international oil, petrochemical and chemical companies. Given today's highly competitive environment, however, the provision of incentives by local, regional and national authorities will not automatically attract foreign investments. The fact that many of the world's leading chemical companies have chosen Singapore as their strategic hub in the Asia-Pacific region points to the competitive position attained by the city-state in embedding these foreign investments. The development of Singapore's chemical industry cluster (SCIC) is closely related to the industrialization phases that Singapore has undergone since the government launched its industrialization programme in the early 1960s. Singapore's role as a centre for petroleum refinery began in 1961 when Royal Dutch Shell opened the country's first refinery on the island of Pulau Bukom (Ng, 1997) (see Figure 11.3). Singapore's geographical location and the worldwide trend by international oil companies to locate refineries near rapidly growing markets were the major reasons determining investments in petroleum refining during this period (Lim and Lloyd, 1986). During the next two decades, the country's refining industry grew significantly to become an important foundation of its economy. More importantly, the refineries served as a launching pad for the future petrochemicals industry (see Table 11.12).

After its separation from the Federation of Malaysia in 1965, Singapore adopted an export-oriented industrialization programme to stimulate industrial growth (Rodan, 1989; Perry et al, 1997; Low, 1998). For the next two decades, Singapore experienced remarkable growth in its manufacturing sector fuelled by increasing foreign investments dominated by petroleum refining and petroleum products. Success in the petroleum refinery sector inspired the Economic Development Board (EDB) to conduct feasibility studies on the potential of developing a petrochemical industry in Singapore (Lee, 1974). Efforts were also made to look for experienced international petrochemical companies to participate in the project. Sumitomo Chemical of Japan eventually assumed the role of a leader and coordinator (Ng, 1997). It was not until 1977 that Sumitomo Chemical and the EDB managed to secure endorsement by the Singapore and the Japanese governments. In July that year, 23 Japanese companies incorporated the Japan–Singapore Petrochemicals Company Limited. Following that in August, the Japan–Singapore Petrochemicals Company and the Singapore government established a 50–50 joint venture, the Petrochemical Corporation of Singapore. The Singapore government offered Pulau Ayer Merbau, an island just off the south coast of mainland Singapore, for the construction of the petrochemical complex (see Figure 11.3). The need for a petrochemical base gained urgency with the launch of Singapore's Second Industrial Revolution in 1979. After nearly two decades of strong growth based on a liberal policy of attracting all kinds of foreign investments, the 'revolution' was aimed at raising the value-added content of the country's economic activity (Rodan, 1989; Ho, 1994; Yeung and Olds, 1998; Yeung, 2002). Given its limited resource base, Singapore now focused on attracting investments and industries that would sustain its long-term economic growth and objectives. As Singapore strove to move up the value-added chain, developing a petrochemical industry was timely to ensure employment in the refining industry. This unemployment issue was further heightened in the early 1980s when Singapore's refining industry was threatened by capacity expansion plans in the Middle East and Indonesia (Ng, 1997).

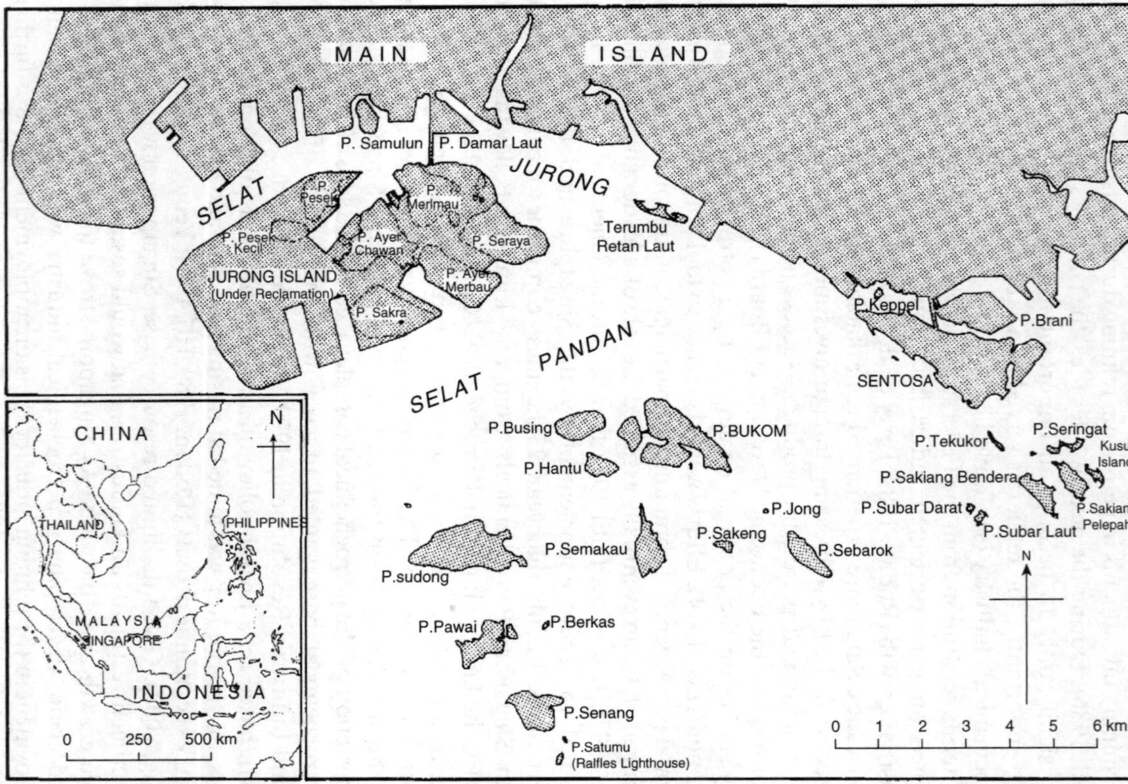

Figure 11.3 The Jurong Island chemical complex in Singapore

Table 11.12 The evolution of Singapore's chemical industry cluster

Phases of industrialization	Landmark events in the history of the Singapore industry chemical cluster	
Initial Phase (1961 to 1965)	1961	Royal Dutch Shell begins operating Singapore's first refinery on Pulau Bukom.
	1962	Japan's Maruzen builds and operates a refinery at Pasir Panjang.
	1963	Mobil begins operating a refinery in Jurong.
	1964	BP buys the Maruzen refinery.
Export-oriented Industrialization (1965 to 1979)	1970	Esso's refinery comes onstream.
	1973	Singapore Petroleum Company (SPC) begins operating a refinery on Pulau Merlimau.
	1977	Incorporation of the main upstream company, Petrochemical Corporation of Singapore (PCS).
Second Industrial Revolution (1979 to 1984)	1980	Incorporation of the first downstream companies: Phillips Petroleum Singapore Chemical (PPSC), The Polyolefin Company Singapore (TPC), Denka Singapore (DSPL).
	1982	Incorporation of Ethylene Glycols Singapore (EGS).
	1984	Start-up of Singapore's first Petrochemical complex – PCS, PPSC, TPC and DSPL plants.
	1991	Proposal of the cluster development strategy in the Strategic Economic Plan.
	1994	Announcement of plans to build PCS II.
	1995	Reclamation work begins under the Jurong Island Project.
	1998	Opening of the $3.4 billion PCS II complex.

Source: Compiled from Ng (1997).

After more than a decade of planning and preparation, Singapore's – and Southeast Asia's – first petrochemical complex (PCS I) finally came on-stream on Pulau Ayer Merbau in February 1984 (see Figure 11.3). This was a milestone in the historical development of Singapore's chemical industry. The establishment of PCS I provided a key link to the integrated process among the sub-sectors of the chemical industry. In the period following the 1985/1986 recession, the Singapore government appointed a ministerial committee to identify the causes of the downturn and to search for 'new directions' for future growth (Ministry of Trade and Industry, 1986). The early 1990s marked the dawn of a new political era in Singapore when Lee Kuan Yew stepped down as Prime Minister and was succeeded by

Goh Chok Tong. The new leadership introduced a portfolio of economic strategies encapsulated in the Strategic Economic Plan (SEP). Under the SEP, an 'Industrial Strategy' was proposed to develop Singapore's own world-class industries (Ministry of Trade and Industry, 1991). To realize this objective, the SEP, in conjunction with the *Manufacturing 2000* (M2000) programme, recommended a cluster approach to industry development. Under this approach, linkages between and within individual industries are analysed to map out the potential opportunities and synergies that can be tapped to enhance overall competitiveness (Economic Development Board, 1996c: 9). Towards this end, a S$1 billion Cluster Development Fund has been set up to enable the EDB to share the risks and costs of strategic investments in Singapore and the region.

Under the M2000 umbrella, the Chemical 2000 (C2000) study was completed with specific recommendations to enhance the chemical cluster. This marks a new phase in the development of the SCIC. Through a cluster development approach, C2000 aims to reinforce Singapore's position as a strategic manufacturing centre of chemicals in the Asia-Pacific region (Economic Development Board, 1995). More importantly, the implementation of the C2000 programme reaffirms the role of the government in developing the SCIC. Committed to nurturing Singapore as a regional chemical hub, the government has invested S$7.2 billion to build a chemical island complex that will rival the world's best. As shown in Figure 11.3, this infrastructural project involves combining seven southern offshore islands of Singapore into a single land mass, known as the Jurong Island Chemical Complex (Economic Development Board, 1995). In March 1994, the buoyant mood in the industry spurred the partners of PCS I to announce a S$3.4 billion investment to build PCS II (Economic Development Board, 1995). A few months later, Hoechst Celanese expressed its intention to build a S$150 million vinyl acetate monomer plant in Singapore, drawing feedstock from the second complex. As related and interdependent activities, these diverse petrochemical investments have contributed to the cluster development strategy by adding strength to a highly integrated industry structure. Such a strategy, however, would not have been realized without the significant role and support of foreign chemical firms (see detailed analysis in Wang and Yeung, 2000).

CONCLUSION

Singapore's experience in chartering its peculiar pathway to high-tech industrialization since the 1980s is unique among newly industrialized economies in Asia. Its entrepôt status and the state's pursuit of an export-oriented industrialization strategy have inevitably enabled the city-state to be a relevant player in the global economy (see also Olds and Yeung, 2004). And yet the state in Singapore has been able to intervene in the market economy to develop a unique repertoire of innovative capacity in various sectors and clusters of the national economy. By carefully managing the development of science parks as a spatial congregation of R&D activities, SME supplier networks as collaborators in high-tech production orchestrated by foreign TNCs, and the chemical clusters as a core pillar of Singapore's manufacturing industries, the developmental state continues to harness global forces to its own advantage (see also Yeung, 2005). To a certain extent, Singapore's national innovation and technological system can be regarded as a highly coordinated and managed system that brings together contributions from the developmental state, foreign TNCs and local enterprises. Such a unique tripartite combination of actors distinguishes Singapore's case from other innovation systems in advanced industrialized economies where local enterprises and state institutions remain the main actors in local and regional economic development.

If high-tech development is impossible in resource-scarce Singapore, had it not been nurtured by the deliberate economic planning undertaken by the state, what then are the specific implications for development strategies in 21st century Singapore (see also Perry et al., 1997; Low, 1998; Yeung, 2002; Rajan, 2003)? In short, the state should continue to adopt an active role in developing specific industrial clusters and TNC–SME supplier networks by removing obstacles, relaxing constraints, eliminating inefficiencies, facilitating collective learning processes, and, ultimately, building institutional capacities. If cluster development and high-tech industrialization are to be reinforced by inbound FDI, efforts should be focused on attracting different companies in the same field, with supporting investments in specialized training, infrastructure and other aspects of the business environment. The ultimate aim is, therefore, to develop a 'critical mass' with *enduring* competitive advantages that both foreign TNCs and local firms can tap to ensure success in an increasingly globalizing world economy.

More specifically, what emerges from the above analysis is that the developmental state in Singapore has always been putting its *political credibility* and *policy consistency* as the top priority in its engagement with global capital and in managing economic forces associated with globalization tendencies. This institutional capacity can be best observed in its labour and financial market governance. In both markets, the state has consistently managed flexibility and domestic interests to attract global capital. Its ability is predicated on the character and legitimacy of domestic institutions, not on the alleged external pressures created by globalization.

While the state and its myriad of associated institutions cannot possibly guarantee the future success of Singapore's high-tech road to economic development, its accumulated capacity to effect changes and transformations can be crucial to the continuous remaking of the Singapore political-economy into something that might just be more resilient and versatile in the face of apparently growing global competition.

NOTES

1. With effect from 1 January 2002, the NSTB has been renamed as the Agency For Science, Technology And Research (A*Star; see http://www.a-star.gov.sg). I would like to thank the organizer, Dr Patarapong Intarakumnerd, for kindly inviting me to participate in the conference where an earlier draft was presented. I am also most grateful to my former students, Yoke Tong Chew, Su-Ann Phillips and Jason Wang, for collecting most of the empirical material for this paper. The reworking of the materials for this chapter was facilitated by the NUS Academic Research Fund (R-109-000-050-112). I am solely responsible for all errors and misinterpretations.
2. The exchange rate in March 2004 was about US$1 to S$1.8.
3. Interview with the NSTB spokesperson, 29 February 2000.
4. A wholly-owned subsidiary of the Jurong Town Corporation since its incorporation in 1990, Arcasia is the developer and manager of the Singapore Science Park. It was renamed as Ascendas on 8 January 2001 following its merger with JTC International, another subsidiary of the JTC.
5. My empirical analysis below results from corporate surveys and interviews with tenants in the Singapore Science Park. A combination of survey and interview method was used. We sent 166 postal survey questionnaires to all tenants of the Singapore Science Park in early 2000. The sampling criterion was that the firm's mailing address had to be located in the Singapore Science Park. Only firms in technical operations were registered. Such facilities as eateries and childcare centres were omitted. The Singapore Science Park directory on the Internet served as the primary list (http://www. sciencepark.com.sg). The Park's tenants were telephoned consequently to ensure a greater survey return, during which 12 firms were no longer in operation, two companies indicated that they were 'dormant', and another two were in the process of moving out of the Park. This reduced the effective survey sample size to 150 firms: 34 (22.7%) responses were received. This survey was followed by 12 in-depth interviews with firms, two with research institutes and two with organizations directly related to the Park. These firms were selected on the basis of their differences in origin, type of firm, employment, industry, main function and proportion of R&D expenditure. All 16 interviews were recorded and transcribed.
6. Primary data on SME suppliers were gathered through a self-administered postal survey targeted at suppliers to foreign TNCs in the manufacturing sector. According to the definition of SME provided by the Economic Development Board (1999), a SME must have at least 30% local equity share, an annual sales revenue of less than S$50 million and fixed assets investments (defined as net book value of factory building, machinery and equipment) of less than S$30 million (see also *The Straits Times*, 6 December 1998). We selected a total of 213 companies that fulfilled these criteria. Out of a total of 220 mailed questionnaires, 68 questionnaires were returned and only 41 were usable because they were SMEs with TNC customers: 32 sample SMEs were wrong samples (e.g. foreign firms) or had ceased operations. The survey thus achieved an effective response rate of 21.8% (41 usable responses from 188 sampled firms). In order to complement the postal survey, we conducted face-to-face interviews with 11 companies.

REFERENCES

Amin, Ash (1999), 'An institutionalist perspective on regional economic development', *International Journal of Urban and Regional Research*, **23** (2), 365–78.

Amin, Ash and Nigel Thrift (eds) (1994), *Globalization, Institutions, and Regional Development in Europe*, Oxford: Oxford University Press.

Arora, A., R. Landau and N. Rosenberg (eds) (1998), *Chemicals and Long-Term Economic Growth: Insights from the Chemical Industry*, New York: John Wiley.

Asheim, Bjørn T. (1996), 'Industrial districts as "learning regions"', *European Planning Studies*, **4** (4), 379–400.

Becattini, Giacomo, Marco Bellandi, Gabi Dei Ottati and Fabio Sforzi (eds) (2003), *From Industrial Districts to Local Development: an Itinerary of Research*, Cheltenham, UK and Northampton, MA, USA: Edward Elgar.

Boekema, Frans, Kevin Morgan, Silvia Bakkers and Roel Rutten (eds) (2000), *Knowledge, Innovation and Economic Growth: The Theory and Practice of Learning Regions*, Cheltenham, UK and Northampton, MA, USA: Edward Elgar.

Braczyk, Hans-Joachim, Philip Cooke and Martin Heidenreich (eds) (1997), *Regional Innovation Systems: The Role of Governances in a Globalized World*, London: UCL Press.

Bunnell, Timothy G. and Neil M. Coe (2001), 'Spaces and scales of innovation', *Progress in Human Geography*, **25** (4), 569–90.

Business Times, various issues, Singapore.

Cantwell, John and Odile Janne (1999), 'Technological globalisation and innovative centres: the role of corporate technological leadership and locational hierarchy', *Research Policy*, **28**, 119–44.

Chew, Yoke-Tong and Henry Wai-chung Yeung (2001), 'The SME advantage: adding local touch to foreign transnational corporations in Singapore', *Regional Studies*, **35** (5), 431–48.

Chia, Siow Yue (1997), 'Singapore: advanced production base and smart hub of the electronics industry', in Wendy Dobson and Chia Siow Yue (eds), *Multinationals and East Asian Integration*, Canada: IDRC, pp. 31–61.

Chiu, Stephen W.K., Kong Chong Ho and Tai-Lok Lui (1997), *City-States in the Global Economy: Industrial Restructuring in Hong Kong and Singapore*, Boulder, CO: Westview.

Clarke, A.E. (1994), 'Spatial linkages and subcontracting relationships among high-technology industries in the Northeast Ohio region', *Environment and Planning A*, **26,** 1579–603.

Coe, Neil M. and Philip F. Kelly (2000), 'Distance and discourse in the local labour market: the case of Singapore', Area, **32** (4), 413–22.

Coe, Neil M. and Philip F. Kelly (2002), 'Languages of labour: representational strategies in Singapore's labour control regime', *Political Geography*, **21** (3), 341–71.

Cooke, Philip N. (2001), *Clusters, Learning and Co-operative Advantage*, London: Routledge.

Dent, Christopher M. (2002), *The Foreign Economic Policies of Singapore, South Korea and Taiwan*, Cheltenham, UK and Northampton, MA, USA: Edward Elgar.

Dent, Christopher M. (2003), 'Transnational capital, the state and foreign economic policy: Singapore, South Korea and Taiwan', *Review of International Political Economy*, **10** (2), 246–77.

Department of Statistics (1992), *Singapore's Corporate Sector: Size, Composition and Financial Structure*, Singapore: DOS.
Department of Statistics (1997), *Profile of Growing Small and Medium Enterprises in Singapore*, Singapore: DOS.
Department of Statistics (2000), *Singapore's Corporate Sector 1996–1997*, Singapore: DOS.
Department of Statistics (various years a), *Statistical Yearbook of Singapore*, Singapore: DOS.
Department of Statistics (various years b), *Foreign Equity Investment in Singapore*, Singapore: DOS.
Department of Statistics (various years c), *Report on the Census of Industrial Production*, Singapore: DOS.
Department of Statistics (various years d), *Report on the Census of Services*, Singapore: DOS.
Department of Statistics (various years e), *Singapore's Investment Abroad*, Singapore: DOS.
Dicken, Peter (2003), *Global Shift: Reshaping the Global Economic Map in the 21st Century*, 4th edn, London: Sage.
Doremus, Paul N., William W. Keller, Louis W. Pauly and Simon Reich (1998), *The Myth of the Global Corporation*, Princeton, NJ: Princeton University Press.
Economic Development Board (1992), *EDB Industry Report: The Disk Drive Industry* **1** (2), Singapore: EDB.
Economic Development Board (1994), *EDB Annual Report*, Singapore: EDB.
Economic Development Board (1995), 'Jurong Island: new-age infrastructure', *Singapore Investment News*, June.
Economic Development Board (1996a), *Singapore Unlimited*, Singapore: EDB.
Economic Development Board (1996b), *Singapore Directory of Chemical Manufacturers and Supporting Services 1996*, Singapore: EDB.
Economic Development Board (1996c), *The Chemistry Industry of Singapore*, Singapore: EDB.
Economic Development Board (1999), http://www.edb.gov.com.sg, accessed in January 1999.
Economic Development Board (various years a), *Report on the Census for Industrial Production, 1994 to 1997*, Singapore: EDB.
Economic Development Board (various years b), *EDB Annual Report*, Singapore: EDB.
Ellison, G. and Edward L. Glaeser (1999), 'The geographic concentration of industry: Does natural advantage explain agglomeration?', *American Economic Review*, **89** (2), 311–16.
Fröbel, Folker, Jurgen Heinrichs and Otto Kreye (1980), *The New International Division of Labour*, Cambridge: Cambridge University Press.
Fujita, Masahisa, Paul Krugman and Anthony J. Venables (1999), *The Spatial Economy: Cities, Regions, and International Trade*, Cambridge, MA: MIT Press.
Gertler, Meric S. (1995), '"Being there": proximity, organization, and culture in the development and adoption of advanced manufacturing technologies', *Economic Geography*, **71** (1), 1–26.
Gertler, Meric S. (2003), 'Tacit knowledge and the economic geography of context', *Journal of Economic Geography*, **3** (1), 75–99.
Gertler, Meric S., David A. Wolfe and David Garkut (2000), 'No place like home?

The embeddedness of innovation in a regional economy', *Review of International Political Economy*, **7** (4), 688–718.
Ghosh, B.C., S.K. Teo and A.M. Low (1993), 'Factors contributing to the success of local SMEs: an insight from Singapore', *Journal of Small Business and Entrepreneurship*, **10**, 33–46.
Goh, C.B. (1998), 'Creating a research and development culture in Southeast Asia: lessons from Singapore's experience', *Southeast Asian Journal of Social Science*, **26** (1), 49–68.
Grundy-Warr, Carl, Karen Peachey and Martin Perry (1999), 'Fragmented integration in the Singapore–Indonesian border zone: Southeast Asia's "growth triangle" against the global economy', *International Journal of Urban and Regional Research*, **23** (2), 304–28.
Henderson, Jeffrey (1989), *The Globalisation of High Technology Production*, London: Routledge.
Hirschman, A. (1958), *The Strategy of Economic Development*, New Haven, CT: Yale University Press.
Ho, Kong Chong (1994), 'Industrial restructuring, the Singapore city-state, and the regional division of labour', *Environment and Planning A*, **26** (1), 33–51.
Hudson, Ray (1999), 'The learning economy, the learning firm and the learning region: a sympathetic critique of the limits to learning', *European Urban and Regional Studies*, **6** (1), 59–72.
Huff, W.G. (1994), *The Economic Growth of Singapore: Trade and Development in the Twentieth Century*, Cambridge: Cambridge University Press.
Huff, W.G. (1995), 'The developmental state, government, and Singapore's economic development since 1960', *World Development*, **23** (8), 1421–38.
Keeble, David E. and Frank Wilkinson (eds) (1999), 'Special issue on regional networking, collective learning and innovation in high technology SMEs in Europe', *Regional Studies*, **33** (4), 295–400.
Lee, Soo Ann (1974), *Industrialization in Singapore*, Sydney: Longman.
Lim, Chong Yah and Peter J. Lloyd (eds) (1986), *Singapore: Resources and Growth*, Singapore: Oxford University Press.
Low, Linda (1998), *The Political Economy of a City-State: Government-Made Singapore*, Singapore: Oxford University Press.
Lundvall, B.-Å. (ed.) (1992), *National Systems of Innovation: Towards a Theory of Innovation and Interactive Learning*, London: Pinter.
MacKinnon, Danny, Andrew Cumbers and Keith Chapman (2002), 'Learning, innovation and regional development: a critical appraisal of recent debates', *Progress in Human Geography*, **26** (3), 293–311.
MacLeod, Gordon (2001), 'Beyond soft institutionalism: accumulation, regulation, and their geographical fixes', *Environment and Planning A*, **33** (7), 1145–67.
Malmberg, Anders and Peter Maskell (1997), 'Toward an explanation of regional specialization and industry agglomeration', *European Planning Studies*, **5**, 25–41.
Malmberg, Anders and Peter Maskell (2002), 'The elusive concept of localization economies: towards a knowledge-based theory of spatial clustering', *Environment and Planning A*, **34** (3), 429–49.
Martin, Ron and Peter Sunley (2003), 'Deconstructing clusters: chaotic concept or policy panacea?', *Journal of Economic Geography*, **3** (1), 5–35.
Maskell, Peter and Anders Malmberg (1999), 'Localised learning and industrial competitiveness', *Cambridge Journal of Economics*, **23** (2), 167–85.

Mathews, John A. (1999), 'A silicon island of the east: creating a semiconductor industry in Singapore', *California Management Review*, **41** (2), 55–78.
McKendrick, David G., Richard F. Doner and Stephan Haggard (2000), *From Silicon Valley to Singapore: Location and Competitive Advantage in the Hard Disk Drive Industry*, Stanford: Stanford University Press.
Ministry of Trade and Industry (1986), *The Singapore Economy: New Directions*, Report of the Economic Committee, Singapore: MTI.
Ministry of Trade and Industry (1991), *Strategic Economic Plan*, Singapore: MTI.
Ministry of Trade and Industry (1996), *The Report on the Census of Industrial Production*, Singapore: MTI.
Mirza, Hafiz (1986), *Multinationals and the Growth of the Singapore Economy*, London: Croom Helm.
Narula, Rajneesh (2003), *Globalization and Technology: Interdependence, Innovation Systems and Industrial Policy*, Cambridge: Polity.
Ng, W.H. (1997), *Guide to the Petrochemical and Chemical Industry in Singapore 1997*, Singapore: The Strategist.
Olds, Kris and Henry Wai-chung Yeung (2004), 'Pathways to global city formation: a view from the developmental city-state of Singapore', *Review of International Political Economy*, **11** (3), 489–521.
Patel, Pari (1995), 'Localised production of technology for global markets', *Cambridge Journal of Economics*, **19**, 141–53.
Patel, Pari and Keith L.R. Pavitt (1991), 'Large firms in the production of the world's technology: an important case of "non-globalisation"', *Journal of International Business Studies*, **22** (1), 1–21.
Pereira, Alexius A. (2000), 'State collaboration with transnational corporations: the case of Singapore's industrial programmes (1965–1999)', *Competition and Change*, **4** (4), 1–29.
Pereira, Alexius A. (2003), *State Collaboration and Development Strategies in China: The Case of the China–Singapore Suzhou Industrial Park*, London: RoutledgeCurzon.
Perry, Martin and Boon Hui Tan (1998), 'Global manufacturing and local linkage in Singapore', *Environment and Planning A*, **30** (9), 1603–624.
Perry, Martin, Lily Kong and Brenda Yeoh (1997), *Singapore: A Developmental City State*, London: John Wiley.
Phelps, Nicholas A. and Philip Raines (eds) (2003), *The New Competition for Inward Investment: Companies, Institutions and Territorial Development*, Cheltenham, UK and Northampton, MA, USA: Edward Elgar.
Phillips, Su-Ann Mae and Henry Wai-chung Yeung (2003), 'A place for R&D? The Singapore Science Park', *Urban Studies*, **40** (4), 707–32.
Porter, Michael E. (1998a), *On Competition*, Boston: Harvard Business School Press.
Porter, Michael E. (1998b), 'Clusters and the new economics of competition', *Harvard Business Review*, **76** (6), 77–90.
Rajan, Ramkishen S. (ed.) (2003), *Sustaining Competitiveness in the New Global Economy: The Experience of Singapore*, Cheltenham, UK and Northampton, MA, USA: Edward Elgar.
Régnier, Philippe (1991), *Singapore: City-State in South-East Asia*, Hawaii: University of Hawaii Press.
Rodan, Garry (1989), *The Political Economy of Singapore's Industralization: National State and International Capital*, London: Macmillan.

Schmitz, Hubert (ed.) (2004), *Local Enterprises in the Global Economy: Issues of Governance and Upgrading*, Cheltenham, UK and Northampton, MA, USA: Edward Elgar.

Schmitz, Hubert and Khalid Nadvi (1999), 'Clustering and industrialization: an introduction', *World Development*, **27** (9), 1503–14.

Scott, Allen J. (1998), *Regions and the World Economy: The Coming Shape of Global Production, Competition and Political Order*, Oxford: Oxford University Press.

Scott, Allen J. and Michael Storper (2003), 'Regions, globalization, development', *Regional Studies*, **37** (6/7), 579–93.

Simmie, James (ed.) (1997), *Innovation, Networks and Learning Regions?*, London: Jessica Kingsley.

Simmie, James (2003), 'Innovation and urban regions as national and international nodes for the transfer and sharing of knowledge', *Regional Studies*, **37** (6/7), 607–20.

Soon, Teck Wong (1995), 'Singapore', in *Industrial Structures and the Development of Small and Medium Enterprise Linkages: Examples from East Asia*, Washington, DC: World Bank, pp. 67–90.

Sternberg, Rolf and Olaf Arndt (2001), 'The firm or the region: what determines the innovation behavior of European firms?', *Economic Geography*, **77** (4), 364–82.

Storper, Michael (1997), *The Regional World: Territorial Development in a Global Economy*, New York: Guilford Press.

The Straits Times, various issues, Singapore.

Tödtling, Franz (1994), 'The uneven landscape of innovation poles: local embeddedness and global networks', in Ash Amin and Nigel Thrift (eds), *Globalization, Institutions, and Regional Development in Europe*, Oxford: Oxford University Press, pp. 68–90.

Wang, Jason H.J. and Henry Wai-chung Yeung (2000), 'Strategies for global competition: transnational chemical firms and Singapore's chemical cluster', *Environment and Planning A*, **32** (5), 847–69.

Wong, Poh Kam (1991), 'Technology development through subcontracting linkages: lessons from Singapore', in *Enhancing Intra-Industry Linkages*, Kuala Lumpur: Institute of Strategic and International Studies, pp. 49–59.

Wong, Poh Kam (1995), 'Competing in the global electronics industry: a comparative study of the innovation networks of Singapore and Taiwan', *Journal of Industry Studies*, **2** (2), 35–62.

Yeung, Henry Wai-chung (1998), 'The political economy of transnational corporations: a study of the regionalisation of Singaporean firms', *Political Geography*, **17** (4), 389–416.

Yeung, Henry Wai-chung (1999), 'Regulating investment abroad? The political economy of the regionalisation of Singaporean firms', *Antipode*, **31** (3), 245–73.

Yeung, Henry Wai-chung (2002), *Entrepreneurship and the Internationalisation of Asian Firms: An Institutional Perspective*, Cheltenham, UK and Northampton, MA, USA: Edward Elgar.

Yeung, Henry Wai-chung (2005), 'Institutional capacity and Singapore's developmental state: managing economic (in)security in the global economy', in Helen E.S. Nesadurai (ed.), *Globalization and Economic Security in East Asia: Governance and Institutions*, London: Routledge, pp. 85–106.

Yeung, Henry Wai-chung and Kris Olds (1998), 'Singapore's global reach: situating the city-state in the global economy', *International Journal of Urban Sciences*, **2** (1), 24–47.

Young, R.C., J.D. Francis and C.H. Young (1994), 'Flexibility in small manufacturing firms and regional industrial formations', *Regional Studies*, **28** (1), 27–38.

Zander, Ivo (1999), 'How do you mean "global"? An empirical investigation of innovation networks in the multinational corporation', *Research Policy*, **28**, 195–213.

12. Policy learning as a key process in the transformation of the Chinese innovation systems

Shulin Gu and Bengt-Åke Lundvall

In this chapter we develop the idea of innovation systems (IS) and link it to policy learning in a developmental context with special reference to the transformation of China's innovation system. We emphasize the evolutionary and systemic foundation of the innovation system's approach and the requirements for an adaptive innovation policy. We present the transformation of China's innovation system from the perspective of policy learning and indicate what further lessons remain to be learnt.

INNOVATION SYSTEMS, ECONOMIC DEVELOPMENT AND POLICY LEARNING

Innovation systems may be defined as sets of institutions, which jointly and individually contribute to the generation, diffusion and use of knowledge for the development, diffusion and application of new technologies. A major point with the system's perspective is that the innovation performance of an economy depends not only on how the individual institutions, such as firms, research institutes and universities, perform in isolation, but especially on how they interact with each other in connection with collective knowledge creation and use (Metcalfe, 1995; Smith, 1996; OECD, 1999a: 24).

The broader definition of innovation systems includes social institutions, macroeconomic regulation, financial systems, education and communication infrastructures and market conditions as far as these have a major impact on innovation. The set of institutions that constitute an innovation system also provides the more or less broad framework in which governments play their role in formulating and implementing policies. As far as the wider set of institutions have a critical impact on innovation performance and as far as they can be shaped by government policy their impact

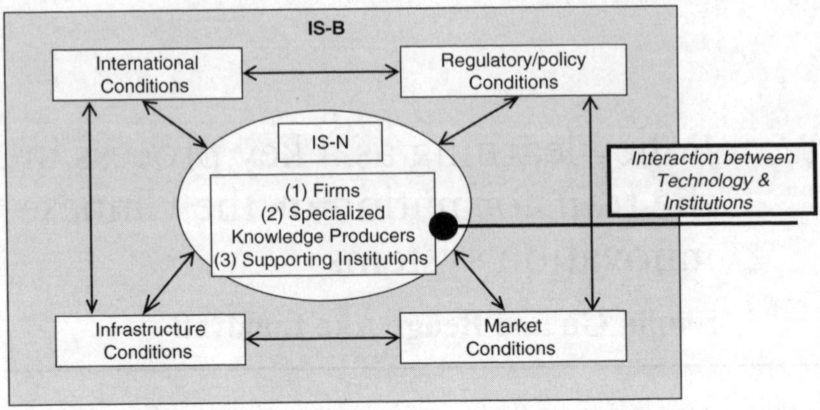

Source: Adapted from OECD (1999a), p 23.

Figure 12.1 A scheme of innovation systems: a broad and a narrow perspective

should be taken into account when designing policies in fields that are normally not included in 'innovation policy' (such as environmental policy, education policy, economic policy and competition policies).

The scheme in Figure 12.1 depicts our definition of innovation systems. Several distinctive policy relevant issues may be raised on the basis of this definition of the innovation system.

1. It is a fundamental assumption behind the innovation system approach that linkages between firms, R&D institutes, and other related components of the system contribute to the generation, dissemination and use of knowledge for innovation. To establish linkages that promote innovation therefore becomes a key issue for policy. This may take at least three different forms:

   a. Increasing absorptive capacity
   b. Increasing transfer capacity
   c. Establishing interconnections between parties

   It is important to note that in the systemic perspective linkages are never seen as going only one way. Where there is a flow of knowledge or information downstream (as from science to technology) there is always a feedback mechanism and a flow the opposite way, and to neglect this will normally result in policy failure. This is one reason why it is fruitful to characterize innovation as a process of 'interactive learning' where both sides are active in the process.

2. The systemic approach to innovation may be applied at different levels – for example a national system, a regional system, a supra-regional (across boundaries of nations) system, a sector system and even a corporate system. Initially, the idea of the innovation system was developed at the national level (Freeman, 1987; Lundvall, 1992; Nelson, 1993) reflecting the constant pressure to improve national competitiveness in the international market. With a delay of less than 10 years, the study of regional innovation systems received rapidly growing attention (Storper, 1995; Braczk et al., 1998) from scholars in economic geography among other disciplines. This reflected the insight that, in spite of the information age, much knowledge remains in a tacit form rooted in localities. With the expansion of international trade, the region increasingly comes to serve as a nexus of international networking, based on local non-traded interdependences. At the same time the study of regional innovation systems opens up an important policy area. This policy area is particularly relevant for developing countries like China, where there has been little policy attention given to regionally differentiated strategies and policies. As will be argued below, this neglect may also be problematic for its negative impact on overall national performance in terms of innovation.
3. Third, as outlined above, the construction and functioning of innovation systems are affected and shaped by the macroeconomic and regulatory environment. The fundamental system's transformation is indispensable when the macroeconomic system undergoes a restructuring, as in the past twenty years in China. In the 1980s and 1990s China initiated market reform programs, which have altered the macroeconomic environment from a centrally planned towards a market-oriented economy. It has changed the institutional framework for innovation and innovation policy radically. The process of innovation was deeply affected and so was the impact of innovation on regional income distribution and economic performance. Not all consequences could be foreseen in advance and there had to be a series of learning-based adaptations of innovation policies. More such adaptations are on the agenda for the future.

THE EVOLUTIONARY AND SYSTEMIC CHARACTER OF INNOVATION AND DEVELOPMENT

Inherent in the innovation system concept is an evolutionary approach to social and economic change, and this helps to understand the importance of policy learning. One important aspect of the system's analysis is the co-evolution of technology and institutions. Innovation and technological progress are regarded as outcomes of a socio-economic system whose structure and internal relationships have been historically created and socially rooted. And, the other way around, it links economic and social development with technological innovation.

Systems are evolutionary in the sense that they can be described in terms of mechanisms for selection, reproduction and creation of variety. They are characterized by *selectiveness* – each specific system has its own selection mechanisms regarding technological change and this will be reflected in the technological specialization of the system. They are also characterized by *path-dependency* – historical patterns of institutions and technologies will be 'remembered' by the system for a long time. Finally they are characterized by *unpredictability and uncertainty* – innovation is a process that permanently introduces *new varieties* that cannot be completely foreseen (Nelson and Winter, 1982; Rosenberg, 1976; 1982; Dosi, 1982; 1985; North, 1990; Kline and Rosenberg, 1986). There are profound implications of such an evolutionary perspective not only for the policy areas we discuss below, but also for the basic understanding of development plans and related policy mechanisms.

The innovation system concept provides a systemic perspective in the sense that it links micro-behavior to the system's level in a two-way direction. It sees change at the system's level as constituted by interactions at the micro-level of the system while it is 'the system' that sets the rules and shapes the patterns of individual innovation players in their selection of technology and in the ways in which they generate and use knowledge. The evolutionary and systemic perspective of innovation and development is in contrast to the conventional economic perspective that basically concerns optimization of resource allocation in a static context. Innovation and especially radical innovation is inherently uncertain and cannot be planned and controlled. The policy aiming at stimulating innovation therefore will involve elements of trial and error, and it needs to be flexible and based upon learning from experience.

THE NEED FOR PARADIGMATIC SHIFT IN POLICY ANALYSIS

Applying the evolutionary perspective of IS in development countries indicates a need for 'paradigm shift' in conceiving development and designing policies. Development strategies based on standard economics will give little attention to institutions and see the major role of policy to be intervention when there is 'market failure'. The ideal state of the economy is one of general equilibrium where all resources have been allocated in such a way that nobody has anything to gain from disturbing the equilibrium. This perspective gives little help when it comes to guiding policy-making for enhancing competences and building technological capabilities.[1]

Economic development in developing countries is, however, by no means a process of reaching or even converging to an equilibrium state. Rather, we find at its very core changes in technology, institutions, structure and attitude, through learning (Amsden, 1989; Kim, 1997; Mytelka, 1998; Johnson et al., 2003). Metcalfe et al. (2003) have characterized an evolutionary view of development in this way: 'Development is about finding new uses for resources, it is a creative process in which qualitative and quantitative changes are intertwined. Innovation redefines the basis for generating profits and this reshapes the evolution of the market structure.' Development would stop if it really fell into an equilibrium state, which would necessarily be a 'low track'-equilibrium.

Although knowledge and institutions have increasingly been accepted as key elements for development (The World Bank, 1999; UNIDO, 2002; UNDP, 2001), the influence of conventional thought, sometimes summarized under the heading 'Washington Consensus' remains strong.[2] Indeed we still need to build a 'Schumpeterian development economics' (Reinert and Reinert, 2003). In this situation the IS-approach may contribute substantially to the development of new perspectives turning development policies towards learning and innovation. This would also improve the policy capacity to integrate science and technology policies with development strategies and macroeconomic policies. It helps to bring into focus institutions at different levels of aggregation that shape, select and diffuse innovations. Innovation policy and policy learning needs to consider and reconsider institutional design in order to make sure that institutions at different levels support each other rather than being in conflict with each other. Historically the most successful innovation systems have been those where institutions match each other and the key technologies in the respective era (Freeman, 1995).

PLURAL PATHS OF INDUSTRIALIZATION AND POLICY LEARNING

One form of policy learning that has become increasingly important is based on international exchange of analysis and policy recommendations. International organizations such as the IMF and the World Bank have been important vehicles for diffusing what they present as 'best-practice' for development. This is not compatible with the understanding of the innovation system approach. It is fundamental that you cannot transplant single elements that work well in one national system to another and expect the same impact on economic performance.

One serious weakness with standard economics and the Washington Consensus is that they tend to see only one institutional best model. The idea is that a market economy with a minimum of public intervention is 'best-practice', perhaps with the appendix that parts of education and research activities are recognized as legitimate fields for governmental activities or support.

This is in contrast to the evolutionary and innovation system perspectives on development. They recognize that there are several different institutional set-ups that, without being 'optimal', work reasonably well in supporting innovation. Development paths and patterns are shaped by processes of selection, reproduction and innovation, which are specific to each country. It is obvious from some large programs of mapping national systems (Nelson, 1993; OECD, 1999a) that national innovation systems differ among OECD countries in terms of institutional set-ups, linkages, their specialized strengths and weaknesses in science and technology as well as their competitive advantage in international trade. Likewise, patterns and paths of development and learning among the Asian Newly Industrializing Economies (NIEs) were quite differentiated, although they had extraordinary growth performance in common.[3]

South Korea and Taiwan (Province of China, hereafter simply Taiwan), for example, have developed technological strengths in rather different areas. Korea was by the end of the 1990s competitive in semiconductors (DRAM), automobiles, electronic home appliances, shipbuilding, and intermediate goods such as iron and steel and petrochemicals (Choi, 2000), while Taiwan was strong in computer peripherals such as mouse technology, image scanners, monitors and keyboards (all took 50–80 per cent of the world market), ASIC (application-specific integrated circuits), and simpler CNC machine tools. The technological strengths of Korea were characteristic of capital-intensive large-scale investments and mass production. Taiwan on the other hand was characterized by smaller and 'lighter' systems and subsystems with less capital intensity.

Why did these differences in technology and sector specialization become established? One major reason was that they were developed within different institutional structures. Korea had developed a rather concentrated industrial structure while in Taiwan small firms dominated the economy.[4] Hence through the co-evolution of technology and institutions, the trajectories of learning diverged between the two cases. Such trajectories were, once established, cumulative in nature. Suh's study (1997) illustrates that such cumulative results can be recognized even in the accustomed firm attitude to decision-making on technology. He reported that in the 1990s Korean firms became rather used to looking for technologies that involved economies of scale.

ADAPTIVE INNOVATION POLICIES AND THE CRUCIAL IMPORTANCE OF POLICY LEARNING

The dynamic nature of development, and the complex co-evolution process of it, requires a new understanding of policy making. Policy operating from the perspective of the innovation system has to be evolutionary and systemic as well, in contrast to traditional static and deterministic approaches. Since there are alternative development paths reflecting interactions between institutions and technology in a particular context, it is not meaningful to specify unique 'best practice' in development policies, even if they have been proved to be successful in their original circumstances. The recent campaign in favor of attracting foreign investment was generalized on the basis of the success of East Asian NIEs. In this context too little attention was given to how the incoming foreign capital needs to be combined with local strengths and needs, and what complementary efforts need to be associated with it. As a result, the 'Asian miracle' has not been replicated elsewhere, and in Africa most countries lagged further behind during the 1990s in spite of their efforts to attract foreign direct investments (UNDP, 2003).

Accepting that there is no best practice policy to be imported from abroad, a serious question arises: how to develop policies that can guide the development process. One answer is that this should be done through *experimentation-based policy learning*.

Learning is a generic concept used with different meanings. In this context of policy learning, it refers to access to *more information* (for instance through access to data on R&D efforts) and a *more adequate understanding* (for instance through access to new models of causality) or *approach* with which policy makers try to manage the formation of policy, as well as to *enhanced competences* (for instance through experimentation and systematic exchange of and reflection on experiences) among individuals and organizations in charge of innovation policy. The rapid rate of change in the learning economy implies that policies and institutions designed and developed in a certain period will become obsolete and that there is a need to adapt and renew innovation policy continuously. Hence there is a need for policy learning and for adaptive innovation policies.

Policies aiming to adjust various aspects of a national innovation system – regional, sectoral or national, are all to be adaptive. In this chapter, we concentrate on policies that are oriented for *fundamental transformation* of national innovation systems. Fundamental transformation of innovation systems is understood in contrast to incremental development of innovation systems; the latter relies more on self-organization of system elements and their relations (OECD, 2002: 17–18).

Our knowledge of such processes of broad and radical institutional change is inadequate. As Ruttan states, thus far improvements in institutions and social systems have resulted primarily through the slow accumulation of successful precedents, a mode of learning that largely rests on trial and error (Ruttan, 2001: 133–4). Advances in social science knowledge may to some extent speed up learning processes and lead to improvement in the knowledge base for policy learning and institutional change (Mjoset, 2002). The innovation systems approach may be seen as one advance toward building such knowledge.

Based on experiences in China, we develop our ideas on system transformation and policy learning under five headings: (1) triggering factors; (2) sources of policy learning and learning ability; (3) mechanisms of transformation: reconstructing agencies and their networks; (4) interest-augmentation versus interest-redistribution; and (5) deepening and broadening the knowledge basis.

Since 1949, the establishment of the People's Republic, China has created a centrally planned innovation system. From the 1980s onwards, a fundamental transformation of the innovation system was undertaken along with market reforms of the overall economic system. With the accession to WTO and with the accumulation of inefficiency over the past twenty years, China is now faced with the challenge of restructuring its development strategy according to changed domestic and international conditions. The innovation system in China needs to deepen, and a further round of transformation is required.

Triggering Factors

While incremental institutional change within innovation systems may proceed through self-organization of system constituents and their relations, a fundamental transformation has to be caused by a major crisis or by higher levels of policy authority. This is because transformation implies setting a new goal and entering a new trajectory. The inertia accumulated in the structure of the system makes it impossible for decentralized agents to take on such a role.

Not only does the system as a whole suffer from a lock-in effect on the eve of transformation, so also do the political institutions. Hence often a change in top leadership, or a shock from external pressure is required to start a radical reform program. The reforms of the economic system and science and technology system in China were decided after Mr Deng Xiaoping took over the political leadership in 1978, even though tensions with the previous centrally planned system had become apparent long before that.[5] Fresh top leaders help policy institutions to overcome barriers to radical reforms, and decisions at the top level are a necessary condition for mobilizing system members in experiences and experimentation on a grand scale.

China has achieved an impressive high growth rate following the reforms of the 1980s and 1990s. The innovation system in China has become more open to an international exchange of knowledge and technology. A large number of previously government-run industrial technology R&D institutes have transformed to be closely associated with industrial production. A host of technology-intensive enterprises has grown by means of the mobility of technologically skilled persons, which has enabled the restructuring of the IT industry in China from a military-oriented to an application-oriented industry with great vitality (Gu, 2003).

However, in many ways, the capacity of China's innovation system remains rather modest. Many firms have not developed an in-house R&D capacity to engage in genuine in-house innovations. They grew successfully in the past two decades on the basis of cheap labor and with an extensive home market where shortage of supply was inherited from the planned economy era. In this context buying 'second-hand' technology from abroad was sufficient and the incentive to develop in-house technology was weak. With the access of China to the WTO, competition from multinational companies is becoming tougher and low-wage exports are confronted with trade disputes not only with advanced countries but also with developing countries. The speed-up of technical change, the new IPR-regime and the growing presence of MNEs in China also undermine a strategy based upon imitation.

The accumulated weakness, and the challenges from the changing environment for development have been increasingly evident since the second half of the 1990s. But the trajectory of development has remained the same. It might be that the succession of Mr Hu Jintao and Win Jiabao for the nation's leadership in 2004 heralds a new period, in which substantial efforts are made to further transform the innovation system. The government's bold decision to promote endogenous innovation, made at the beginning of 2005, may be seen as a possible signal of such a shift.

Sources for Policy Learning and Learning Ability

Two kinds of sources for policy learning can be identified. One comes from international demonstration effects while the other is internal to the national system. The impressive achievements in OECD countries and of China's Asian neighbors, and the surge of liberal ideology in political and academic spheres since the 1980s, have exerted a strong influence on the initiation and orientation of the reform programs in China. This illustrates how international demonstration may affect the goal-setting for fundamental reform programs. In China specific goals for the science and technology system reform were set forth clearly from the beginning. Inspired by international comparison and debate, the former Prime Minister Mr Zhao Ziyang put it in the following words:

> The current science and technology institution in our country has evolved over the years under special historical situations. The advantages embodied in this system manifested themselves in concerted efforts to tackle major scientific and technological projects, which were achieved with great success. However, there is growing evidence to show that the system can no longer accommodate the situation in the four modernizations program, which depends heavily on scientific and technological progress. One of the glaring drawbacks of this system is the disconnection of science and technology from production, a problem, which is a source of great concern for all of us. . . .
>
> By their very nature, there is an organic linkage between scientific research and production. For this linkage a horizontal, regular, many-leveled and many-sided channel should be provided. The management system as practiced until now has actually clogged this direct linkage, so that research institutes were only responsible to the leading departments above, in a vertical relationship, with no channels for interaction with the society as a whole or for providing consultancy services to production units. This is the root cause of the inability of our scientific research to meet our production needs over the years. . . . This state of affairs can hardly be altered if we confine ourselves to the beaten track. The way out lies in a reform (Zhao Ziyang, 1985).

The other major source of policy learning is the ongoing experimentation within the system. In the reform in China, the policy measure of assigning autonomy to R&D institutes was in a sense a transposition of ideas based upon experience developed in rural reforms. These proceeded by assigning autonomy to a farmer's family in order to move out of the previous commune system, originally created by a group of farmers in Anhui Province. The approach of 'One institute two mandates' was initially developed in the Chinese Academy of Sciences before it spread to R&D institutes as well as to universities (Liu and White, 2001). 'One institute two mandates' refers to a situation where two operational rules with different orientations co-exist in a certain research institute; one is academic and the other commercial. This might not be optimal for creating a stable framework for research since it produces enormous conflicts in the organization. Nonetheless, it was instrumental in breaking up institutional rigidity and created the flexibility, which made more experimentation possible.

In contrast to international sources, domestic sources serve as constructive means for effective system restructuring; in other words, knowledge and information from experimentation form indispensable inputs in the implementation of the transformation program. This is indispensable not only because followers experience proximity to the experimenting pioneer, but more importantly because such creation results from interaction within the system; hence it may also enhance the quality of system-fitness.

Evolutionary economists contend that from a systemic and evolutionary view, a policy initiative may be seen as the introduction of a social innovation into the system. Similar to that of a firm's manager initiating technological change, the policy maker is faced with a dynamic world, information is imperfect and outcomes are characterized by fundamental uncertainty. He must adjust the policy over time by learning about the reactions of the system, or from feedback – often both unexpected and undesired (Kline and Rosenberg, 1986; Metcalfe and Georghiou, 1998: 80).

The 'learning ability' of policy making refers to the capability to develop competences to launch and implement reform programs. It includes the ability of goal-setting of a reform program, the capacity to mobilize actions taken by system members, and the effective creation, dissemination and incorporation of practice-based experiences. The central government in China has the power to initiate ambitious reforms and to coordinate participatory actions at various levels of the system, as was done in the 1980s and 1990s reforms. Information flows from the bottom to the top policy centers were intense, especially in the first years of the science and technology systems reforms. These were important conditions for the successful reforms at the time. On the other hand, the horizontal links or 'learning networks' across interfaces between different policy areas or across geographical regions are more weakly developed in China. One result is that the science and technology policy has not been adequately integrated in the economic policy and also that reforms in the S&T system were inadequately attuned to the reforms in industrial sectors. Especially in the current context of a rapidly changing environment, hierarchical organizations of policy institution divided into highly specialized branches do not support the comprehensive policy learning necessary for system transformation and knowledge-based development strategies.

Mechanisms of Transformation: Reconstructing Agencies and their Networks

Even though decision-making at the top level is crucial for systems transformation, the call for decisions by the top authority by no means implies a 'shock-therapy' type of reform program. A 'shock-therapy' program assumes that transformation proceeds in a deterministic manner and leaves no space for experimentation and policy learning (Rodrik, 1999). Both top-down initiation and bottom-up feedbacks are necessary components for successful transformation.

Why is bottom-up feedback necessary? In a complex and highly interdependent system, any major reform initiative forces the various institutional units to make adjustments to cope with the disturbance and to reshape their interrelations. By observing radical systems' transformation associated with the development process in developing countries, Aoki and his associates argue that a learning policy process is necessary because of the interrelationships between parts and components of a complex system (Aoki, 1996; Aoki et al., 1996). Such institutional adjustments take place according to their own logic, often with unexpected outcomes (Aoki calls it 'unexpected fit'). It is impossible for any ambitious reform program to be designed perfectly in advance – surprises are doomed to appear. We might refer to this as involving 'radical' fundamental uncertainty in institutional innovation since we know *for certain* that the unknown and unforeseen will appear. The only way to handle this kind of uncertainty is to build 'learning' into the process, for instance through regular evaluation and adaptation combined with policy organizations that take on the characteristics of learning organizations and become members of learning networks.

So the essence of systems transformation is institutional restructuring through fitting and refitting elements/subsystems to the goal and to each other. The initial reform will indicate a set of new rules, norms and incentives to induce institutional changes but it cannot possibly shape in detail the network of organizations and subsystems. In China the policies for the reforms were on the one hand, to withdraw the control by the government over the operation of R&D institutes by cutting regular funds appropriation, and on the other, to force them to earn revenue through selling their knowledge (Liu and White, 2001; Suttmeier, 2000). This went hand in hand with the opening of 'technology markets' as intermediation mechanisms between R&D institutes and industry. In order to give legal support to the reform, an intellectual property rights system including patent law and national patent office were established.

Soon after the reform was initiated, it was found that promoting transactions at the technology market could not by itself fulfill the goal for effectively transferring achievements of the R&D institutes to industrial users. Nor could these institutes survive exclusively on the basis of such transfers. It became clear that the initial design of the policies was incomplete and simplistic regarding institutional restructuring and creation. Thereafter frequent modifications of policy measures occurred in the first four to five years after 1985 when the reform was initiated. The policies were subjected to a series of revisions and additions. Gradually the scope of institutional instruments was expanded to encompass various approaches, ranging from transactions of technologies at the technology market, contract research and development, creation of a supply network, creation of affiliated commercial units, mergers with industrial firms, and transforming industrial technology research institutes as a whole to become technology-intensive enterprises. Many of the approaches were initially tested by individual institutes or in one specific industrial sector. Institutional fitting and refitting in effect resulted in changes in operational rules, external links and internal deployment of assets of research institutes at the very grass-roots level, as well as in the incumbent firm structure and supply–demand structure of technology at the sector level (Gu, 1999b).

This means that any central initiative for introducing radical change to a system would be only a starting point for a period of economy-wide adjustment and re-adjustment in the operation, organization and interactive patterns of the system's units. A myriad of organizational responses by entities at lower levels are necessary to make the new elements fit into the new system. Radical institutional innovation is followed by incremental innovation that reduces uncertainty and the original mismatch caused by radical reform. Such mechanisms are fundamental aspects of the adaptation and learning that need to be at the core of a radical reform policy.

Interest-augmentation versus Interest-redistribution Transformations

Systems transformation is not to be seen as a purely techno-economic process aiming at higher efficiency. It is certainly a process with wide social and political implications as well. In this context it is useful to distinguish between interest-augmentation and interest-redistribution transformations. This helps us understand what assistance or resistance can be expected from different interest groups.

The reforms in the 1980s and 1990s basically moved towards interest-augmentation, in the sense that the central government decentralized decision-making autonomy to local (provincial, municipal and county) administration bodies, and to state-owned enterprises and R&D institutes. At the same time it opened the way for private initiatives in a variety of economic activities. Various potential actors – local administrators and managers for enterprise and R&D-institutes as well as individual entrepreneurs – welcomed the reforms. They were stimulated and mobilized when offered more autonomy and stronger incentives. In the first ten or so years of reform, interest-augmentation effects were reflected in widening opportunities for participation in, as well as benefiting from, economic and innovative activities. The 'recombination learning', as Gu and Steinmueller reported (Gu and Steimueller, 2000), illustrates the learning process underpinning the performance of the IT industry which was achieved through combining locally cumulated knowledge and skills with inflows of technology at the marketplace.

Currently, the most needed reforms for China might turn out to be of the interest-redistribution kind. This is for two reasons. First, as mentioned earlier, the easy extensive ways to growth may now have been exhausted. Future growth needs to be based not only on cheap labor, nor on a domestic market characterized by supply shortage of supply and the transfer of export-oriented manufacturing from Hong Kong and Taiwan. As competition is intensified and customers become more demanding, firms will have to invest more revenue in intangible assets. Second, the direct engagement of local and central governments in commercial activities creates anomalies and problems that need to be tackled. Local government bodies increasingly behave like entrepreneurs. While this was a result of the decentralization reform programs, it has become a problem for local governance. The combination of economic and political power at the local level now constitutes unchecked power. Local administrators have become a group with vested interests, and many of them have gathered extraordinary personal wealth.

Would they be supportive of the next series of reforms, which certainly must discipline them and which might reduce their expectations for further power and monetary benefits? To which extent will decisions made at the top become as effective as in the 1980s when the reforms could count on the mobilization of these groups in favor of the reforms? Are they prepared to learn from growing internal tensions and international competition so that they take a rational attitude to necessary reforms? At present the answers to these questions are difficult to predict.

Deepening and Broadening the Knowledge Base

We have discussed different sources of knowledge and information relevant for policy learning. Among those are 'learning-by-comparing' through international comparison, domestic transfer of institutional forms from one sector to another and institutional fitting and refitting in response to new rules and incentives set by the transformation program. Apart from these, 'the absorptive capacity' of the policy institution and policy learning relies increasingly and intensely on a supportive knowledge base.

In 1983–4 in order to formulate the reform program for the science and technology system that was put in action in 1985, China introduced the Frascati-indicators of R&D and carried out a national S&T survey. That was the first time that China employed internationally comparable S&T

statistical methods and moved away from the Soviet statistical system. For the introduction and implementation of the survey, huge human and capital resources were mobilized, and as a result, thousands of S&T statisticians and analysts were trained and became professional in the field.

One of the areas given little attention in the 1985 reform program was the national technological infrastructure. All R&D institutes with technology development (that is, experimental development in Frascati-terms) as the major activity were transformed into organizations that operate for commercial profits. One result has been that industries with a low degree of concentration constituted by small and medium sized firms, often weak in terms of technology and management competence, suffer from the weakening of supportive services. This was partly a reflection of the fact that 'technological infrastructure' was not a key concept in S&T policy under a centrally planned economy, whereas the international comparison made for the program was partially focused on formal R&D and science-based activities (Jensen et al., 2004).[6] The outcome was a biased understanding, paying little attention to market failure in providing technological services to SMEs. The lesson is that in order to utilize international experiences effectively, there is a need for deep knowledge and analytic capacity. A full analysis of the innovation system – both its science base and its wider knowledge base – is necessary in order to support policy learning.

Now China is faced with the challenge to move further ahead in terms of knowledge- and learning-based development. As mentioned, firms, regions and public agents, including government bodies, have to become effective learners. Linkages between knowledge centers and productive enterprises have to be enhanced to improve the system's efficiency. Reform programs and policies must become more specific and more responsive to feed-back from emerging changes and experiences. In comparison, the tasks of the reforms in the 1980s and 1990s were relatively simple. The knowledge base for future reform is insufficient in different respects. There is a need for research on China's innovation system but also for strengthening the competence of policy makers.

For example, regional administrators in China have acquired more autonomy and resources for regional development since the 1980s (Suttmeier, 2000). But regional governments still lack policy capacity. One sees a lot of duplication in policy language in different regions' development plans. With inadequate capacity to analyze specific conditions and opportunities, regional policy agents have also borrowed concepts and ideas from central plans and policies in cases where they are not applicable.

China has had a unique experience in developing and implementing a long-term science and technology plan. Improving the organization and methods of the plan in such a way that the need for policy learning is taken explicitly into account can play an important role in making the next major transformation reform a success. Research on the socio-economics of the national system of innovation, including the regional dynamics, is also necessary in order to understand better the prerequisites for building endogenous innovation capabilities both in science-based and experience-based activities (Jensen et al., 2004).

CONCLUSIONS

We have discussed implications of the idea of innovation systems for development policies in developing countries. If anything, policy learning is more demanding in development countries than it is in the rich part of the world. It is a characteristic of the economic development process that from time to time a new trajectory needs to be opened up.

We have used the experiences from the transformation of China's innovation system to demonstrate the usefulness of a systemic and evolutionary perspective on adaptive policy and policy learning. Specifically we have argued that the innovation systems approach is helpful in supporting the management of the endogenous process of policy and institutional evolution. This is in sharp contrast to some conventional understandings of policy, where it is assumed that the policy maker is in a position to once and for all design a master plan that defines the optimal solutions to all problems. At least implicitly this latter approach to policy still has some influence.

This calls for initiatives at the central level, but in order to be successful there must follow a period of adaptive learning at all levels of the innovation system. Neither over-centralized systems that leave no autonomy for the lower level of policy learning nor decentralized ones that lack the central governance mechanism necessary to initiate and coordinate change will be able to cope with this double challenge.

NOTES

1. The trade liberalization (or neo-classical) view is currently prevailing with its hallmark being the aggressive World Bank and IMF Structural Adjustment Programmes (SAP). Theoretical reasoning for SAP is centered on the role of market in efficient resources allocation, or, 'getting the prices right', although the balance-of-payment problems were one of the major triggering factors to the SAP which were faced by Latin American countries (Kruger, 1995). A review on SAP Programmes concludes that 'getting prices systematically and significantly wrong that import-substituting countries have done in the past has been a costly mistake. But few would disagree that getting prices right, in itself and of itself, will be sufficient to make Bolivia or Ghana grow at Korean rates.' (Rodrik, 1995: 2971–2).

2. With reference to the current situation of development thoughts: 'The seeming disappearance of development economics as a separate discipline some quarter century ago could not have come at a more inopportune time. Some of the criticisms . . . are valid . . .But their argument that developing countries are just like more developed countries, only lacking as much physical (and later,. . . human) capital and their assumption that competitive equilibrium theorem can be applied in a straightforward way is, if anything, even more misguided.'. . .

   'In the last two decades, there has been a growing awareness of the limitations of the competitive paradigm . . . Yet, in this same period, the reigning paradigm in development economics was the Washington consensus, which ignored these considerations, despite the fact that they are even more important to developing countries.' . . . 'A new development agenda thus must center around (i) identifying and explaining key characteristics of developing countries, . . .and exploring the macro-economic implications . . .; (ii) describing the process of *change*, how institutions and economic structures are altered in the process of development. . . . It must do so in light of changes in the global economy . . .'(Stiglitz, 2001).
3. Aoki and his colleagues contend for plural paths of development, based on game theory, that plural states for systems are the norm because various institutional, political, cultural and historical factors differ greatly among economic systems.
4. It is reported that in the mid-1990s and in the 'engineering' industries (namely machinery and electronics sectors), firms' concentration degree in Taiwan is considerably lower than that in Korea. In the industrial machinery sector the 96 largest Taiwanese business groups produced 9.8% of the total sales; in comparison, the 50 largest Korean *chaebols* made up 34.9% of the total. We have the data for more pairs: for the electronics sector it was 22.7 % versus 50.9%; for precision instruments, nil versus 14%; and for transportation equipment, 39% versus 79%. Source: Working Paper 5887, National Bureau of Economic Research, Washington DC, cited from Juana Kuramoto, INTECH mimeo, 1998.
5. Similarly, current reforms in Brazil can be traced to the event when civilian presidents took over the Federal Government from the military authorities in the mid-1980s (Velho and Saenz, 2002); and reforms in Thailand, related both to external shocks which came from the financial crisis in 1997 and to Thaksin Shinawatra taking over the Thai administration in 2001 (Intarakumnerd, 2004).
6. It is shown on the basis of empirical data and statistical tests that both science-based and experience-based learning are important for innovation in the learning economy. Most important for the innovative capacity of the firm is to combine science- and experience-based learning. It is argued that innovation policy is biased toward promoting STI-learning in science-based sectors and industries. Policy should also focus on experience-based learning in these industries, as well as on the need for firms in low- and medium-tech industries to get support for science-based learning. See Jensen et al., 2004.

REFERENCES

Amsden, A.H. (1989), *Asia's Next Giant: South Korea and Late Industrialization*, New York, Oxford: Oxford University Press.

Aoki, Masahiko (1996), 'Unintended fit: organizational evolution and government design of institutions in Japan', in Kim Aoki and Masahiko Okuno-Fujiwara (eds), *The Role of Government in East Asian Economic Development, Comparative Institutional Analysis*, Oxford: Clarendon Press, pp. 233–53.

Aoki, Masahiko, Hyung-ki Kim and Masahiro Okuno-Fujiwara (eds) (1996), *The Role of Government in East Asian Economic Development*, Oxford: Clarendon Press.

Braczk, H.-J., H. Cooke and M. Heidenreich (eds) (1998), *Regional Innovation Systems: The Role of Governments in a Globalized World*, London: UCL Press.

Choi, Y. (2000), 'Paradigm shift in Korea's science and technology policy', presentation to the international High-Level Seminar on Technological Innovation, co-sponsored by *United Nations University and the Ministry of Science and Technology of China*, Beijing, China, 5–7 September.

David, P.A. (1975), *Technical Choice, Innovation and Economic Growth*, Cambridge: Cambridge University Press.

David, P.A. (1985), 'Clio and the economics of QWERTY', *Economic History*, **75**, 332–7.

Dosi, G. (1982), 'Technological paradigm and technological trajectory', *Research Policy*, **11**, 147–62.

Freeman, C. (1987), *Technology Policy and Economy Performance: Lessons from Japan*, London: Pinter.

Freeman, C. (1995), 'The National Innovation Systems in historical perspective', in *Cambridge Journal of Economics*, **19** (1).

Gu, S. (1999a), *China's Industrial Technology, Market Reform and Organizational Change*, London and New York: Routledge in association with the UNU Press.

Gu, S. (1999b), 'Implications of National Innovation Systems for developing countries: managing change and complexity in economic development', *UNU/INTECH Discussion Paper #9903*, Maastricht, The Netherlands.

Gu, S. (2003), 'NIS transformation and recombination learning in China', paper prepared for the First International GLOBELICS Conference, Rio de Janeiro, 2–6 November.

Gu, S. and W.E. Steinmueller (2000), 'National innovation systems and the innovative recombination of technological capability in economic transition in China', UNU/INTECH Discussion Paper #2000–2c.

Intarakumnerd, Patarapong (2004), 'Thailand's National Innovation System in transition', paper presented at the First ASIALICS International Conference on Innovation Systems and Clusters in Asia: Challenges and Regional Integration, 1–2 April, Bangkok, Thailand.

Jensen, M.B., B. Johnson, N. Lorenz and B.-Å. Lundvall (2004), 'Forms of knowledge, modes of learning, and economic development', Paper presented at the Second GLOBELICS Conference, October, Beijing, in *IMPP* 2005.

Johnson, Björn, Charles Edquist and Bengt-Åke Lundvall (2003), 'Economic development and the national system of innovation approach', paper presented at the First International GLOBELICS Conference, Rio de Janeiro, 2–6 November.

Kim, L. (1997), *Imitation to Innovation: The Dynamics of Korea's Technological Learning*, Boston, MA: Harvard Business School Press.

Kline, Steven J. and Nathan Rosenberg (1986), 'An overview of innovation', in Ralph Landau and Nathan Rosenberg (eds), *The Positive Sum Strategy, Harnessing Technology for Economic Growth*, Washington, DC: National Academy Press.

Kruger, A.O. (1995), 'Policy lessons from development experience since the second world war', in J. Behrman and T.N. Srinivasan (eds), *Handbook of Development Economics*, Amsterdam: North-Holland, pp. 2497–550.

Liu, X. and S. White (2001), 'Comparing innovation systems: a framework and application to China's transition context', *Research Policy*, **30**, 1091–114.

Liu, Xielin and Steven White (2002), 'Institutions, evolution and performance in China's national innovation system', paper presented at the International High-level Seminar on Technological Innovation, co-sponsored by UNU and the Ministry of Science and Technology of China, Beijing, 5–7 September.

Lundvall, B.-Å. (ed.) (1992), *National Systems of Innovation*, London: Pinter.
Metcalfe, J.S. (1995), 'The economic foundations of technology policy: equilibrium and evolutionary perspectives', in Paul Stoneman (ed.), *Handbook of the Economics of Innovation and Technological Change*, Oxford, UK and Cambridge, USA: Blackwell, pp. 409–512.
Metcalfe, J.S. and L. Georghiou (1998), 'Equilibrium and evolutionary foundations of technology policy', in OECD *Science, Technology Industry Review* No. 22, *Special Issue on New Rationale and Approaches in Technology and Innovation Policy*, Paris: OECD, pp. 75–136.
Metcalfe, J.S., R. Ramlogan and E. Uyarra (2003), 'Economic development and the competitive process', paper prepared for the First International GLOBELICS Conference, Rio de Janeiro, 2–6 November.
Mjoset, Lars (2002), 'An essay on the foundations of comparative historical social science', ARENA and The Department of Sociology and Human Geography, University of Oslo, Working Paper No. 22, August.
Mytelka, L.K. (1998), 'Learning, innovation, and industrial policy, some lessons from Korea', in M. Storper, S.B. Thomadakis and L. Tsipouri, (eds), *Latecomers in the Global Economy*, London and New York: Routledge.
Nelson, R.R. (ed.) (1993), *National Innovation Systems: A Comparative Analysis*, New York and Oxford: Oxford University Press.
Nelson, R.R. and S.G. Winter (1982), *An Evolutionary Theory of Economic Change*, Cambridge, MA and London: The Belknap Press of Harvard University Press.
North, D.C. (1990), *Institutions, Institutional Changes and Economic Performance*, Cambridge: Cambridge University Press.
OECD (1999a), *Managing National Innovation Systems*, Paris: OECD.
OECD (1999b), *Boosting Innovation, The Cluster Approach*, Paris: OECD.
OECD (2002), *Dynamising National Innovation Systems*, Paris: OECD.
Reinert, Erik S. and Sophus A. Reinert (2003), 'Innovation systems of the past: modern nation-states in a historical perspective, the role of innovations and of systemic effects in economic thought and policy', paper presented at the First International GLOBELICS Conference, Rio de Janeiro, 2–6 November.
Rodrik, D. (1995), 'Trade and industrial policy reform', in J. Behrman and J.N. Srinivasan (eds), *Handbook of Development Economics*, Amsterdam: North-Holland, pp. 2925–82.
Rodrik, D. (1999), 'The new global economy and developing countries: making openness work', Washington, DC: Overseas Development Council.
Rosenberg, N. (1976), *Perspectives on Technology*, Cambridge: Cambridge University Press.
Rosenberg, N. (1982), *Inside the Black Box: Technology and Economics*, Cambridge: Cambridge University Press.
Ruttan, V.W. (2001), *Technology, Growth and Development, An Induced Innovation Perspective*, New York, Oxford: Oxford University Press.
Smith, K. (1996), *Systems Approaches to Innovation Policy: Some Policy Issues*, paper for the ISE-project, Oslo: Stepgroup, mimeo.
Smith, K. (1997), 'Economic infrastructure and innovation systems', in C. Edquist (ed.), *Systems of Innovation, Technology, Institutions, and Organization*, London and Washington: Pinter, pp. 86–106.
Stiglitz, Joseph E. (2001), *An Agenda for the New Development Economics*, draft paper prepared for the discussion at the UNRISD meeting on 'The Need to

Rethink Development Economics', 7–8 September, Cape Town, South Africa, Geneva: United Nations Research Institute for Social Development, http://www.unrisd.org.

Storper, M. (1995), 'The resurgence of regional economics, ten years later: the region as nexus of untraded interdependences', in *European Urban and Regional Studies*, **2** (3), 191–221.

Suh, J. (1997), 'Outward orientation, technology transfer and firm dynamics: the case of Korea', a paper presented at International Symposium *Innovation and Competitiveness in Newly Industrializing Economies*, Seoul, Korea, 26–27 May.

Suttmeier, Richard P. (2000), 'Globalization, structural change, and the role of government in China's search for a national innovation strategy', paper presented at the International High-level Seminar on Technological Innovation, co-sponsored by UNU and the Ministry of Science and Technology of China, Beijing, 5–7 September.

UNDP (2001), *Human Development Report 2001: Making New Technologies for Human Development*, New York and Oxford: Oxford University Press.

UNDP (2003), *Human Development Report 2003: Millennium Development Goals: A Compact Among Nations to End Human Poverty*, New York and Oxford: Oxford University Press.

UNIDO (2002), *World Industrial Development Report 2002/3 – Competing Through Innovation and Learning*, Vienna: UNIDO.

Velho, Lea and T.V. Saenz (2002), 'R&D in the public and private sector in Brazil: complements or substitutes?', *UNU/INTECH Discussion Paper* 2002–8.

The World Bank (1999), *Knowledge for Development, World Development Report 1998/1999*, Washington, DC: World Bank.

Zhao, Ziyang (1985), 'Speech to the national working conference of science and technology', 6 March, in SSTC White Paper No. 1, 293–7.

Zysman, J. (1994), 'How institutions create historically rooted trajectories of growth', in *Industry and Corporate Change*, **3** (1), 243–83.

Index